Making the Connection
Research and Teaching in Undergraduate Mathematics Education

© 2008 by
The Mathematical Association of America (Incorporated)

Library of Congress Catalog Card Number 2008926251

ISBN 978-0-88385-183-8

Printed in the United States of America

Current Printing (last digit):
10 9 8 7 6 5 4 3 2 1

Making the Connection

Research and Teaching in Undergraduate Mathematics Education

Edited by

Marilyn P. Carlson
Arizona State University

and

Chris Rasmussen
San Diego State University

Published and Distributed by
Mathematical Association of America

The MAA Notes Series, started in 1982, addresses a broad range of topics and themes of interest to all who are involved with undergraduate mathematics. The volumes in this series are readable, informative, and useful, and help the mathematical community keep up with developments of importance to mathematics.

MAA Notes

MAA Service Center
P.O. Box 91112
Washington, DC 20090-1112
1-800-331-1MAA   FAX: 1-301-206-9789

Preface

This volume is intended for mathematicians and mathematics instructors who want to enhance the learning and achievement of students in their undergraduate mathematics courses. The chapters in this volume are based on research that closely examines the cognitive and social complexities of how learners build mathematical ideas.

Academic mathematicians spend much of their time teaching, preparing to teach, or thinking about how to improve their teaching. Their conversations with colleagues and personal reflections often raise difficult questions about teaching and student learning. All the while the undergraduate mathematics education research community produces theories, models, curricula, and learning materials that speak to the questions mathematics instructors ask. To date, however, few vehicles have been available to assist instructors in using this research knowledge to better understand students' conceptual growth and to facilitate their reflection on teaching practice.

This volume is intended to bring some of the knowledge created by mathematics education researchers to the attention and service of mathematics instructors. The 23 chapters in the volume are divided into two sections. In Part 1, "Student Thinking, " the chapters describe perspectives and findings derived from investigations about how people learn central ideas in the undergraduate mathematics curriculum. Part 2, "Cross-Cutting Themes," contains chapters that focus on the teaching of mathematics and various ways to frame issues that are inextricably related to the art of teaching. In the table of contents, for each article there is a brief annotation describing what issues the article addresses.

In both sections authors synthesize and describe research findings from multiple studies in ways that they believe will make the findings useful for classroom instructors. When chapters do include data from particular studies, the authors provide useful details on methodology and references for further elaboration.

The authors in Part 1 each describe some facet of the growing body of knowledge regarding how students develop increasingly sophisticated ways of reasoning about particular mathematical ideas. In Part 2 the chapters focus on what research is revealing about the fundamental processes of teaching mathematics—including teacher interactions with students, teacher strategies for promoting student learning, teacher knowledge, and classroom and institutional values and norms.

Readers will find that mathematics education research, like any other field, has developed terminology to help systematize the collection of data and the interpretation of results. In this regard, authors have attempted to define terms likely to be new to readers outside of mathematics education.

The variety of research methods used in mathematics education research and described in this volume may also be novel to some readers. Unlike mathematics, our field is empirical rather than deductive. It employs both quantitative methods—such as the administration of assessments to gauge learning and statistical analyses of a population's performance, and qualitative methods—such as close studies of individual students or teachers or groups using data gathered through clinical interviews, observation of classrooms, coding and interpretation of videos, and examination of student work.

All of the chapters in this volume aim to reveal fundamental aspects of learning and teaching while simultaneously working toward solving pressing problems of practice. The chapters in this volume typically do not identify particular teaching methods that are "proven" to be effective – these types of studies would be "pure applied" studies, in that they would only attempt to show what works without proposing why. Such studies would employ tightly controlled experimental designs, defining and manipulating variables, then analyzing the variables' impact.

In our view, two factors limit the usefulness of experimental studies in the early stages of mathematics education research. First, it is difficult to control a classroom environment sufficiently so as to isolate variables and determine cause and effect. Second, classical experimental design is used to test and verify theories. But mathematics education research is a young field, and for many research questions there is no established theory to test.

At this stage of the field's development, a central goal of our research is to develop explanatory models for patterns in observations. Thus, readers will notice that many of the chapters in this volume speak about frameworks or theoretical constructs that have emerged by systematically analyzing and searching for patterns in qualitative data. From these patterns researchers can begin to build theories about learning and teaching that are useful for guiding the development and refinement of curriculum and other instructional tools. Once theories about knowing or learning specific content become stable, researchers may also develop quantitative tools to continue testing and expanding the theories.

Readers will also notice that transcripts of students' speaking are included in many chapters. Such transcripts of students "reasoning aloud" are a source from which we draw data and make observations. We believe that reviewing these transcripts will also provide readers with greater insights into how students think and how the authors derived their conclusions.

Above all, we hope that reading the chapters in this volume will encourage all mathematics teachers to become more reflective about their craft. In this need for reflection, teaching is much like problem solving. Teachers and problem solvers both become more effective as they acquire knowledge and skills; they become most effective when they can draw on their knowledge and skills to reflect on what works, why it works, what does not work, and why it does not work, in solving the problems upon which they are focused. May it also provide useful knowledge for the mathematics community as a whole, populated as it is by teacher-scholars who care deeply about doing the best possible job of assisting their students to develop as strong mathematical thinkers.

Acknowledgements

We wish to thank the MAA Notes Series Board, headed by Barbara Reynolds and then Stephen Maurer, for their assistance and support in preparing this volume. We also thank SIGMAA on RUME Publication Committee members Anne Brown, Barbara Edwards, Annie Selden, and Rina Zazkis for valuable assistance in reviewing manuscripts, and Janet Bowers for her editorial assistance.

During the 2004 Joint Meetings, the Special Interest Group of the MAA on Research in Undergraduate Mathematics Education decided to take on the challenge of preparing a publication to communicate knowledge and perspectives of the undergraduate mathematics education research community to mathematicians and mathematics instructors. The SIGMAA on RUME Publications Committee came to consensus on major areas to include and authors who had made significant contributions in these areas were identified and invited to submit a manuscript for consideration. Each submitted chapter received three reviews. The SIGMAA on RUME Publication Committee then met to select chapters to carry forward. The authors revised their chapters based on the three reviews and on editors' recommendations. Chapters in which the authors had adequately addressed the concerns of the reviewers were submitted to the MAA Notes Committee for consideration. The MAA Notes Committee then carried out its own review, with each chapter being reviewed by at least two reviewers. Recommendations were addressed, and the final chapters are included in this volume.

Contents

Part I
Student Thinking

a. Foundations for Beginning Calculus

1

On Developing a Rich Conception of Variable

María Trigueros, *Instituto Tecnológico Autónomo de México*
Sally Jacobs, *Scottsdale Community College*

Introduction

Have you ever considered that what mathematicians call 'variable' is not a mathematically well-defined concept? And that variable can have different meanings in different settings? Unlike the concept of function, for example, variable has no precise mathematical definition. It has come to be a "catch all" term to cover a variety of uses of letters in expressions and equations. As a result, students are often unclear about the different ways letters are used in mathematics.

Later in this chapter, we provide practical suggestions to help students develop a rich conception of variable as called for by the National Council of Teachers of Mathematics (NCTM) in *Principles and Standards for School Mathematics* (NCTM, 2000). For now, though, let's begin this chapter with some traditional problems that students typically encounter in their high school or college math courses. Work through each problem, and pay attention to the roles that your symbols play during the process of solving them.

Problem 1. Laureen trained for a bicycle race by repeatedly going up and down a hill near her house. Every time she went up the hill, she rode her bike at an average speed of 8 km/h and she rode back down the hill at an average speed of 17 km/h ending at the same spot where she started. One day, she went up and down the hill repeatedly for two and a half hours. How long did it take Laureen to go up the hill each time? What is the total distance she travelled that day?

Problem 2. Find the family of lines that pass through the point (–2, 3). What is the slope of the line that goes through (7, 4)?

Problem 3. Find the values of a so that the function given by $f(x) = \begin{cases} 3x^2 - 2, & x \le a \\ 3x + 4, & x > a \end{cases}$ is continuous over its entire domain.

As you solved these problems, did you notice the complexity of demands for thinking about variable? Did your symbols take on different roles during the problem solving process?

All of these problems are straightforward for most advanced students. But for many average students, it is difficult to understand the role of the symbols at each different step in the solution process. It is hard for them to consider symbols sometimes as variables related in a function, at other times as unknown numbers to be found, and at still other times as general numbers. (By "general number" we mean a symbol whose value is neither assigned throughout the solution process nor is to be determined at the final stage of the solution process.) Yet many teachers say 'variable' for all these different instances of symbol. Using the same name for a symbol that plays a variety of roles is very confusing to students. Most of them think of symbols as unknowns that have to be found no matter where they appear; a few consider them as standing for any number and they know how to operate on them procedurally; but all students have difficulty integrating all the different meanings of what mathematicians call 'variable.' Because the term 'variable' covers a variety of uses of letters and lacks an analytic definition, it should be no surprise that attaining a well-developed robust variable conception is problematic for students.

To elaborate more fully, we turn now to a discussion of the complexity of the demands for thinking about variable in each of the problems above.

Analysis of Problem 1

The solution to the first part of the problem requires dividing the distance travelled by Laureen into two parts: the way up and the way down the hill. It is necessary to consider the time Laureen took while going uphill as a general number. That is, it can be represented with a letter, say t, that can take any value in a set and on which one can operate. The time to go downhill can be symbolized as $T - t$, where T is another general number. In order to find the total distance, one has to consider the distance up the hill the same as the distance down, since Laureen ended at the same spot where she started. Using the fact that the distance travelled is proportional to the time spent travelling, and that the constant of proportionality is the average speed, it is necessary to introduce functional relationships between the distance travelled each part of the trip and time spent in each of them: $v_1 t = d_1$ and $v_2 (T - t) = d_2$. In these relationships, it is important to note that t can be considered as the independent variable in the function, d_1 and d_2 as the dependent variables, and v and T as parameters. Using these two relationships, one gets a new functional relationship, $v_1 t + v_2 (T - t) = d_1 + d_2$. Since the values of the parameters are given, $d_1 = d_2$ results in the equation, $8t = 17(2.5 - t)$, where the symbol t is a specific unknown that must be determined. Once the value of t is obtained, it is necessary to go back to the function that relates time and distance, substitute the value for t and solve a new equation for d_1. The total distance can then be calculated by multiplying the uphill distance by 2.

Did you notice how the symbols take on different meanings at different times during the solution process? Initially, t and T were used as general numbers. Then t became an independent variable coordinating with the dependent variables d_1 and d_2. At that same time, T changed from a general number to a parameter. Near the end of the process, the roles of t and d_1 changed to unknown from independent variable and dependent variable, respectively.

Analysis of Problem 2

The solution to this problem requires students to find an expression for all the possible lines that pass through the given point. They need to generalize what they know about a line to consider a family of lines each with different slopes. They must thus conceptualize slope as a varying quantity; but then they must introduce a parameter (a symbol used to generalize an algebraic statement so that the expression covers a family of cases) when they express the functional relationship between the independent variable x and the dependent variable y: $y - 3 = m(x + 2)$. Students need to distinguish between the meaning of the symbol m, as a parameter, and the meaning of the symbols x and y, as covariates. This distinction is required in order for them to recognize that m (not x!) now becomes the unknown to be determined. Take note of the complexity of understanding required here: the student must move flexibly from variable as varying quantity to variable as parameter, as independent/dependent variable, and as unknown.

Analysis of Problem 3

To solve this problem the student needs to understand the functional relationship between x and $f(x)$, where x can take any real value. It also helps if the student considers this relationship as a dynamic covariation between the independent and dependent variables. Furthermore, the ability to distinguish the different roles played by the variables becomes crucial when applying the definition of continuity for a real function. In the expression, $\lim_{x \to a} f(x) = f(a)$, the student must realize that the independent variable x varies dynamically with the dependent variable $f(x)$ as it nears the value of the parameter a. At the moment when x takes the value of a, the student must recognize that the role of $f(x)$ changes from dependent variable to a specific unknown that must be found.

Following in the spirit of the analysis of these three problems, you may now have a more acute awareness of the different uses and the changes in meaning of the term 'variable.' Several questions may come to mind: When is the last time you discussed the concept of variable with your students? How often do you point out the different ways in which variables are used in mathematics? Are you satisfied with the treatment given to variable in the textbooks? Do you assume, perhaps, that students just "know" variable and that it's not necessary to spend much class time talking about it?

Mainstream mathematics textbooks used in the high school, community college and first-year mathematics university programs (including the calculus sequence) generally give a cursory treatment to the concept of variable. The concept is not well developed and it is not given much attention. Thus it is not surprising that high school and college students possess variable conceptions that in some cases are narrow, limiting, and generally underdeveloped. In the next section, we present literature regarding the impoverished nature of some commonly held variable notions. Moreover, the research literature suggests that these limited conceptions of variable can pose obstacles for students when they advance to calculus and higher levels. This chapter discusses some of this research and provides practical suggestions.

Research Literature About Students' Conceptions of Variable

Research on both high school and beginning college math students suggests that their conceptions of variable are often superficial and lacking in richness. A rich variable conception should include the notion of changing quantity and joint variation. Also, dynamic imagery plays a key role in a well-developed variable conception. In this section, we summarize several studies related to these aspects of a rich variable conception.

Superficial Conceptions

In an investigation involving 167 beginning college students, Ursini and Trigueros (1997) concluded that these students had generally superficial conceptions of variable. The majority were able to interpret, symbolize and manipulate variables as specific unknowns only at a very elementary level. For example, they could interpret the symbol x in $x + 3 = 7$ as representing an unknown value, and they knew that x in $x + 3$ is not an unknown to be determined. Furthermore, they could symbolize the relationship $y = 2x$ from the data of a given table, and could correctly manipulate the symbols in expressions such as $2x - 5 = 4x + 3$. They could recognize the presence of an unknown number in a problem; but their ability to process other given information was limited (e.g., the ability to use contextual data to symbolize an equation). In particular, they often had difficulty discriminating between variable as *unknown* and variable as *general number* in fairly simple expressions. For example, they had trouble seeing how variable is used differently in equations such as $4x^2 - 12x + 9 = 5$ and $4x^2 - 12x + 9 = (2x - 3)^2$. Further, they consistently avoided manipulation of variables. Additionally, situations involving related variables posed difficulties for these students. While they could adequately handle correspondence between specific numbers, they had trouble with the notion of related variation. Their difficulties seemed to stem from the absence of a conception of relation as a transformation process or a dynamical process of variation.

The researchers noticed, also, that while these students were capable of recognizing the role played by the variable in very simple expressions and problems, any small increase in complexity provoked inadequate generalizations. Overall, students' understanding of the concept of variable lacked the flexibility that is expected at the college level. In a later study, Trigueros and Ursini (2003) concluded that when students engaged in more complex problems in which different uses of variable are involved (for example, where they must pose and solve an equation or set up and work with a specific functional relationship), they were unable to differentiate among the different uses of variable and integrate them successfully.

Varying Uses of Variable

Based on a seminal large scale British study, Küchemann (1980) developed a model for describing and classifying different ways secondary students use algebraic letters. This classification model delineates six levels of interpretation and use of letters: Letter Evaluated; Letter Ignored; Letter as Object; Letter as Specific Unknown; Letter as Generalized Number; and Letter as Variable. Building on the work of Küchemann and others, Trigueros and Ursini (1999, 2001, 2003) developed the "3 Uses of Variable Model" (3UV model) — not for classification purposes — but for the purpose of analyzing student difficulties, textbook treatment, and classroom observations. In addition, they use the 3UV model in the development and testing of instructional design. Their work analyzes the different uses of variable and the different aspects involved in its use when solving elementary algebra problems.

The 3UV model identifies variable as (a) *unknown*, sometimes called 'indeterminate' in older textbooks, (b) *general number*, and (c) *related variables*, such as those found when working with functions or curves. In this model, the term *general number* refers to the meaning associated with symbols in general expressions in which it is necessary to perform algebraic operations; or when symbolizing generalizations (for example, those found in problems where it is needed to find the next term in a sequence of numbers, or the number of points or lines that a specific geometric pattern will have after a certain number of iterations). To be fluent with variables, students need to be able to interpret all three uses of variable in different parts of a multi-step problem. They also need to be able to symbolize a quantity with a variable and manipulate variables. In the case of functions and curves, they need to be able to construct graphs of related variables and interpret them. According to Trigueros and Ursini (1999, 2001, 2003), a well-developed understanding of algebra necessitates the ability to differentiate among the three uses of variable and to flexibly integrate their uses during the solution of any problem.

Parameters

The 3UV model has also been used to analyze problems that include the use of parameters. Ursini and Trigueros (2004) consider parameters as a particular use of general numbers since they are needed to generalize expressions that already include variables. But in their study with 62 undergraduates, they found that students think of parameters as variables. During interviews, most students responded, "this letter stands for a constant that can change, it is another variable." These students interpreted parameters as general numbers and did not differentiate them from other variables unless the problem they were confronted with provided them with a concrete referent where the parameter acquires a specific meaning (for example, when parameters appear in the equation for a line or within a well-known formula such as the quadratic formula). Students showed difficulties manipulating the parameters, and on many occasions they tended to ignore them.

To illustrate, students solved the equation without taking the parameter into account when they were presented with the following problem: *Given the equation $3x^2 + px + 7 = 0$, for which values of p does the equation have only one solution? What are the roles of p and x in this equation?* When asked to explain their answer, they could not attach any meaning to p. They did not see p as the coefficient part of the linear term; they ignored it and were able to solve only a particular case. Most of the time, when these students were asked to symbolize a generalization, they either ignored the parameter or identified some general elements from the problem, but ultimately could not write an appropriate expression or equation.

Changing Quantity, Dynamic versus Static Imagery, and Joint Variation

Using variables to represent changing quantities and express relationships is particularly problematic for students, as substantiated by numerous reports (e.g., Küchemann, 1980; Kieran, 1992; Ursini & Trigueros, 1997, 2001; Trigueros & Ursini, 1999, 2001, 2003; Jacobs, 2002). Studies show that high school and beginning college students can work appropriately with correspondence between numbers, but the idea of joint variation is not easy for them. They can plug in a value of one variable into a functional relationship, but they are unable to determine variation intervals (Kieran, 1992; English & Sharry, 1996) or think about this relationship in a more dynamic way (Ursini & Trigueros, 1997).

In exploring this difficulty, a few studies have revealed the role played by a covariational view of function that is supported by dynamic imagery in the formation of student conceptions of variables (e.g., Cottrill et al., 1996; Jacobs, 2002; see also Oehrtman, Carlson, and Thompson, this volume). In the context of limit, investigations of calculus students' views about variable have uncovered interesting results. Cottrill et al. (1996) found that, among first semester calculus students, mental construction of the domain process ($x \rightarrow a$) was dynamic, whereas thinking about the range entailed the static image of considering only a single value, $f(a)$. A similar finding was reported by Jacobs (2002) in her exploratory study of Advanced Placement BC (AP/BC) calculus students' notions about variable. When these students were asked to discuss the meaning of 'x' in the expression $\lim_{x \rightarrow a} f(x) = L$, they used dynamic imagery as they referred to x occurring in $x \rightarrow a$; at the same time, however, they held a static view when they talked about x

occurring in $f(x)$ as a single value that is 'plugged in.' In other words, they did not conceive of the two instances of x within the same equation in quite the same way (dynamic image in one instance, static in the other).

Moreover, Jacobs found that when these same students discussed derivative, the notion of continuous variation in a changing quantity seemed to be largely absent from their thinking. They gave no indication of holding an image of one variable changing in tandem with another variable (a dynamic covariational view) in the context of derivative. Also, they tended not to mention changing rate in their discussion of derivative. Jacobs concluded that, in general, the ability to view variable as capable of having changing values seems to play an important role in conceptualizing changing rate. Recognition that something is *changing* is an essential underpinning to understanding the key ideas in calculus such as derivative, changing rate, integration and the Fundamental Theorem.

White and Mitchelmore (1996) revealed serious deficiencies in first semester calculus students' conceptions of variable that affected their ability to represent changing quantities. Their study involved four versions of four problems (each problem having to do with application of the first derivative), where the most contextual presentation required more translation than the purely symbolic presentation. Most of the students had difficulty using variables to represent changing quantities; few were able to correctly symbolize. The investigators noted that "... defining and using new variables is qualitatively different from relating explicitly given variables in symbolic form... [it] involves forming relationships at a higher level of abstraction than relating those already given" (p. 91). In particular, students' written work and their follow-up interviews demonstrated that many of them were confused about how to define appropriate variables and whether letters represented changing or constant values. Moreover, they tended to think about two or more variables as things to be manipulated rather than as representations of quantities having a relationship. The researchers noted that these students seemed to view variables as literal symbols detached from any concrete meaning. That is, their conception of variable was limited to algebraic symbols. The study concluded that an underdeveloped conception of variable is a major obstacle to applying calculus successfully.

Examining variable understanding in the context of linear inequality, Sokolowski (2000) found that a conception of variable as a varying quantity is linked with the ability to model, solve, and interpret solutions to linear inequality problems. In addition, she concluded that most students lacked a deep and robust understanding of variable and the ability to use variables flexibly. Other authors have called attention to the role that covariational reasoning plays in students' understanding of variable quantities changing in tandem with each other (e.g., Carlson, Jacobs, Coe, Larsen, & Hsu, 2002).

Possible Explanations

Why do secondary and post-secondary students have such difficulty with variable? The roots of the problem are indeed complex. Certainly over the past several centuries, variable has conveyed different meanings at different times. A retrospective look at these shifts in meaning provides some insight into why students may have difficulties.

Historical Development: Shifts Over The Ages

Variable is a concept for which conventional meanings and published definitions have varied considerably since its early origins. The meaning of variable has changed in emphasis at different times. In the 16th century, François Vieta (1540–1603) (or Viète) proposed a general method to solve problems. He referred to his method as the "analytical art" (Klein, 1968). Vieta's greatest innovation was his conception of (a) a general object which could be introduced in the general method and which could be operated on by well-defined rules and (b) a notational scheme: "a vowel to represent the quantity in algebra that was assumed to be unknown or undetermined and a consonant to represent a magnitude or number assumed to be known or given" (see Boyer & Merzbach, 1989, p. 341). He did not, however, apply his method to problems involving relationships between variables. The notion of related variables was advanced some years later. Working independently, both René Descartes (1596–1650) and Pierre Fermat (1601–1665) applied Vieta's ideas to geometry; Descartes, in particular, considered general objects as a useful tool to model and think about problems, to introduce dependence between symbols, and to calculate the value of one of the objects when the value of the other was known (Youschkevitch, 1976).

In the 18th century the mathematical term 'variable' conveyed the meaning of something that actually varies, such as time that passes, temperature that oscillates, days that lengthen, mortality rate that decreases, etc. (see Freudenthal,

1983). The earlier meaning is imbued with a kinesthetic quality, as evidenced by mathematical expressions such as 'ε converges to 0,' 'x runs through the set S,' or 'n approaches infinity.' According to Hamley (1934), that kind of interpretation began to give way in the late 1800s to the idea of variable as an abstract concept relating to 'pure' number values disassociated from any concrete embodiment in physical quantity. This latter notion of variable, which Freudenthal (1983) calls 'polyvalent name,' has persisted throughout the last century. The modern mathematical practice, however, is to mix polyvalent names and variable mathematical objects into one term, 'variable.'

A look at definitions used for the term 'variable' over the last few centuries reveals qualitative shifts in emphasis. Schoenfeld and Arcavi (1988) cite 10 different definitions of variable in various technical publications printed between 1710 and 1984. These definitions differ according to the importance they attribute to notions such as domain (modern emphasis) or variable quantity (earlier emphasis). Mainstream mathematics curricula in several countries, for example México and the United States, tend to introduce variables as symbols (usually letters) that stand for numbers and whose value is changeable. Contrast this approach with the notion of changing amounts of some measurable quantity like amount of time. Appreciation of this distinction invites the question: Which is the variable, 'time' or 't'? Adding to the complexity of variable for the uninitiated student, phrases such as 'x varies,' 'as x gets closer and closer to,' and 'let number of hours be the independent variable' are often heard.

Some researchers have expressed their concern about the polyvalence of meaning of variable. Thompson (1994b) theorizes that Newton's insight leading him to the Fundamental Theorem of Calculus was supported by a mental image of dynamic quantities. Similarly, the development of an image of rate in 7th graders, he found, begins with an image of change in some quantity. Thus, according to Thompson (1994a), a conception of variable as changing magnitude is important for developing a mature image of rate. He asserts:

> In today's K–14 mathematics curriculum there is no emphasis on function as covariation. In fact, there is no emphasis on variation. . . . This is in stark contrast to the Japanese elementary curriculum which repeatedly provokes students to conceptualize literal notations as representing a continuum of states in dynamic situations. . . . It seems, to me anyway, that a progressively more abstract notion of covariation rests upon a progressively more abstract image of variable magnitude. (p. 29)

Even earlier, Menger (1956) had expressed dismay over the blending of meanings into one term because it resulted in loss of preciseness in mathematical language. Freudenthal (1983) also disagreed with this convention on both pedagogical and mathematical grounds, since it obscures the important aspect of kinesthetics. Janvier (1996) notes the difference between magnitudes and numbers. He contends that the essence of the term 'magnitude' captures the idea of a measuring number as opposed to a 'pure' number. For him, this double meaning has serious implications for curriculum design, particularly in the areas of modeling and function approaches to algebra.

Calculational Attitude

Another possible explanation for student difficulties may relate to the tendency for both teachers and students to focus more on calculations and less on concepts involving variable. The role of calculational versus conceptual orientation to variable surfaced in the Jacobs' (2002) study. Her investigation of AP/BC high school calculus students revealed that when they oriented themselves to a given task in a calculational manner (i.e., they talked about the task in terms of 'find,' 'solve,' 'answer' or 'plug in'), they viewed a variable as an unknown, a letter that stands for a number or general number, an input to a function, or as a receptacle that receives a number in specific cases or parameter. These views about variable contrasted sharply with the views of students who exhibited a conceptual orientation and who talked about the task in terms of dependency relationships and varying values. In the latter case, students tended to view variables as tools for expressing mathematical relationships; their variable conception was characterized by a concern for how a variable relates to its domain and how two or more variables relate to each other.

In the context of function, these AP/BC students tended to focus on the independent variable and the process of evaluating the function at a particular input value. The independent variable was in the foreground of their discussions while the dependent variable remained in the background. In other words, they were not mentally coordinating two changing quantities in a covarying relationship.

In the context of limit, students' calculational attitudes were associated with their tendency to view limit (or the variable L) as an unknown value to be procedurally determined. The limit L was never discussed as a variable. Also, a mental coordination of L with a (as in 'x approaches a') was seldom observed in these students.

In the context of derivative, a calculational attitude was associated with students' inability to view variable as a changing quantity. Whether they talked about slope, difference quotient, rate, velocity or derivative, the matter of two simultaneously changing quantities was never mentioned. What they almost always alluded to, however, was the procedure for calculating slope/rate/velocity or finding the derivative function.

White and Mitchelmore (1996) reported similar findings regarding students' tendency to approach a derivative problem with a view toward using variables to perform a calculation or follow a procedure. As previously mentioned, they found that university calculus students' difficulties in applying calculus were directly related to their tendency to view variables merely as algebraic symbols that are to be manipulated.

Lack of Rich Curricular Material

Current school mathematics curricula often fail to account for the conceptual complexity of variable. Not only do curriculum materials generally fall short of scaffolding a robust variable understanding, but instructional practices often neglect developing a rich conception. Textbook analysis and classroom observations indicate that curriculum and instruction at the secondary level concentrate mainly on the use of variable as unknown. Teachers seem to expect that as students encounter algebraic expressions, word problems, and problem-solving exercises, they will construct (all by themselves!) a robust, flexible and coherent conception of variable as a mathematical entity. However, the large body of research findings clearly indicates that this expectation has no grounding in empirical studies. Teachers need to become aware that something more explicit is needed to ensure that students develop the ability to differentiate general numbers from specific unknowns and work with variables as changing entities that can be related in a dynamic way. They need to concentrate on helping students work with the different uses of variable in a flexible way, to foster students' ability to overcome a merely computational way of thinking about symbols, and to ensure that students achieve a stronger conceptual understanding.

Simply having students take more math courses does not improve their variable understanding. Results from a study by Trigueros and Ursini (1999) suggest that, as students continue to take algebra courses through high school, their conception of variable shows little sign of substantial and sustained improvement between middle school and university levels. The researchers examined interpretation, symbolization and manipulation of variable in a variety of representations (verbal statements, tables, graphs, algebraic expressions). They found that even though students are exposed to different uses of variable in algebra courses, they nonetheless fail to comprehend variable as a multifaceted entity. Surprisingly, on tasks involving variables in a functional relationship or identification and interpretation of the unknown, students who had not yet taken algebra performed better than students who had already studied algebra. This finding indicates that student difficulties with variable conception are probably attributable to current didactical approaches.

Interventions and Promising Results

Attempts to address the potential of students to understand the concept of variable after suitable intervention are reported by Ursini and Trigueros (2001). They designed a teaching approach that follows a spiral path where at each stage students are introduced first to a problem requiring only one use of variable at a time (variable as specific unknown, variable as general number, or variables in functional relationship) and then to a problem of the same level of difficulty but requiring integration of the different uses of variable. This procedure is repeated numerous times with ever increasing levels of task difficulty. This teaching design was tested in a study with 12- and 13-year-old public school students in México (Ursini, Trigueros, Escareño, & González, 2002). The analysis of these students' work showed that after 15 sessions they had acquired the ability to recognize the different uses of variable and use them appropriately when working on simple algebraic problems. Also, they were able to differentiate among the uses of variable and integrate the different uses throughout the process of solving the problem. Overall, they demonstrated gains in their ability to shift between the different uses of variable in a flexible way. The next section proposes specific suggestions for the practitioner.

A Call To Action

The National Council of Teachers of Mathematics (NCTM) has designated specific goals relating to variable understanding in children from pre-K through grade 12. In the *Principles and Standards for School Mathematics* (NCTM, 2000), a comprehensive resource guide and set of recommendations for mathematics curriculum, strategies for developing student conceptions of variable are explicitly recommended for each grade band (Grades preK–2, Grades 3–5, Grades 6–8, Grades 9–12). One of the NCTM's guiding philosophies for developing variable conception is expressed in this document in a quote by Anna Sfard: "A thorough understanding of variable develops over a long time, and it needs to be grounded in extensive experience" (p. 39).

It is interesting to note that in the NCTM (2000) document, most of the curricular recommendations addressing variable concept development target the middle grades. There is very little discussion of variable at the secondary level. The same is true of the programs for the three compulsory mathematics courses for middle school education published by the Ministry of Education in México, as well as the most widespread programs for high school mathematics in México. A careful examination of all these curricula leaves us with the distinct impression that a robust variable understanding is tacitly assumed to be firmly established by the time students reach high school. The literature in mathematics education, however, strongly suggests otherwise. In fact, as we have documented throughout this chapter, high school algebra students tend to have under-developed conceptions of variable and this impoverishment holds true for college students as well.

Practical Suggestions

The teaching of variable at all levels before calculus must develop the capacity of the learner to think of this concept as an entity and to understand its different uses as diverse facets of the same concept. Students need opportunities to discuss the differences in variables that appear in a problem so that they can become acquainted with different uses and can integrate and interiorize them into that which teachers and mathematicians call 'variable.' Students also need opportunities to reflect on the roles that symbols play at different moments in the solution of specific problems, as demonstrated in the analysis of problem solutions discussed at the beginning of this chapter.

When students arrive at the calculus level, their conception of variable is far from complete. They still need to be given opportunities to revisit variables and think about them in the differentiated way presented in these pages. Students are capable of developing a rich conception of variable, but they need the full complexity of the concept to be addressed in a more explicit manner in their courses. And since many of the conceptual obstacles facing calculus students are related to variables playing a dynamic role in functional relationships, mathematics educators must give increased curricular attention to variables in such roles. We must be overt and intentional as we redirect our efforts toward fostering well-connected understandings.

Specifically, it is important that teachers realize the need (a) to become aware of students' difficulties with variables and (b) to address them explicitly in the classroom. In particular, teachers can incorporate into their classroom practice the following suggestions:

- Encourage students to choose meaningful symbols for variables in a contextual setting (not always x and y), and take care that they are able to change the name of the variable when needed to be sure they are not using the variable only as the replacement of the name of an object, but as a symbol to operate with. In expressions such as $T - t$ in the bicycle problem presented at the beginning of this chapter, it is helpful to ask students to write 1–2 sentences to explain the difference in meaning between T and t.

- Ask students to identify all of their symbolic expressions. For example, have them write in words what each expression represents and discuss similarities and differences. Engage them in explaining the difference between equations and open expressions. Have them write in words what each open expression represents. For their equations, have them explain when the symbols stand for a number or numbers they need to find and when they represent any number in a certain domain. In the case where there is a relationship between variables, have them explain in words which variables are related and how.

- Discuss with students the different roles (unknown, general number, related variables) played by each of the symbols in specific problems. For instance, have students work in groups to construct a table that shows, for each successive step of the solution, what role each symbol plays in that particular step. Allow each group to present their table to the whole class and discuss the different roles played by their symbols with particular attention to when the roles change. Discuss similarities and differences between the tables presented.

- Request that students state the appropriate unit of measure that goes with each variable in a contextual setting. For example, when students identify variables and their relationship, have them write the unit of measure that goes with each variable. Also, when they identify their open expressions, have them write appropriate units.

- Provide opportunities for students to identify related variables and to explain how these quantities vary in different problems. Encourage them to imagine the variation and describe it in their own words. For example, when working with a problem such as the bicycle problem presented at the beginning of this chapter (e.g., number of times that Laureen rides up the hill, down the hill; distance travelled; duration of training that day, etc.), talk with students about all the quantities that vary. Ask students questions such as: What is the smallest and largest value that you can imagine for each quantity identified? What intermediate values do you imagine would occur between the smallest and largest values?

- Place increased emphasis on the covariation aspect of related variables. Have students work in groups to construct a 2-column table for each pair of variables that co-vary. Using their tables, have them construct graphs where they label and scale both axes. Have them select two points on their graph and discuss what each point represents (in terms of coordinated variables). Also have them explain what changes occur when moving from one point to the next. Then have them construct 3-column tables to show how the values of each variable change in tandem with the values of related variables.

- Help students develop dynamic imagery in connection with the concept of variable. Design classroom activities where students kinesthetically act out variation in some designated quantity over time (by walking, by hand movement along an axis on a graph, or by using physical models). Have them attend to aspects of the variation (for example, magnitude, amount of change, increasing/decreasing magnitudes, rate of change and changing rate). Then create for them a function setting where the varying quantities do not explicitly involve time and have them attend to aspects of variation in each quantity as it relates to the other.

- Reinforce all of these suggestions by including items on tests, quizzes, and other assessments that engage students in sense-making. For example, a problem like the bicycle problem (discussed earlier) lends itself to asking questions such as the following: What does the equation $f(t) = d$ mean? Explain each letter. What is the unit of measure for $f(2)$? What does the equation $f(t) = 1$ mean? What does $f(t)$ mean?

Classroom time spent on developing a rich conception of variable is time well spent. With appropriate curriculum and instruction, students can achieve gains in their confidence and proficiency in working with variables. A better knowledge of variable can help students overcome a merely computational orientation and achieve rich conceptual thinking. When students possess rich variable conceptions, they are better prepared to advance to their next higher course in mathematics. Our goal for students is that they will no longer struggle with the complexities of variable in the ways that they are struggling today.

Concluding Remarks

The research findings reported in this chapter make it clear that variable is a very complex concept and that students need help in developing a rich conception. The current body of research alerts us to specific difficulties that students face regarding the concept of variable and provides useful information to guide the design of interventions for helping students work more fluently with variables. The way is open for new curricular design; the way is open for new instructional methods. With research guiding the way, mathematics educators are now obligated to direct their efforts toward ensuring that students advance their conceptual understanding of variable.

References

Boyer, C., & Merzbach, U. (1989). *A history of mathematics* (2nd ed.). New York: Wiley.

Carlson, M., Jacobs, S., Coe, E., Larsen, S., & Hsu, E. (2002). Applying covariational reasoning while modeling dynamic events: A framework and a study. *Journal for Research in Mathematics Education, 33*(5), 352–378.

Cottrill, J., Dubinsky, E., Nichols, D., Schwingendorf, K., Thomas, K., & Vidakovic, D. (1996). Understanding the limit concept: Beginning with a coordinated process schema. *Journal of Mathematical Behavior*, *15*(2), 167–192.

English, L. D., & Sharry, P.V. (1996). Analogical reasoning and the development of algebraic abstraction. *Educational Studies in Mathematics*, *30*, 135–157.

Freudenthal, H. (1983). *Didactical phenomenology of mathematical structures*. Boston: Reidel.

Hamley, H. R. (1934). *Relational and functional thinking in mathematics* (The National Council of Teachers of Mathematics Ninth Yearbook). New York: Bureau of Publications, Teachers College, Columbia University.

Jacobs, S. (2002). *Advanced placement BC calculus students' ways of thinking about variable*. Unpublished doctoral dissertation, Arizona State University, Arizona, USA.

Janvier, C. (1996). Modeling and the initiation into algebra. In N. Bednarz, C. Kieran, & L. Lee (Eds.), *Approaches to algebra: Perspectives for research and teaching*. (pp. 225–236). Boston, MA: Kluwer.

Kieran, C. (1992). *Multiple solutions to problems: The role of non-linear functions and graphical representations as catalysts in changing students' beliefs*. Seminar presented at CINVESTAV, México.

Klein, J. (1968). *Greek mathematical thought and the origin of algebra*. Cambridge, MA: The MIT Press.

Küchemann D., (1980). *The understanding of generalised arithmetic (algebra) by secondary school children*. PhD thesis, University of London, GB.

Menger, K. (1956). What are x and y? *The Mathematical Gazette*, *40*, 246–55.

National Council of Teachers of Mathematics (2000). *Principles and standards for school mathematics*. Reston, VA: National Council of Teachers of Mathematics.

Schoenfeld, A. H., & Arcavi, A. (1988). On the meaning of variable. *Mathematics Teacher*, *81*(6), 420–427.

Sokolowski, C. (2000). The variable in linear inequality: College students' understandings. In M. Fernandez (Ed.), *Proceedings of the 22nd Annual Meeting of the North American Chapter of the International Group for the Psychology of Mathematics Education* (pp. 141–146). Columbus, OH: ERIC Clearinghouse for Science, Mathematics, and Environmental Education.

Thompson, P. W. (1994a). Students, functions, and the undergraduate curriculum. *Research in Collegiate Mathematics Education. I. CBMS Issues in Mathematics Education, 4,* 21-44.

—— (1994b). Images of rate and operational understanding of the fundamental theorem of calculus. *Educational Studies in Mathematics, 26,* 229–274.

Trigueros, M., & Ursini, S. (1999). Does the understanding of variable evolve through schooling? In O. Zaslavsky (Ed.), *Proceedings of the 23rd International Conference for the Psychology of Mathematics Education* (*Vol. 4,* pp. 273–280). Haifa, Israel.

—— (2001). Approaching the study of algebra through the concept of variable. In H. Chick, K. Stacey, Jill Vincent, & John Vincent (Eds.), *The future of the teaching and learning of algebra, Proceedings of the 12th ICMI Study Conference*. The University of Melbourne, Australia.

—— (2003). Starting college students' difficulties in working with different uses of variable. *Research in Collegiate Mathematics Education. CBMS Issues in Mathematics Education* (Vol. 5, pp. 1–29). Providence, RI. American Mathematical Society.

Ursini, S., & Trigueros, M. (1997). Understanding of different uses of variable: A study with starting college students. In Pehkonen, E, (Ed.). *Proceedings of the 21st International Conference for the Psychology of Mathematics Education* (*Vol. 4,* pp. 254–261). Lahti, Finland.

—— (2001). A model for the uses of variable in elementary algebra. In M. Van den Heuvel-Panhuizen (Ed.), *Proceedings of the 25th International Conference for the Psychology of Mathematics Education* (*Vol. 4,* pp. 327–334). Utrecht, Netherlands.

—— (2004). How do high school students interpret parameters in algebra? In M. Johnsen & A. Berit (Eds.), *Proceedings of the 28th Conference of the International Group for the Psychology in Mathematics Education* (*Vol. 4*, pp. 361–369). Bergen, Norway.

Ursini, S., Trigueros, M., Escareño, F., & González, D. (2002). Teaching algebra using the 3UV model. In M. Van den Heuvel-Panhuizen (Ed.), *Proceedings of the 25th International Conference for the Psychology of Mathematics Education* (*Vol. 4*, pp. 327–334). Utrecht, Netherlands.

White, P., & Mitchelmore, M. (1996). Conceptual knowledge in introductory calculus. *Journal for Research in Mathematics Education, 27*(1), 7–95.

Youschkevitch, A. P. (1976). The concept of function up to the middle of the 19th century. *Archives of History of Exact Science, 16*(1), 37–85.

Rethinking Change

Bob Speiser and Chuck Walter
Brigham Young University

A cat springs from a walk into a run. A spiral shell takes form through growth. With suitable mathematics, what might we learn about the motion or the form? Calculus is said to help with questions about movement, growth and change. How, exactly, does it help us? What might we need to understand? How could we gain such understanding? In this paper we explore two tasks that we designed for students, one about a moving cat and one about a spiral shell. We have written extensively about these tasks (Speiser & Walter, 1994, 1996, 2004; Speiser, Walter & Maher, 2003), as well as several related tasks (Speiser, Walter & Glaze, 2005; Speiser, 2004). We revisit this research briefly, partly to introduce it to new readers, partly to suggest further implications.

For concreteness, consider a moving cat. *Based on information to be gathered from the photographs in Figure 1, how fast might the cat be moving in Frame 10?* We will consider this task in more detail below, but we encourage readers to tackle it right away, and to reflect on the kinds of thinking needed to resolve it.[1]

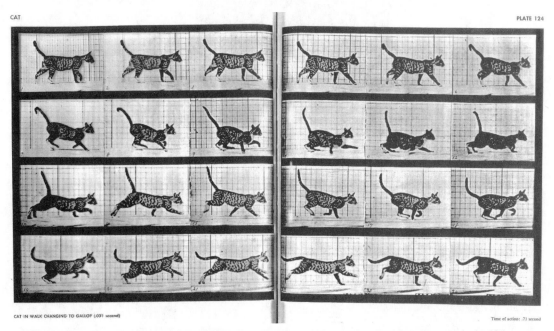

Figure 1. The cat (Muybridge, 1887). Note: The time interval between successive photographs is .031 sec. The distance between lines in the background grid is 5 cm. Every tenth vertical black line is emphasized.

Under typical instruction, far too many students fail to build sufficient understanding of the central mathematical ideas. In response, we would like to learn much more about the way *successful* learning might take place and about conditions that support such learning. On the one hand, a researcher might respond (as some have, especially in the

---

[1] It would certainly help to enlarge the figure given here, or else to copy from Muybridge (1887), which is still in print and legal to copy.

quite extensive misconceptions literature) by documenting and systematizing the surprising range of difficulties that so many students have. While such research has made us more aware of the specific difficulties many students have, it has not, by itself, at least in our view, led to much progress in developing instruction—until quite recently.[2] In general, the tendency has been to avoid pitfalls rather than to understand what *successful* learning might be like and how to support it. The work we survey here, on the other hand, takes a very different path: to study attentively, in detail, specific instances where students (typical students, not just a favored few) do learn successfully. In our analyses, we try to clarify what that learning might consist of, and to identify conditions that support it.

Our research is anchored in case studies where groups of learners, with little teacher intervention, tackle challenging investigations. We focus on how groups of students reason, based on information that they build together in complex, realistic problem situations. We are especially interested in task designs that invite students to tackle fundamental issues about change and motion. In the cases we consider here, students can discover *how to reason* about change. For us such cases, studied in detail, offer chances to clarify important theoretical and practical issues about student learning.

In our data, students build and advocate for *models*, that is, mathematical structures to describe or predict data. In the next section, we consider what students can learn from building models to fit or predict data in rich, realistic contexts that present important challenges. In later sections, we build from this basic theme.

Perspectives on Models

For clarity, we first consider building models in the context of a spiral shell (Speiser & Walter, 2004). Specifically, give students photographs of the fossil shell, as shown in Figure 2, and simple metric rulers.

Figure 2. The *Placenticeras* fossil. Age: 170 million years. Found near Glendive, Montana. The students' photocopies were about twice this size. (*Photo: Two Samurai Graphics.*)

To treat the spiral mathematically, it is natural to introduce polar coordinates (r, θ) centered at the spiral's origin. What can you say about r as a function of θ?

This task was designed by the first author for first-semester calculus students, to invite them to look deeply at how models might be built in realistic situations. In *Placenticeras*, by design, students must first construct the data they will try to model. Hence students must decide where to locate the polar origin and axis, which angles θ they will sample, and then make whatever measurements they need. The shell has visible irregularities. Thus, in addition to measurement discrepancies, students may decide to face important questions of interpretation.

---

[2] Awareness of the diversity of student difficulties, however, has begun to trigger research, by a number of workers, where epistemological and pedagogical issues have been explored productively. See the work of Carlson, Rasmussen, Selden & Selden, Zaskis, and others in this volume.

For many students in an initial trial,[3] an exponential model of the form

$$(1) \qquad r = r_0 a^\theta$$

emerged quickly. Several students, however, made a cogent observation: when they plotted a model of the form (1) against a scatter plot of their data, they found that the data points, as one student put it "sort of snake back and forth" across the exponential. (Different groups of students, however, obtained different amplitudes and phases for their wobbles.) Leading the class, the first author asked for the period of the observed oscillations. Several students, although using different models, soon estimated a period of 2π. This estimate led several students to propose models of the form

$$(2) \qquad r = r_0 a^\theta + b\sin(\theta + c).$$

These models, based on students' visual estimates, offered much closer agreement with the data.[4] If mere agreement with one's data were the standard, we could have stopped right there. But discrepancies for b and c across several student groups raised a further issue.

Let's take stock for a moment. Equations (1) and (2) present models for the spiral of *Placenticeras*. To move from (1) to (2), students reasoned from a graph of (1) together with the data points, in order to obtain good values for the constants b and c. That a model of the form (2) fits data better than a model of form (1) might indicate that (2) describes the shell more accurately. But the discrepancies for b and c suggest that the sine term in (2) might not describe the shell.

One class session later, several students suggested that the sine terms reflected different choices for the polar origin. To support this guess, they used their values for b and c to relocate the polar origin. Measuring again obtained new models of the form (1), that is, with different values for r_0 and a, which gave still better fits. In other words, they used the sine term in model (2) to reconstruct their data, and then returned to models of the form (1) without the controversial sine term.

If we think of data as inhabiting one world, and models inhabiting another, then all we need to do is find a formula to fit the data. As we have seen, it's not that simple. When we reason carefully in realistic situations, building models and obtaining data often interact. Such interaction can impel students to reexamine fundamental mathematics, as they did above when they interpreted the sine term. The data (in this case measurements) present the shell, and so reflect decisions we have made about the shell's geometry.

Returning to a model of form (1) for the *Placenticeras*, we see that r can be interpreted as growing exponentially as a function of θ. In other words, the *Placenticeras* grew proportionally, and in particular (neglecting a few obvious dents) it must therefore be self-similar. In this way our model tells us something *new* about *Placenticeras*.

The spiral's self-similar geometry can also be the basis for constructing models. In 2002 the first author taught an experimental mathematics class for students in the arts. In November, Sara Godfrey, a ballet major, noticed a copy of Figure 2 on a bulletin board and sketched it in her notebook. In a conversation several days later, Sara proposed to base her senior choreography project on the *Placenticeras* form, and asked the first author to help her and her dancers with the mathematics. A few days later, Sara recorded measurements of r at θ-increments of 90 degrees into her journal next to a second rough sketch of the spiral.[5]

In Sara's words: "When I see [the shell], I see movement. I can start from something small and investigate evolving into something larger, almost an *explosion*. Or, going backwards, an engulfment." The shell's chambers, "how each one relates to the one before," challenged her to investigate the underlying geometric structure. She wanted her choreography to reflect the unifying rigor of the mathematics. "I see the space [that the dancers] are working in, and how it's utilized, in a mathematical sense. I see *failures* in that, too. I want to try some things here, to best represent [the shell's] shape in choreography that is anchored in the mathematical concepts."[6]

At one point in Sara's choreography, six dancers formed a spiral on the floor of the stage. First, a lone male dancer

---

[3] September 18, 1991.

[4] The students' software could compute a least-squares estimate, but no student reported using such an estimate.

[5] This and a related choreographic project, and their documentation, was supported by a modest grant from BYU's Office of Research and Creative Activities. We are very grateful for this help.

[6] Conversation with Speiser, November 19, 2002.

Figure 3. The spiral on the floor. Here five women build the spiral form. Note: Their backs
also faced the center when they selected their locations. (Photo: *Two Samurai Graphics*.)

ran in from the wings and sat down on the stage to mark the center. Next, five women entered in succession. The first
woman took a spot not far from the center, to locate a first point on the spiral. The next dancer then took her place to
give a second spiral point. Following her, the three remaining dancers quickly found appropriate locations. Figure 3,
from a dress rehearsal on March 20, 2003, shows the resulting spiral just as the sixth and final dancer took her place.

At first the dancers found the spiral difficult to build and understand. When they tried to form it, they found
themselves too tightly wound around the center. For them, as they posed it, the issue was to grasp more clearly how
they should relate, in space, to one another.[7] Sara then asked Speiser, observing from one side, to help the dancers
grasp better how to build the form. With a rough sketch of Figure 4 in mind, he formed the angle α with his arms.

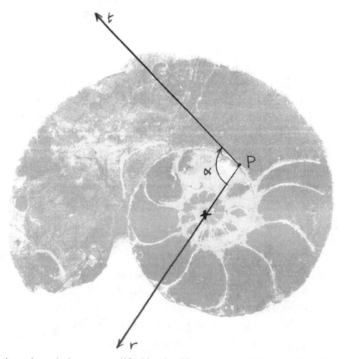

Figure 4. The equiangular spiral, as exemplified by the *Placenticeras*. (*Montage*: *Two Samurai Graphics*.)

---

[7] Sara explained to us in conversations that dancers in classical ballet typically travel along straight lines, especially the *corps*. Hence her dancers,
as they explained to us in interviews, saw spiral motions as challenging and new, both geometrically (the dancers tended toward circular arcs) and
also physically, in that the acceleration needed to produce tight curves required surprising force and preparation. Sara sought here, quite consciously,
as she explained in interviews, to widen the foundations of ballet.

Then he moved outward, holding the angle of his arms fixed, using the ray t (given by his right arm and hand) to determine his direction. Because the spiral is self-similar, the angle α, giving the tangent direction, remains constant through the motion. Sara had worked through this construction before, but her dancers had not. Immediately they returned to their rehearsal. This time, as they acknowledged right away, the spiral they were seeking soon emerged as they had hoped.

Based on videotaped data from rehearsals, Sara's dancers built the spiral form on stage *on the basis of a mathematical model*. Their model, shown in Figure 4, is both geometric and dynamic, in that it centers on developing the spiral curve through time. Each dancer found her place according to this model.[8] In interviews the dancers emphasized the local and dynamic nature of their thinking. They never visualized a static spiral form projected on the stage, but rather each dancer sensed, from the positions of the dancers just ahead, where she had to land. In this way, we could say the model (equiangular growth) gave each dancer a tool to *build* or *find* a data point (a place on stage) to fit her model. Again, as we reflected, the data cannot simply be given (they vary from performance to performance), and the models were not built even to fit pre-given data, but rather to help the dancers to *construct* those data.

In the next section, we explore how students reason.

Rates of Change in Realistic Contexts

One message of this chapter is that if we take realistic modeling seriously (c.f., Speiser & Walter, 1994), then we need to treat the basic concepts of calculus with much more subtlety and depth than standard pedagogy does. In this section, we focus on the interplay between data and models in situations where the data come from motion. We anchor this discussion to the case of a specific moving cat, through 24 historic time-lapse photographs by Eadweard Muybridge (1887). These photographs[9] appear in Figure 1 above.

In September 1993 we gave full-sized copies of the photographs in Figure 1 to first-semester calculus students, reviewed briefly how to find average speeds, and then posed two questions.

Based on information to be gathered from these photographs: (1) How fast is the cat running in Frame 10? (2) How fast is the cat running in Frame 20?

Because cats pursue their prey by sight, they should tend to move with their heads steady. Hence we suggested that students begin by taking the cat's position to be given by the tip of the cat's nose. In these photographs the cat shifts from a walk to a full gallop. The shift seems to be right around Frame 10, so we expected the average velocities for intervals before and after Frame 10 to be significantly different. From the photographs, students could calculate the average velocity between each pair of successive frames. Around frame 20, they found that the average speed between successive frames was nearly constant, but as the cat moved through frame 10, its average speed approximately *tripled*. What would the students do with this?

We posed this task (Speiser & Walter, 1994) to motivate discussion of the derivative as giving, at each moment, an instantaneous rate of change. We chose Muybridge's photographs because we wanted students to confront such rates of change in a complex, realistic setting. We especially wanted students to appreciate how much the concept of an instantaneous rate of change can be an idealization, a kind of *postulate* that might reach notably beyond the discrete data that one might observe. Students argued after some discussion that one cannot *know* (in particular, cannot convincingly defend a value for) the instantaneous speed at Frame 10 based on the evidence at hand, because the average speed varied too drastically.

Well and good, we thought. But a surprise came next, when we attempted to define the derivative formally, as the limit of a difference quotient.[10] Our students, thinking of the cat, raised several very fundamental questions. In our view, as mathematicians, we might seem to gain precision when we introduce a function $s = f(t)$ to represent the position of the cat's nose at time t. Hence, in this kind of setting, we would typically presume that the function $f(t)$

---

[8] Our use of the word *model* here seems quite similar to the way models derived from data are said in practice to offer predictions.

[9] We learned of these photographs from David Lomen and David Lovelock (Cushing, Gay, Grove, Lomen, & Lovelock, 1992) who introduced them to illustrate potential uses of technology in teaching calculus.

[10] We will try here to explain what happened very briefly, but we urge readers to consult our first paper on the cat task (Speiser & Walter, 1994) for full details, interpretation and discussion.

has been defined for every *t*, from Frame 1 to 24, even though we *know* values of $f(t)$ only for those 24 particular *t* at which a camera fired. We might further assume that the function $f(t)$ is continuous, or even smooth. In this way, we imagine that an average speed exists for *any* interval [*t, t + h*] in our range, with *h* nonzero. Each such average speed can then be represented by the classical difference quotient

$$Q_h = \frac{f(t+h) - f(t)}{h},$$

whether or not we can provide numerical values, based on measurements, for the two terms in the numerator. Of course we can provide such values for each pair of successive frames, as we did earlier. The 23 average speeds that we obtain in this way give useful information about the motion of the cat, even though they do not seem to tell us much about what may have happened at Frame 10. In practice, position measurements are *necessarily* discrete, so there seems to be no way to get around the lack of local information.

Based on this analysis, we enter suspect terrain the moment we enquire about a possible *limit* $Q_h \rightarrow L$ as $h \rightarrow 0$. This limit, should it exist, would be the derivative $f'(t)$, here the instantaneous speed at time *t*. But, as our analysis suggests, this number would be simply a *postulate* if all we know for sure are discrete data. Further, asserting (or assuming) the derivative's *existence* for a given *t* might seem like quite a drastic move for someone who has kept the data from the photographs clearly in mind.

In a typical calculus course, perhaps by force of habit, we interpret the difference quotient geometrically as the slope of a secant, under the additional assumption that the function *f* is not just defined for every *t* in a suitable interval, but also that the graph of *f* is smooth. Thus we might explain the limit of Q_h as $h \rightarrow 0$ dynamically, as the point of the graph given by *t + h* is imagined to approach the point given by *t*. In this way, imagining the tangent line and perhaps sketching it, we would argue for the tangent as a limit of secants. Of course, when *h* is very small, one cannot really see what happens, so some texts and instructors adopt the further strategy of zooming, progressively magnifying the graph of $f(t)$, perhaps using a graphing calculator, until the curve looks linear. In this way a situation like that of Frame 10 can come to resemble that of Frame 20, which, upon reflection, might seem strange. Indeed, when the second author tried this strategy (also in 1993), he found himself surprised by how his students understood the function $f(t)$ in question. (Speiser & Walter, 1994, p. 146).

> Chuck asked the students to imagine what happens to the secant when one of the two points approaches the other. Their response was that they could not do so. Chuck… then asked his students to imagine that the picture… instead of being on the blackboard, was now on their calculator screens, and asked them to imagine zooming in on the two points as they come together. Wouldn't the curve look straighter after zooming in? Here is the subsequent exchange, as Chuck remembers it:
>
> "No," said the students, "the curve looks straighter, but we still can't tell you what will happen, because the curve has *thickness*."
>
> "Thickness?" Chuck asked.
>
> "Yes, thickness," the class responded.
>
> "Is this thickness due to uncertainties?" Chuck asked.
>
> "Yes."
>
> "As in the motion of the *cat*?" Chuck asked.
>
> "Yes," said the students, "we really don't know where the cat is, so we can't say how the *points* will come together."

Through further discussions with our students, it became quite clear that the difference quotient and derivative (and later, the definite integral) made quite good sense when the function $f(t)$ is well defined and smooth, as would be the case if *f* were built up in standard ways from elementary functions. As a result, we felt, we and our students needed to make clear to what extent a *postulated model* (to which calculus, in general, applies) might be taken to describe a given set of *data* (to which calculus may often not apply, as we have seen above). In other words, we chose to make the selection and design of models more visible, more open to inspection and discussion with the learners, in relation to a given set of data.

A further point is that the heuristic discussion of the tangent as limit of secants assumes directly that the tangent *needs to be there*. Psychologically at least, this makes good sense. For example, if we imagine driving through the *x, y* plane along a road, say, given by a smooth function $y = f(x)$, a driver or passenger would tend to look directly forward, along a straight line tangent to the road, our path of motion (Speiser & Walter, 1994, p. 148). Secants might seem at least as natural (as lines determined by two points) but the central issue, as we've seen, is to *relate* the secant to the tangent as the two points come together. These observations led the authors (loc. cit.) to discover a straightforward geometric argument, using the half-planes given by a moving *tangent*, that the secant limits to the tangent when the two points coincide. In this argument no zooms are needed, and there is no temptation to make use of one. Note, however that what seems natural certainly depends on the observer's viewpoint. To understand a tangent as above, we must imagine moving with a point along a road; in contrast, for a secant, we would very likely need to stand outside the plane, to draw a line between two given points.

At the very least, the cat task can raise important questions about continuity, and perhaps especially discontinuity (as at Frame 10) in cases where no evidence could dismiss the possibility that there might have been an *instantaneous* change of speed—as if a quantum of momentum suddenly arrived. As a result, in 1994 (ibid.) we urged including discontinuous functions as potential models early. By the time we wrote a second paper on the cat task (Speiser & Walter, 1996) our rationale seemed even stronger.

Making Sense of Realistic Data

The initial studies (Speiser & Walter, 1994, 1996) were to some extent exploratory, in the sense that student data came from field notes and informal interviews. In 1999, as part of Carolyn Maher's Kenilworth longitudinal study, we posed the cat task to a group of third-year high-school students[11] in a nearly ideal research setting (Speiser, Walter & Maher, 2003). We obtained student data of unprecedented richness, and hence could study learners' choices, arguments and presentations in extraordinary depth.

In the Kenilworth data, linear models for the cat's motion took on particular importance, once the students came to recognize the potential of such models to help them verify and then make sense of standard graphs they had initially constructed for position and velocity as functions of time. They had built these graphs, as graphing calculator plots, from data they had found by measuring.[12] When they found themselves unable to connect position to velocity as they discussed these graphs, several students began checking independently to be sure the data in their tables made sense to them numerically. Magda and Aquisha, for example, measured distances directly on the photographs, rather than by means of Muybridge's squared background. To do so, they superimposed two overhead transparencies, as shown in Figure 5, to compare the cat's position in each pair of successive frames.

Figure 5. Superimposed, transparent cats, by Magda and Aquisha.
(This image, from videotape, appears here for the first time in print.)

---

[11] These students, about 17 years old, had recently completed a precaculus course together, but had not yet begun to study calculus.

[12] Graphing calculators were quite helpful for the students here, especially for building certain graphs from data tables. Once built, these graphs could be projected onto whiteboards for further layers of elaboration, the latter done entirely by hand.

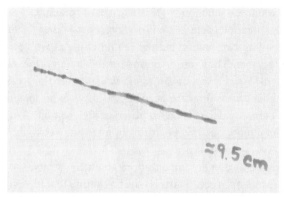

Figure 6. Aquisha's line (Speiser, Walter & Maher, 2003).

Then Aquisha, working alone, took an important further step, eliminating measurement entirely: she recorded each change of position as a *line segment*, traced directly from each pair of superimposed, successive frames. She then assembled the 23 resulting segments, in alternating colors, to form a line that showed the cat's position as it moved from frame to frame, as shown in Figure 6.[13] Read from left to right, Aquisha's line progresses through an initial series of short segments and then suddenly (at Frame 10) continues with drastically longer segments.[14]

We had seen line models earlier (Speiser & Walter, 1996), in a series of discussions about change and motion with art and dance faculty, for whom standard graphs made little sense initially. For these colleagues, line models seemed much more readable. Sara Lee Gibb, who teaches modern dance, suggested that the authors join her in a rehearsal studio, build a line model on the floor, and then run it as she beat a steady pulse. We were startled by the force and preparation needed to step out beyond frame ten, after a sequence of tight steps, and continue moving at the pace required. Hence, at Kenilworth (Speiser, Walter & Maher, 2003), we asked the students (many of them competition athletes) to build a version of Aquisha's line that they could run. Their model, built in a hallway of their high school, was slightly more than 65 meters long, scaling the cat's motion upward by a factor of 50. These students, too, were startled by the sudden burst of energy they needed to go past Frame 10.

Many learners find mathematics difficult when its constructs strike them as artificial or fictitious. Running physically through the hallway model, as several students argued later, helped them to connect the tables in their calculators to the immediate, detailed experience of physically moving. In effect, the act of running gave them opportunities, not to replicate the motion of the cat, but instead, as several students pointed out, to build a *new model* for it: their own motion. This experiential model, as student discussion soon made quite explicit, helped several students to connect not just the graphs, but indeed nearly every model they had built, to key features of their own experience of moving, and then, by extension, to the motion of the cat. To achieve this, they will need to reason carefully, and most likely collaborate in detail. In particular, they would need sufficient time and freedom to build convincing arguments based on the range of models then in play, as well as clear encouragement to do so.

Making Realistic Sense of Graphs

To find the average speed between two frames, we can compare, as Magda and Aquisha did, two successive snapshots of the cat in motion, by setting one photograph on top of another. We might be doing something similar, but with numbers, when we calculate the change between two successive measurements to form the numerator of the corresponding difference quotient. When we run the hallway model, however, it seems that we would experience our motion more as an evolving whole than as a mere aggregate of discrete pieces set out side by side for subsequent comparison.

---

[13] Magda and another student, Romina, then measured Aquisha's segments and rescaled them to obtain "cat distances" that allowed an independent check of their group's data tables.

[14] This idea looks very classical. Recall that Euclid (Heath, 1956) did not work with measurements, but rather used implicitly a kind of algebra for line segments (Hartshorne, 2000) in which lengths of segments could be added and compared. Hartshorne, in particular, drew his figures by hand, using the classical tools (compass and straightedge), and encouraged students to do the same, in this way gaining physical experience that many teachers, perhaps over centuries, have seen as helpful.

Similarly, in Section 2, to unpack the relation between tangents and nearby secant lines, we found it helpful to imagine looking forward (on a line of sight) as we drove along a given curve—again seen as a single, unimpeded motion. In both cases, from the viewpoint of our own participation in the motion, what happens when we divide time into discrete intervals would need to be related, through detailed reflection, to our own sense of what happened when we moved. In this way, successive snapshots or successive measurements might come to tell a story—although, strange to say, a set of snapshots and resulting measurements helped anchor the actions we performed. In this sense, we can say that we have *superposed* our personal experience of moving across key information from successive photographs. The personal experience allows us to connect distinct events to form a narrative.

In Figure 7, one student, Matt, has drawn a graph by hand across a projected image (on a whiteboard) of the graph of velocity with respect to time stored in Romina's calculator (Speiser, Walter & Maher, 2003). Romina's graph, a scatter plot, suggests a step function. The scatter points display a rapid increase in velocity at Frame 10, followed by a further rise at Frame 18. Elsewhere, the plot looks roughly constant. Matt's graph, in effect, interprets the scatter points by suggesting possible velocities between them. Based on his experience of running, he emphasizes the *second* rise especially in his discussion. This further acceleration, after the cat has launched into its run, does not seem obvious from Aquisha's line model (Figure 6), and may not have been clear to Matt until he ran the model in the hallway.[15] Matt conjectured that both rises correspond to moments when the cat, running, had pushed with its hind legs. Matt's graph, built from his own experience of running, allowed him to conjecture a connection between that graph and how the *cat* had moved.

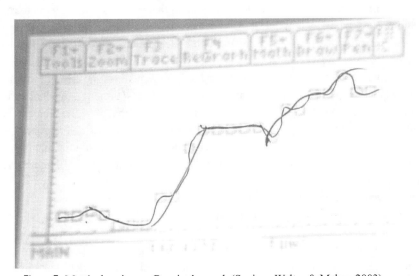

Figure 7. Matt's drawing on Romina's graph (Speiser, Walter & Maher, 2003).

In a few minutes, Matt and a second student, Benny, photographs in hand, proved that conjecture right. In this way, the following presentations were connected: the photographs, the students' tables, the calculator graph of velocity with respect to time, Matt's drawing on the projection of this graph, Aquisha's line model, its elaboration on the hallway floor, and students' detailed analyses of the experience of running. A key observation here is that of these seven presentations, only two (the photos and the calculator graph) were in some sense standard, and all but one (the photos) were constructed fully by the students. In effect, each presentation has come to be superposed on every other, through the ways the students have connected them. Further, based on our analysis (Speiser, Walter & Maher, 2003), if these students' emerging understanding could be seen as anchored to any one of the available array of presentations they have built, it might be anchored perhaps just as usefully in any other. In this way, using the full variety of presentations, the students argued strongly that the speed at Frame 10 could not be known, based on the data from the photographs, while at Frame 20 a close estimate could be made for the instantaneous speed. These students had not yet studied calculus.

---

[15] In other words, when they ran the line model, students discovered further detail, significant detail, that they had not noticed when they simply looked at Aquisha's presentation, or even at the marked floor of the hallway before running.

The Kenilworth experiment strongly suggests that students have much to gain from work on tasks that give them opportunities to look closely at fundamental mathematical issues, in settings where they have sufficient time and freedom to build and then work through a range of standard and (as needed) original, new forms of presentation to support their reasoning.

Connecting Teaching and Research

In an important sense, our research approach develops from a cognitive tradition, in that we focus strongly on how ideas and information are represented. In this sense, our work builds on Bob Davis' later work (1984), most especially his strongly cognitive approach to problem solving. For us, therefore, a representation is a *presentation*, either to oneself as part of an ongoing thought, or to others, as a tool for communication (Speiser, Walter & Maher, 2003). Put simply, whether to persuade or understand, we first build presentations and then reason from them. Such presentations might, for example, take the form of words, symbolic expressions, graphs, diagrams or models.

Our work also reflects, perhaps equally fundamentally, research traditions that stress the social sharing and negotiation of ideas. As in Maher's work (Maher & Martino, 1996), we pay close attention to how ideas and arguments are built and move through groups of learners, as they build and reason from key presentations.

As researchers in *mathematics* education, we are especially concerned with understanding how best to help students to reason mathematically. In particular, in our own teaching, we place strong emphasis on explanation, argument and, when appropriate, on formal proof. While it is widely understood that students seldom learn to reason mathematically simply by being told or shown how one might do it, nonetheless, as we have seen, powerful and fully mathematical thinking can emerge, both individually and socially, through collective task investigations that have raised important issues for the learners.

In this connection, we should emphasize how central *listening* has become for us, both as researchers and as teachers. In order to respond helpfully, we need to understand our students' thinking, in detail and depth. This need has led to a quite specific research methodology: (1) the development of tasks that raise fundamental issues for the learners and are rich enough to trigger extensive exploration and discussion of alternatives by learners; (2) close study, in this context, of how specific learners build, select, connect and restructure the presentations that they reason from; and (3) detailed attention to the way ideas and information move through the community of learners, and to the conditions that support such interchange.

We cannot emphasize too strongly how our research methods and our teaching practice, in related ways, entail a central emphasis on task selection and design. Our teaching, and hence our classroom research, might take place either within a standard mathematics class (as with our first work with the cat task), or perhaps (as, later, at Kenilworth) outside of one. In the work we have described, regardless of the setting, learners addressed important issues (such as the complexity of making sense of realistic data) that many course designs are structured to avoid.

Perhaps the largest difference between our work and what might be called the "standard model" for instruction stems from our strong disagreement with the widespread postulate, within that model, that students will learn best when they begin with simple, pre-structured situations that might be taken as convenient points of entry for more complex situations to be treated later. To provide the structures ready-made, however, can deprive students of the opportunity to learn, from hard experience, how to structure situations by and for themselves. Instead, we offer students problem situations to explore that, in effect, *require* the learners to propose and test potential structures and possible action plans against specific goals they see as pertinent. In this way, starting with rich situations and proceeding later to more simple cases, we have found that many learners build robust conceptions, lines of reasoning and basic skills that they can adapt directly to the simpler special cases. This view of learning, quite explicitly, informed the task designs we have described.

In both *Placenticeras* and the cat task, to facilitate our teaching, we designed rich situations for the learners to explore and structure.[16] In *Placenticeras*, student explorations often center on how to make good choices for a model,

---

[16] This view reflects a long tradition, which contrasts strongly with the "standard model." Consider Freudenthal (1991, p. 28 ff.), who strongly criticized the widespread use of logical or attribute blocks in elementary instruction: "Who is not familiar with the so-called logical blocks, 24 in number, available in a rich variety of models, for instance, red/blue, circle/square/triangle, big/small, thick/thin? They are a paragon of an entirely pre-structured world: one piece for each combination of the four criteria. Abstract operations on sets can be concretized marvelously by means of such a model. I chose this example not because I may believe in classifying as an important cognitive activity, but rather an example that, thanks to its simplicity, sharply features the difference between poor and structured on the one hand, and rich and to be structured, on the other."

because no preferred, pre-structured choice had been concealed in given data for the students to "discover."[17] Or, thinking of the dance students, our analyses suggest that the learners' thinking centered strongly on how to generate the form, or as one student put it, to *create* the spiral. In the first case, the students even had to build the data for themselves, before they could consider structures for them. In the second case, the data could only take form once a structure had been chosen for them. Further, in the cat task, while standard graphs of position and velocity were constructed early, these did not, by themselves, help to resolve the questions that the students sought to answer. Such questions, indeed, led the learners to invent and build from new presentations, like Aquisha's line, the line model on the hallway floor, and the physical experience of running. Only then, we feel, could truly useful structuring begin, as it did in Matt's discussion, when *still further presentations*, such as drawing on projected images, provided means for students to connect and to contrast what different prior presentations might have made available. In this way, the students structured the cat's motion, and, through the way they built and thought about the structures they considered, deeply enriched their understanding.

References

Cushing, J., Gay, D., Grove, L., Lomen, D., & Lovelock, D. (1992). The Arizona experience: Software development and use. In F. Demana, B. K. Waits & J. Harvey (Eds.), *Proceedings of the Third International Conference on Technology in Collegiate Mathematics* (pp. 41–47): Reading, MA: Addison-Wesley.

Davis, R. B. (1984). *Learning mathematics: The cognitive science approach to mathematics education.* Norwood, NJ: Ablex.

Freudenthal, H. (1991). *Revisiting mathematics education.* Dordrecht: Kluwer Academic Publishers.

Hartshorne, R. (2000). *Geometry: Euclid and Beyond.* New York, Berlin, Heidelberg, Tokyo, Moab: Springer-Verlag.

Heath, T. L. (1956). *The Thirteen Books of Euclid's Elements.* New York: Dover.

Maher, C. A., & Martino, A. M. (1996). The development of the idea of mathematical proof: A 5-year case study. *Journal for Research in Mathematics Education, 27*(2), 194–214.

Muybridge, E. (1887). *Animals in motion* (Reprint, 1957 ed.). New York: Dover.

Speiser, B. (2004). Experimental Teaching as a Way of Building Bridges. *Proceedings of the Twenty-Sixth annual meeting of the International Group for the Psychology of Mathematics Education, North American Chapter (PME-NA 26)* (pp. 21–36). Toronto.

Speiser, B., & Walter, C. (1994). Catwalk: first-semester calculus. *Journal of Mathematical Behavior, 13,* 135–152.

Speiser, B., & Walter, C. (1996). Second Catwalk: narrative, context, embodiment. *Journal of Mathematical Behavior, 15,* 351–371.

Speiser, B., & Walter, C. (2003). Getting at the mathematics: Sara's journal. In N. A. Pateman, B. J. Dougherty & J. T. Zilliox (Eds.), *International Group for the Psychology of Mathematics Education (joint meeting of PME and PME-NA)* (Vol. 4, pp. 245–249). Honolulu.

Speiser, B., & Walter, C. (2004). Placenticeras: Evolution of a task. *Journal of Mathematical Behavior, 23,* 259–269.

Speiser, B., Walter, C., & Glaze, T. (2005). Getting at the mathematics: Sara's Journal. *Educational Studies in Mathematics, 58,* 189–207.

Speiser, B., Walter, C., & Maher, C. (2003). Representing motion: an experiment in learning. *Journal of Mathematical Behavior, 22,* 1–35.

---

[17] Many students, as they began Placenticeras, also needed, in effect, to reinvent polar coordinates. Even the use of angles greater than 2π, which might seem very natural here, came as a surprise for most.

3

Foundational Reasoning Abilities that Promote Coherence in Students' Function Understanding

Michael Oehrtman, Marilyn Carlson and Patrick W. Thompson
Arizona State University

The concept of function is central to undergraduate mathematics, foundational to modern mathematics, and essential in related areas of the sciences. A strong understanding of the function concept is also essential for any student hoping to understand calculus — a critical course for the development of future scientists, engineers, and mathematicians.

Since 1888, there have been repeated calls for school curricula to place greater emphasis on functions (College Entrance Examination Board, 1959; Hamley, 1934; Hedrick, 1922; Klein, 1883; National Council of Teachers of Mathematics, 1934, 1989, 2000). Despite these and other calls, students continue to emerge from high school and freshman college courses with a weak understanding of this important concept (Carlson, 1998; Carlson, Jacobs, Coe, Larsen & Hsu, 2002; Cooney & Wilson, 1996; Monk, 1992; Monk & Nemirovsky, 1994; Thompson, 1994a). This impoverished understanding of a central concept of secondary and undergraduate mathematics likely results in many students discontinuing their study of mathematics. The primarily procedural orientation to using functions to solve specific problems is absent of meaning and coherence for students and has been observed to cause them frustration (Carlson, 1998). We advocate that instructional shifts that promote rich conceptions and powerful reasoning abilities may generate students' curiosity and interest in mathematics, and subsequently lead to increases in the number of students who continue their study of mathematics.

This article provides an overview of essential processes involved in knowing and learning the function concept. We have included discussions of the reasoning abilities involved in understanding and using functions, including the dynamic conceptualizations needed for understanding major concepts of calculus, parametric functions, functions of several variables, and differential equations. Our discussion also provides information about common conceptual obstacles to knowing and learning the function concept that students have been observed encountering. We make frequent use of examples to illustrate the 'ways of thinking' and major understandings that research suggests are essential for students' effective use of functions during problem solving, and that are needed for students' continued mathematics learning. We also provide some suggestions for promising approaches for developing a deep and coherent view of the concept of function.

Why Is The Function Concept So Important?

Studies have revealed that learning the function concept is complex, with many high performing undergraduates (e.g., students receiving course grades of A in calculus) possessing weak function understandings (Breidenbach, Dubinsky, Hawks, & Nichols, 1992; Carlson, 1998; Thompson, 1994a). We are beginning to understand that the conceptions and reasoning patterns needed for a strong and flexible understanding of functions are more complex than is typically assumed by designers of curriculum and instruction (Breidenbach et al., 1992; Carlson, 1998; Thompson, 1994a). Students who think about functions only in terms of symbolic manipulations and procedural techniques are unable to comprehend a more general mapping of a set of input values to a set of output values; they also lack the conceptual structures for modeling function relationships in which the function value (output variable) changes continuously in tandem with continuous changes in the input variable (Carlson, 1998; Monk & Nemirovsky, 1994; Thompson, 1994a). These reasoning abilities have been shown to be essential for representing and interpreting the changing nature of a wide array of function situations (Carlson, Jacobs, Coe, Larsen, & Hsu, 2002; Thompson, 1994a); they are also foundational for understanding major concepts in advanced mathematics (Carlson, Smith & Persson, 2003; Cottrill et al., 1996; Kaput, 1992; Rasmussen, 2000; Thompson, 1994a; Zandieh, 2000).

It is noteworthy that many of the reform calculus texts of the early 1990s, e.g., Ostabee & Zorn (1997), Harvard Calculus (Hughes-Hallet & Gleason, 1994), and C4L (Dubinsky, Schwingendorf, & Mathews, 1994), included a stronger conceptual orientation to learning functions. Such past curriculum development projects and the educational research literature are pointing the way for future curricular interventions to assist students in developing a robust function conception — a conception that begins with a view of function as an entity that accepts input and produces output, and progresses to a conception that enables reasoning about dynamic mathematical content and scientific contexts. Research suggests that the predominant approach to calculus instruction is not achieving the foundational understandings and problem solving behaviors that are needed for students' continued mathematical development and course taking (Carlson, 1998, 1999, 2003; Oehrtman, 2002). It is our view that the mathematics community is ready for a careful rethinking of the precalculus and calculus curriculum — one that is driven by past work of mathematicians, as well as the broad body of research on knowing and learning function and major concepts of calculus. It is also our view that if algebraic and procedural methods were more connected to conceptual learning, students would be better equipped to apply their algebraic techniques appropriately in solving novel problems and tasks.

Why is the Function Concept So Difficult for Students to Understand?

As students move through their school and undergraduate mathematics curricula, they are frequently asked to manipulate algebraic equations and compute answers to specific types of questions. This strong emphasis on procedures without accompanying activities to develop deep understanding of the concept has not been effective for building foundational function conceptions—ones that allow for meaningful interpretation and use of functions in various representational and novel settings. Even understanding functions in terms of input and output can be a major challenge for many students. As one example, 43% of A-students at the completion of college algebra attempted to find $f(x + a)$ by adding a onto the end of the expression for f rather than substituting $x + a$ into the function (Carlson, 1998). When probed to explain their thinking, they typically provided some memorized rule or procedure to support their answers. Clearly these students were not thinking of $x + a$ as a value of the function's argument at which the function is being evaluated. Another misconception is thinking that constant functions (e.g., $y = 5$) are not functions because they do not vary. Not viewing $y = 5$ as an example of a function can become problematic for students; as one example, when considering equilibrium solution functions for differential equations such as $dy/dt = 2y(y - 5)$ (Rasmussen, 2000). In one study, only 7% of A-students in a college algebra course could produce a correct example of a function all of whose output values are equal to each other, while 25% of A-students in second semester calculus produced $y = x$ as an example (Carlson, 1998). Even more problematic, students often view functions simply as two expressions separated by an equal sign (Thompson, 1994b). Such an impoverished understanding of functions is insufficient to serve as a base for a rich understanding of more advanced mathematics.

It is also common for developing students to have difficulty distinguishing between an algebraically defined function and an equation (Carlson, 1998). This is not surprising if one considers the various uses of the equal sign and the fact that many instructors refer to a formula as an equation. For the student, this ambiguous use of the word equation appears to cause difficulty for them in distinguishing between the use of the equal sign as a means of defining a relationship between two varying quantities and a statement of equality of two expressions. Our recent work has shown that students benefit from an explicit effort to help them distinguish between functions and equations. The first two authors have developed instructional interventions that promote students' thinking about an equation as a means of equating the output values of two functions, and the act of solving an equation as finding the input value(s) where the output values of these functions are equal.

Many students also tend to believe that all functions should be definable by a single algebraic formula. This focus often hinders flexible thinking about function situations and can lead to erroneous conclusions, such as thinking that all functions must always behave "nicely" in some sense (Breidenbach et al., 1992). For example, many students tend to argue that a piecewise defined function like

$$f(x) = \begin{cases} 0, & x \le 0; \\ e^{-1/x^2}, & x > 0, \end{cases}$$

is actually two separate functions or that a function such as Dirichlet's example,

$$g(x) = \begin{cases} 1, & x \text{ is rational,} \\ 0, & x \text{ is irrational,} \end{cases}$$

is not even a function at all because it "behaves badly." Similarly, many students have difficulty conceiving of different formulas representing the same function, as in the examples

$$f_1(n) = n^2 \quad \text{and} \quad f_2(n) = \sum_{k=1}^{n} [2k - 1],$$

which define the same function on the natural numbers, albeit through very different algebraic operations. Many students also tend to assume that functions are linear or quadratic in cases where this assumption is unwarranted, expecting for example, that any "u-shaped" graph is a parabola (Schwarz & Hershkowitz, 1999). These tendencies are perhaps not so surprising when we consider that functions are typically introduced in the school curriculum through specific function types. As such, a working definition in which functions are equated with formulas is perfectly reasonable, and even mirrors the historical understanding of mathematicians like Euler, Bernoulli, Lagrange, and d'Alembert (Kleiner, 1989; Sierpinska, 1992). It is not, however, the view that Euler himself, and subsequently the mathematics community in general, ultimately found to be most useful. The modern definition of function was motivated largely by debates between d'Alembert and Euler on the nature of a solution to the vibrating string differential equation (Luzin, 1998a, 1998b) and by Cauchy's and others' attempts to decide the conditions under which a limit of a sequence of continuous functions is a continuous function (Boyer, 1968; Lakatos, 1976). Thus, to use the modern definition of function in an introduction to the function concept is to present students with a solution to problems of which they cannot conceive. We recommend that school curricula and instruction include a greater focus on understanding ideas of covariation and multiple representations of covariation (e.g., using different coordinate systems), and that more opportunities be provided for students to experience diverse function types emphasizing multiple representations of the same functions. College curricula could then build on this foundation. This would promote a more flexible and robust view of functions — one that does not lead to inadvertently equating functions and formulas.

Another common difficulty for students is distinguishing between visual attributes of a physical situation and similar perceptual attributes of the graph of a function that models the situation. When dealing with functions as models of concrete situations, there are often topographical structures within the real-world setting itself (e.g., the curves of a racetrack, the elevation of a road traveling across hilly terrain, or the shape of a container being filled with liquid) that students see as being reflected in the function's graph. The considerable salience of these physical features often creates confusion, even for students with a strong understanding of function. Several types of errors can be traced to conflating the shape of a graph with visual attributes of the situation (Carlson, 1998; Monk, 1992; Monk & Nemirovsky, 1994). Consider the following problem:

The following diagram is the side-view of a person cycling up and over a hill. Draw a graph of speed vs. position along the path.

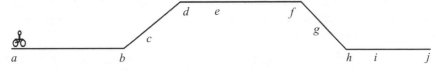

Figure 1. A problem in which students must distinguish between visual features of a situation and representational features of a graph. (From Monk, 1992).

In response to this problem, many students tend to copy features directly from the diagram into their graph (Monk, 1992). Correctly interpreting the situation is not a conceptually trivial task. A student must ignore the fact that the picture looks like a graph, think of how riding uphill (for example) affects the speed of the cyclist, then, while ignoring the shape of the hill in the picture, determine how to represent the result graphically.

When interpreting graphs such as the one in Figure 2, students often confuse velocity for position (Monk, 1992) since the curves are laid out spatially, and position refers to a spatial property. This confusion leads to erroneous claims such as: the two cars collide at $t = 1$ hour or that Car B is catching up to Car A between $t = .75$ hour and $t = 1$ hours. In one study, 88% of students who had earned an A in college algebra made such mistakes, as did 63% of students earning an A in second semester calculus, and 42% of students earning an A in their first graduate mathematics course (Carlson, 1998).

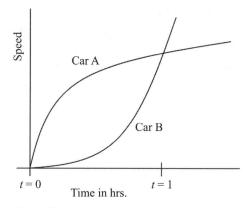

Figure 2. Students fail to interpret the function information conveyed by the graph.

In both these examples, students are thinking of the graph of a function as a picture of a physical situation rather than as a mapping from a set of input values to a set of output values. Developing an understanding of function in such real-world situations that model dynamic change is an important bridge for success in advanced mathematics.

Students' weak understandings of functions have also been observed in their inability to express function relationships using function notation. When asked to express s as a function of t, many high performing precalculus students did not know that their objective was to write a formula in the form of "$s = $ <some expression containing a t>." Some students have also exhibited weaknesses in knowing what each symbol in an algebraically defined function means. Even in the case of a simple function such as $f(x) = 3x$, many students are unaware that the parentheses serves as a marker for the input, that $f(x)$ represents the output values, that f is the name of the function, and that $3x$ specifies how the input x is mapped to the output $f(x)$. Such weak understandings and highly procedural orientations appear to contribute to students' inability to move fluidly between various function representations, such as the inability to construct a formula given a function situation described in words (Carlson, 1998).

Dynamic Conceptualizations Needed for Understanding and Using Functions

In our work to develop and validate the *Precalculus Concept Assessment Instrument*[1] (Carlson, Oehrtman, & Engelke, submitted), the first two authors found that students' ability to respond correctly to a diverse set of function-focused tasks is tightly linked to two types of dynamic reasoning abilities. First, as mentioned above, students must develop an understanding of functions as general processes that accept input and produce output. Second, they must be able to attend to both the changing value of the output and rate of its change as the independent variable is varied through an interval in the domain.

Understanding limits and continuity requires one to make judgments about the behavior of a function over intervals of arbitrarily small sizes. Conceptualizations based on "holes," "poles," and "jumps" as gestalt topographical features (corresponding to removable discontinuity, vertical asymptotes, and jump or one-sided discontinuity, respectively) can lead to misconceptions in more complex limiting situations, such as the definitions of the derivative and definite integral. For example, students can develop an intuitive understanding of the Fundamental Theorem of Calculus with which they explain why the derivative of the volume of a sphere ($v = \frac{4}{3}\pi r^3$) with respect to the length of its radius is its surface area. However, most of these students cannot explain why the same is not true for the volume of a cube ($v = s^3$) with respect to the length of its side (Oehrtman, 2002). In order to resolve such results conceptually, one must be able to coordinate images of changes in the "radius" with the corresponding changes in the volume over a range of small variations. For such variations, students must then be able to imagine the computation of rate of change of volume and see its connection to the computation of surface area.

[1] The Precalculus Concept Assessment Instrument is a 25-item multiple choice instrument for assessing students' understanding of the major aspects of the function concept that are foundational for success in beginning calculus. The answer choices include the correct answer and the common misconceptions that have been expressed by students in research studies (e.g., interviews that have probed students' thinking when providing specific responses to conceptually based tasks).

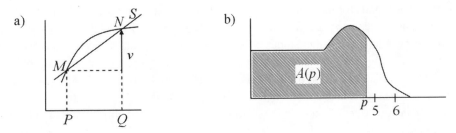

Figure 3. Foundational images for the definitions of a) the derivative and b) the definite integral.

To understand the relationship between average and instantaneous rates and the graphical analog between secant and tangent lines, a student must first conceive of an image as in Figure 3a (Monk, 1987). By employing *covariational reasoning* (e.g., coordinating an image of two varying quantities and attending to how they change in relation to each other), the student is able to transform the image and reason about values of various parameters as the configuration changes. Being able to answer questions that require covarying two quantities, i.e., "When point Q moves toward P, does the slope of S increase or decrease?", is significantly more difficult than being able to answer questions about the value of a function at a single point.

Analyzing the changing nature of an instantaneous rate also requires the ability to conceive of functional situations dynamically. Consider the following question based on a classic related rates problem in calculus:

From a vertical position against a wall, the bottom of a ladder is pulled away at a constant rate. Describe the speed of the top of the ladder as it slides down the wall.

Reasoning about this situation conceptually is difficult for calculus students even when they are given a physical model and scaffolding questions (Monk, 1992) and is similarly challenging for beginning graduate students in mathematics (Carlson, 1999). The standard calculus curriculum presents accumulation in terms of methods of determining static quantities — such as the area of an irregular region of the plane, or the distance traveled over a fixed amount of time given a changing velocity. Students imagine themselves approximating an area of a region. The area happens to be defined by a graph, but the task, to them, is essentially the same as approximating the area of a circle with triangles emanating from the circle's center. Equally important, however, is a dynamic view in which an accumulated total is changing through continual accruals (Kaput, 1994; Thompson, 1994a). For example, in a typical "area so far" function as in Figure 3b, this involves being able to mentally imagine the point p moving to the right by adding slices of area at a rate proportional to the height of the graph. This requires students to engage in covariational reasoning (Carlson, Smith & Persson, 2003) and is significantly more difficult for students than evaluating and even comparing areas at given points (Monk, 1987), for instead of asking them to conceptualize $m = \int_a^b f(x)\,dx$, we are asking them to conceptualize $F(x) = \int_a^x f(t)\,dt$.

In interviews with over 40 precalculus level students, the first two authors found that students who consistently verbalized a view of function as an entity that accepts input and produces output were able to reason effectively through a variety of function-related tasks. For example, these students, when asked to find $f(g(x))$ for specific values of x, given in either a table or words that defined the functions f and g, described a process of inputting a value into g, with the output of g becoming the input of f, and this output providing an output for the composite function $f \circ g$. However, students who provided an incorrect answer to this question were typically attempting to employ some memorized procedure. Without understanding, they invariably made a crucial mistake along the way such as interpreting $f(g(3))$ as meaning "the value of f when g is three," and by mistaking the *output* of g to be 3, arriving at $f(3)$ as an answer. As another example, when asked to solve the equation, $f(x) = 7$, given the graph of f, students who viewed this problem as a request to reverse the function process to determine the input associated with an output of f, had no difficulty responding to this task. Surprisingly, only 38% of 1196 students (550 college algebra and 646 precalculus) provided a correct answer at the completion of their courses. Those unable to provide a correct answer appeared to be applying memorized procedures – they did not speak about a function as a more general mapping of a set of input values to a set of output values. Their impoverished function view was also revealed by their inability to explain the meaning of function composition and function inverse in other settings and their inability to apply function composition to define an algebraic formula for a function situation (e.g., to define area as a function of time for a circle whose radius is expanding at 7 cm per second).

According to several studies, calculus students are slow to develop an ability to interpret varying *rates of change* over intervals of a function's domain. (Carlson, 1998; Kaput, 1992; Monk, 1992; Monk & Nemirovsky, 1994; Nemirovsky, 1996; Tall, 1992; Thompson, 1994a). According to Thompson (1994a), once students are adept at imagining expressions being evaluated continually as they "run rapidly" over a continuum, the groundwork has been laid for them to reflect on a set of possible inputs in relation to the set of corresponding outputs (p. 27). Such a *covariation view* of function has also been found to be essential for understanding central concepts of calculus (Cottrill et al., 1996; Kaput, 1992; Thompson, 1994b; Zandieh, 2000) and for reasoning about average and instantaneous rates of change, concavity, inflection points, and their real-world interpretations (Carlson, 1998; Monk, 1992).

The following section provides additional elaboration of these essential process and covariational understandings of functions.

The Action and Process Views of Functions — A More Formal Examination

Developmental research has provided insights about the reasoning patterns essential for success in collegiate mathematics. As we have previously discussed, investigations of students' function knowledge have consistently revealed that students' underlying conceptual view is important. Researchers have formalized these consistent observations by introducing terms for referencing specific types of conceptual views and their development. Specifically, students must move from what is called an *action view of functions* to what is called a *process view of functions*.

According to Dubinsky & Harel (1992),

> An *action* conception of function would involve the ability to plug numbers into an algebraic expression and calculate. It is a static conception in that the subject will tend to think about it one step at a time (e.g., one evaluation of an expression). A student whose function conception is limited to actions might be able to form the composition of two functions, defined by algebraic expressions, by replacing each occurrence of the variable in one expression by the other expression and then simplifying; however, the students would probably be unable to compose two functions that are defined by tables or graphs.   (pp. 85)

Students whose understanding is limited to an action view of function experience several difficulties. For example, an inability to interpret functions more broadly than by the computations involved in a specific formula results in misconceptions such as believing that a piecewise function is actually several distinct functions, or that different algorithms must produce different functions. More importantly, reasoning dynamically is difficult because it requires one to be able to disregard specific computations and to be able to imagine running through all input-output pairs simultaneously. This ability is not possible with an action view in which each individual computation must be explicitly performed or imagined. Furthermore, from an action view, input and output are not conceived except as a result of values considered one at a time, so the student cannot reason about a function acting on entire intervals. Thus, not only is the complex reasoning required for calculus out of reach for these students, but even simple tasks like conceiving of domain and range as entire sets of inputs and outputs is difficult.

Without a generalized view of inputs and outputs, students cannot think of a function as a process that may be reversed (to obtain the inverse of a function) but are limited to understanding only the related procedural tasks such as switching x and y and solving for y or reflecting the graph of f across the line $y = x$ (Figures 4a and 4b). This procedural approach to determining "an answer" has little or no real meaning for the student unless he or she also possesses an understanding as to why the procedure works. Students with an action view often think of a function's graph as being only a curve (or fixed object) in the plane; they do not view the graph as defining a general mapping of a set of input values to a set of output values. As such, the location of points, the vertical line test, and the "up and over" evaluation of functions on a graph are concepts only about the geometry of the graph, not about the more general mapping that is conveyed by the function, or the meaning that is conveyed by inverting the process for a function that represents a real-world situation. Similarly, with an action view, composition is generally seen simply as an algebra problem in which the task is to substitute one expression for every instance of x into some other expression. An understanding of why these procedures work or how they are related to composing or reversing functions is generally absent.

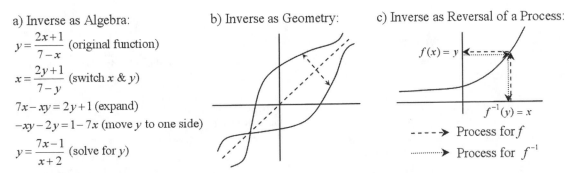

a) Inverse as Algebra:

$y = \dfrac{2x+1}{7-x}$ (original function)

$x = \dfrac{2y+1}{7-y}$ (switch x & y)

$7x - xy = 2y + 1$ (expand)

$-xy - 2y = 1 - 7x$ (move y to one side)

$y = \dfrac{7x-1}{x+2}$ (solve for y)

b) Inverse as Geometry:

c) Inverse as Reversal of a Process:

$f(x) = y$

$f^{-1}(y) = x$

----→ Process for f

.........→ Process for f^{-1}

Figure 4. Various conceptions of the inverse of a function as a) an algebra problem, b) a geometry problem, and c) the reversal of a process. The first two of these are common among students but, in isolation, do not facilitate flexible and powerful reasoning about functional situations.

Students who possess only the procedural orientations of Figures 4a and b, without understanding why the procedures work, are not likely to recognize even simple situations in which these procedures should be applied. Curriculum and instruction has not been broadly effective in building these connections in students' understanding. A recent study of over 2000 precalculus students at the end of the semester (Carlson et al., submitted) showed that only 17% of these students correctly determined the inverse of a function for a specific value, given a table of function values.

In contrast to the conceptual limitations of an action view, Dubinsky and Harel (1992) state:

A *process* conception of function involves a dynamic transformation of quantities according to some repeatable means that, given the same original quantity, will always produce the same transformed quantity. The subject is able to think about the transformation as a complete activity beginning with objects of some kind, doing something to these objects, and obtaining new objects as a result of what was done. When the subject has a process conception, he or she will be able, for example, to combine it with other processes, or even reverse it. Notions such as 1-1 or onto become more accessible as the student's process conception strengthens. (p. 85)

With such a process view, students are freed from having to imagine each individual operation for an algebraically defined function. For example, given the function on the real numbers defined by $f(x) = x^2 + 1$, the student can imagine a set of input values that are mapped to a set of output values by the defining expression for f. In contrast, students with an action view see the defining formula as a procedure for finding an answer for a specific value of x; they view the formula as a set of directions: square the value for x then add one to get the answer. A student with a process view can conceive of the entire process as happening to all values at once, and is able to conceptually run through a continuum of input values while attending to the resulting impact on output values. This is precisely the ability required for covariational reasoning introduced above and discussed more fully in the following section. In Table 1 we provide a characterization of "action views" of functions and their corresponding "process views."

Understanding even the basic idea of equality of two functions requires a generalization of the input-output process, (i.e., the ability to imagine the pairing of inputs to unique outputs without having to perform or even consider the means by which this is done). Students may then come to understand that any means of defining the same relation is the same function. That is, a function is not tied to specific computations or rules that define how to determine the output from a given input. For example, the rules

$$n \overset{f}{\mapsto} n^2 \quad \text{vs.} \quad n \overset{g}{\mapsto} \sum_{k=1}^{n} [2k-1]$$

look different; yet produce the same results (and thus define the same function) on the natural numbers.

Students with a process view are also better able to understand aspects of functions such as composition and inverses. They are consistently able to correctly answer conceptual and computational questions about composition in a variety of representations by coordinating output of one process as the input for a second process. Similarly, students conceiving of an inverse as reversing the function process so that the old outputs become the new inputs and vice-versa (Figure 4c), or by asking "What does one have to do to get back to the original values?" were able to correctly answer a wide variety of questions about inverse functions (Carlson et al., submitted).

Table 1. Action and Process Views of Functions

Action View	Process View
A function is tied to a specific rule, formula, or computation and requires the completion of specific computations and/or steps.	A function is a generalized input-output process that defines a mapping of a set of input values to a set of output values.
A student must perform or imagine *each action*.	A student can imagine the *entire process* without having to perform each action.
The "answer" depends on the formula.	The process is independent of the formula.
A student can only imagine a single value at a time as input or output (e.g., x stands for a specific number).	A student can imagine all input at once or "run through" a continuum of inputs. A function is a transformation of entire spaces.
Composition is substituting a formula or expression for x.	Composition is a *coordination* of two input-output processes; input is processed by one function and its output is processed by a second function.
Inverse is about algebra (switch y and x then solve) or geometry (reflect across $y = x$).	Inverse is the *reversal of a process* that defines a mapping from a set of output values to a set of input values.
Finding domain and range is conceived at most as an algebra problem (e.g., the denominator cannot be zero, and the radicand cannot be negative).	Domain and range are produced by operating and reflecting on the set of all possible inputs and outputs.
Functions are conceived as static.	Functions are conceived as dynamic.
A function's graph is a geometric figure	A function's graph defines a specific mapping of a set of input values to a set of output values.

A process view of function is crucial to understanding the main conceptual strands of calculus (Breidenbach et al., 1992; Monk, 1987; Thompson, 1994a). For example, the ability to coordinate function inputs and outputs dynamically is an essential reasoning ability for limits, derivatives and definite integrals. In order to understand the definition of a limit, a student must coordinate an entire interval of output values, imagine reversing the function process, and determine the corresponding region of input values. The action of a function on these values must be considered simultaneously since another process (one of reducing the size of the neighborhood in the range) must be applied while coordinating the results. Unfortunately, most pre-calculus students do not develop beyond an action view, and even strong calculus students have a poorly developed process view that often leads only to computational proficiency (Carlson, 1998). With intentional instruction, however, students can develop a more robust process view of function (Carlson et al., submitted; Dubinsky, 1991; Sfard, 1991).

Certainly not every aspect of an action view of functions is detrimental to students' understanding, just as the acquisition of a process view does not ensure success with all functional reasoning. However, a process view of functions is crucial for developing rich conceptual understandings of the content in an introductory calculus course. The promotion of the more general 'ways of thinking' that we have advocated should result in producing curricula that are more effective for promoting conceptual structures for students' continued mathematical development.

Fostering a Process View of Functions

We offer the following general recommendations for promoting students' development of a process view of functions:

Ask students to explain basic function facts in terms of input and output. For example, ask students to determine whether $(f \circ g)^{-1}$ is $f^{-1} \circ g^{-1}$ or $g^{-1} \circ f^{-1}$ and explain their reasoning. In the process, most will initially struggle to decide which of the diagrams in Figure 5 represents $(f \circ g)$. Determining both the correct diagram and the correct formula for the inverse encourages students to think in terms of a general input-output process. As another example, students typically learn to carry out rote procedures when asked to solve equations such as $f(x) = 6$ for some specified function f; but asking them to find the input value(s) for which the function's output is 6

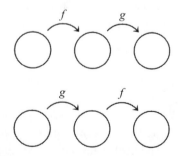

Figure 5. Which diagram represents $f \circ g$? What is its inverse?

(both algebraically and graphically) promotes an understanding that solving an equation can be seen as the reversal of a function process. As yet another example, students typically memorize (without understanding) that the graph of a function g given by $g(x) = f(x + a)$ is shifted to the left of the graph of f, but asking them to discover or interpret this statement as meaning "the output of g at every x is the same as the output of f at every $x + a$" will give them a more powerful way to understand this idea and reinforce a process view of functions. Ask students to determine the domain and range of functions based on the problem context, and relate this to answers (possibly different) derived from algebraic constraints alone. Other possibilities include asking students to explain why composition is associative, to develop the definition of a periodic function on their own, or to graph and explain the results of simple function arithmetic.

Ask about the behavior of functions on entire intervals in addition to single points. Focusing on the image of a function applied to an infinite set also encourages students to think in terms of a general process. Students should be asked to coordinate such judgments with basic compositions and inverses, asking, for example, for the length of an interval after being transformed by two linear functions. Similarly, ask students to find preimages of intervals as in the definition of limit or continuity and to reverse the process of a function even if it is not invertible (e.g., find the preimage of 1 under $f(x) = x^2$).

Ask students to make and compare judgments about functions across multiple representations. Such questions should include multiple algebraic representations to reinforce the independence from a formula as well as the standard representations of graphs, tables, and verbal descriptions. Students should make such determinations; then compare the results for consistency, justifying or discovering why they are the same. For example, asking how the various techniques of inverting a function are related reinforces seeing a reflection across the line $y = x$ as switching the roles of independent and dependent variable, of input and output. Also helpful are predictions about how a graph will look based on how a real-world quantity is changing across its domain, requiring simultaneous attention to multiple input-output pairs and translation between representations. As an example, when asking students to solve standard problems such as 'define the area as a function of time for a circle whose radius is expanding at 7 cm per second,' ask students to begin by constructing a dynamic image of the situation via a computer program or by drawing a picture; then ask them to label using algebraic symbols the varying quantities in the situation. After recognizing the 7cm change in the radius per second can be represented algebraically by labeling the varying length of the radius with the formula $r = 7s$ on the picture, prompt them to determine how to relate area to time in seconds. You could also ask your students to graph the resulting function, $A = \pi(7s)^2$, and determine the average rate of change of the circle as s changes from 3 to 4; then as s changes from 4 to 5; then as s changes from 5 to 6; then as s changes from 6 to 7; then ask them to explain in the context of the growing circle what these average rates imply about how the area of the circle is growing over the time interval from $s = 3$ to $s = 7$. (For additional discussion of the complexities involved in acquiring a flexible view of variable as an unknown, varying quantity, placeholder, etc., see the Jacobs and Trigueros chapter in this volume).

Building on the Process View of Function: Applying Covariational Reasoning

As students begin to explore dynamic function relationships such as how speed varies with time or how the height of water in a bottle varies with volume, they will need to begin considering *how* one variable (often the dependent variable) changes while imagining changes in the other (the independent variable). When coordinating such changes, one must be able to represent and interpret important features in the shape of a graph of a dynamic function event. As a very simple example, a student who has a strong process view of function might see the algebraic formula $A(s) = s^2$ as a means of determining the area of a square for a set of possible input values. She would be viewing the function as an entity that accepts any side length s as input, to produce an output value for the area A. She would have no difficulty determining the side of the square for given values of the area (reversing the process) or with using any particular representation of this function situation (algebraic, tabular, graphical). In this context, the student may begin to notice that as the value of s increases, the value of A increases. By exploring numerical patterns and/or constructing a graph of this function, the student may also observe that as one steps through positive integer values for s, the amount of increase of A is getting larger and larger. He or she may also notice that as s increases continuously, the area is growing faster and faster. By constructing a graph to represent this function relationship, the student may observe that, when s is greater than 0, the slope of the graph gets steeper as s increases. When asked to explain why the graph gets steeper, the student would also be able to unpack the notion of slope (steepness) by describing the relative change of the input (side) and output (area), while stepping through values of s.

The Covariation Framework

Our work to characterize the thinking involved in reasoning flexibly about dynamically changing events has led to our decomposing *covariational reasoning* into five distinct mental actions (Carlson et al., 2002). This decomposition has been useful for guiding the development of curricular modules to promote covariational reasoning in students. These five categories of mental actions (Table 2) describe the reasoning abilities involved in meaningful representation and interpretation of a graphical model of a dynamic function situation. In our work, the first two authors have developed beginning calculus modules that include tasks and prompts to promote these ways of thinking in students. After three iterations of refining these modules (based on our analysis of data of students' reasoning when working through these modules), we are observing dramatic gains in beginning calculus students' *covariational reasoning* abilities over the course of one semester.

The initial image described in the framework for covariational reasoning is one of two variables changing simultaneously. This loose association undergoes multiple refinements as the student moves toward an image of increasing and decreasing rate over the entire domain of the function (Table 2).

Table 2. Mental Actions of the Covariation Framework

Mental Action	*Description of Mental Action*	*Behaviors*
Mental Action 1 (MA1)	Coordinating the dependence of one variable on another variable	• Labeling the axes with verbal indications of coordinating the two variables (e.g., y changes with changes in x)
Mental Action 2 (MA2)	Coordinating the direction of change of one variable with changes in the other variable	• Constructing a monotonic straight line • Verbalizing an awareness of the direction of change of the output while considering changes in the input
Mental Action 3 (MA3)	Coordinating the amount of change of one variable with changes in the other variable	• Plotting points/constructing secant lines • Verbalizing an awareness of the amount of change of the output while considering changes in the input
Mental Action 4 (MA4)	Coordinating the average rate of change of the function with uniform increments of change in the input variable	• Constructing secant lines for contiguous intervals in the domain • Verbalizing an awareness of the rate of change of the output (with respect to the input) while considering uniform increments of the input
Mental Action 5 (MA5)	Coordinating the instantaneous rate of change of the function with continuous changes in the independent variable for the entire domain of the function	• Constructing a smooth curve with clear indications of concavity changes • Verbalizing an awareness of the instantaneous changes in the rate of change for the entire domain of the function (direction of concavities and inflection points are correct)

In our work to study and promote students' emerging covariational reasoning abilities, we have found that the ability to move flexibly between mental actions 3, 4 and 5 is not trivial for students. We have also observed that many precalculus level students only employ Mental Action 1 and Mental Action 2 when asked to construct the graph of a dynamic function situation.

When prompting students to construct the graph of the height as a function of the amount of water in a bottle (Figure 6), the first two authors found that many precalculus students appropriately labeled the axes (MA1) and then constructed an increasing straight line (MA2). When prompted to explain their reasoning, they frequently indicated that "as more water is put into the bottle, the height of the water rises (MA2)." These students were clearly not attending to the amount of change of the height of the water level or the rate at which the water was rising.

We have observed that calculus students frequently provided a strictly concave up graph in response to this question (Carlson, 1998; Carlson et al., 2002). When probed to explain their reasoning, a common type of justification

Imagine this bottle filling with water. Sketch a graph of the height as a function of the amount of water that's in the bottle.

Figure 6. The bottle problem.

was, "as the water is poured in it gets higher and higher on the bottle (MA2)." In contrast, other students who were starting to be able to construct an appropriate graph began coordinating the magnitude of changes in the height with changes in the volume (MA3). This is exemplified in the strategy of imagining pouring in one cup of water at a time and coordinating the resulting change in height based on how "spread out" that layer of water is.

Other students have demonstrated the ability to speak about the average rate of change locally for a specific interval of a function's domain (MA4) but were unable to explain how the rate changes over the domain of the function. Even when calculus students produced a graph that was correct, they commonly had difficulty explaining what was conveyed by the inflection point and why the graph was "smooth" (in particular, why it is C^1 rather than piecewise linear). Students frequently exhibited behaviors that gave the appearance of engaging in Mental Action 5 (e.g., construction of a smooth curve with the correct shape), however when prompted to explain their reasoning, they expressed that they had relied on memorized facts to guide their constructions. They were relying on apparent facts such as "faster means steeper" and "slower means less steep," but they were unable to explain why this was true.

Engaging Covariational Reasoning Through Analysis Of Function Situations

We offer the following suggestions for strengthening students' covariational reasoning abilities:

Generally, ask questions associated with each of the mental actions. For orientation to any problem, MA1 skills and basic function awareness can be addressed by asking what values are changing and what variable(s) influence the quantity of interest (i.e., the dependent variable). Is there a single variable that determines that quantity's values? How are the variables related and in what representations can this relationship be expressed? For MA2, ask whether a function increases or decreases if the independent variable is increased (or decreased). Expect students to make such judgments from multiple representations. At an MA3 level, ask students to make judgments about amounts of change in the function for constant increments of the independent variable. For a dynamic situation, have students draw diagrams representing changes from one output variable to the other for each of two nearby intervals of the input variable, and represent these changes pictorially and algebraically. Ask students to interpret these representations in terms of rate of change in the problem context. To foster MA4 thinking, have students compute several average rates using various representations and find various interpretations for these values and explicitly discuss the meaning of units such as meters per second and even non-temporal rates such as square inches per inch or degrees Kelvin per meter. For MA5, ask students to anticipate second derivative information based on the problem context, e.g., whether the force of gravity between two celestial objects will increase at an increasing rate or at a decreasing rate with respect to a decreasing distance between them. Ask students to describe the rate of change of a function event as the independent variable continuously and dynamically varies through the domain. Ask where inflection points are, what events they correspond to in real-world situations, and how these points are interpreted in terms of changing rate of change.

Ask for clarification of rate of change information in various contexts and representations. Expect students to explain statements about rates in real-world contexts from algebraic or graphical information, e.g., why does a steeper graph mean the quantity represented by the function is increasing faster? Push beyond students' initial, simplified statements such as "the rate of change of position" that ignore the role of time. Require explication of *both* variables involved and relationships about changes in both quantities. Finally, a student may be able to make statements indicative of a Mental Action 5 by attending only to the geometry of the curve and associated phrases such as "increasing at a decreasing rate." Ask them to unpack such statements in terms of the underlying mental actions, in this case perhaps prompts that reveal if they understand what they mean by the phrase "increasing at a decreasing rate." Unpacking what may be pseudo-conceptual knowledge (that is, knowledge that has been memorized and is not based on an underlying conceptual structure and understanding) can be achieved by posing pointed questions that prompt students to reveal their underlying conceptions (e.g., why is the graph concave up or why is the curve "smooth" rather than piecewise linear?) Such questions typically reveal if the student is merely spouting a memorized rule or fact, or if the statement is supported by an understanding of why the rule or statement is true.

Extending Ideas of Covariation To Higher Dimensions

The idea of covariation is fundamentally that of parametric functions. As one imagines scanning through values of one variable and keeping track of values of another variable, one is essentially imagining the parametric function $(x, y) = (t, f(t))$. Once students have developed the ability to reason covariationally, it is a natural (but not small) step to reason about functions defined parametrically by $(x, y) = (f(t), g(t))$. For example, the graph in Figure 7 is $(x, y) = (\sin 10t, \cos 20t)$, $0 \le t \le 1$. Students can conceptualize this graph by generating the graphs of $f(t) = \sin 10t$ and of $g(t) = \cos 20t$ separately and then tracking the values of $x = \sin 10t$ and $y = \cos 20t$ as t varies. This same technique can be used to conceptualize graphs of phase space, $(x, y) = (f(u), f'(u))$, in differential equations.

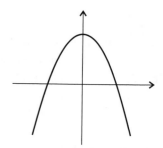

Figure 7. Graph of $(x, y) =$ $(\sin 10t, \cos 20t)$, $0 \le t \le 1$.

Covariation can also support thinking about curves in space. To continue the previous example, imagine that t in Figure 7 is actually an axis, coming straight at your eyes. If you now keep track of t as well as x and y, you get a sense that each point on the graph in Figure 7 is actually some distance toward you from the page. If you rotate your position relative to the graph so that you can see an axis that is perpendicular to x-y, then you have engendered an image like that in Figure 8.

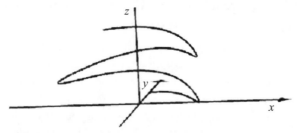

Figure 8. Graph of $(x, y, z) = (\sin 10t, \cos 20t, t)$, $0 \le t \le 1$. As t varies, points in Figure 7 with coordinates $(\sin 10t, \cos 20t)$ are projected t units perpendicularly from the x-y plane.

Finally, ideas of covariation can help students visualize functions of more than one variable. For example, we can envision the behavior of $z = f(x, y)$ in a multitude of ways, such as thinking of y (or x) as a parameter. The graph of f, then, can be visualized as being generated by a family of functions $z = f_y(x)$ as y varies. Figure 9 shows three successive graphs corresponding to $z = x^3 + yx$ at $y = -2$, at $y = -1$, and at $y = 1$, where in each graph, x varies from -2 to 2. Figure 10 shows the surface swept out by $f_y(x) = x^3 + yx$ as y varies continuously from -2 to 2, thus generating the graph of $f(x, y) = x^3 + yx$, $-2 \le x \le 2$, $-2 \le y \le 2$.

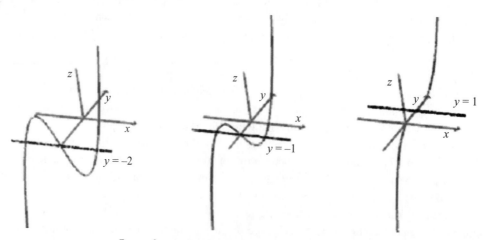

Figure 9. Graphs of $z = f_y(x)$ for $y = -2$, $y = -1$, and $y = 1$.

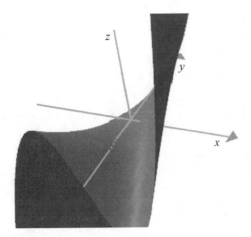

Figure 10. Graphs of $z = f(x, y)$, $-2 \leq x \leq 2, -2 \leq y \leq 2$.

Concluding Remarks

A mature function understanding that is revealed by students' using functions fluidly, flexibly, and powerfully is typically associated with strong conceptual underpinnings. Promoting this conceptual structure in students' understanding may be achieved through both curriculum and instruction including tasks, prompts, and projects that promote and assess the development of these "ways of thinking" in students. We advocate for greater emphasis on enculturating students into using the language of function in order to develop facility in speaking about functions as entities that accept input and produce output, a more conceptual orientation to teaching function inverse and composition, the inclusion of tasks requiring simultaneous judgments about entire intervals of input or output values, and the development of students' ability to mentally run through a continuum of input values while imagining the changes in the output values, with explicit efforts to also promote, at the developmentally appropriate time, the covariational reasoning abilities described in this chapter. Our work also suggests that students would benefit from explicit efforts to promote their understanding of function notation. Additionally, we call for evaluations of students' mathematical development and readiness to include assessments that measure the foundational reasoning abilities needed for a robust function conception. As one example, when teaching calculus I, we begin the semester by assessing students' function understanding. This provides useful knowledge for our selecting and creating tasks to address their misconceptions and promote the reasoning abilities and understandings that we have described in this chapter. We have found that the time spent at the beginning of calculus to strengthen students' function conceptions is crucial for their understanding the major ideas of calculus (For further reading on how the covariation perspective to teaching functions influences students' understanding of ideas of calculus, see Thompson and Silverman, this volume). (Note that the precalculus concept assessment instrument (PCA) and calculus modules mentioned in this chapter can be acquired by contacting the second author at marilyn.carlson@asu.edu.) You may also find it useful to assess your students' thinking and reasoning on tasks that we have discussed in this chapter. Lastly, we advocate that you regularly pose questions and engage your students in tasks that will allow you to gauge their development in understanding major ideas of your courses. This instructional perspective will require that you have clarity on the mathematical thinking, understandings, and problem solving behaviors that your students need to acquire to advance their mathematical development. It also sets up a challenge for you to scaffold your instruction based on what your students know and understand, but should in turn lead to greater success for your students and a more rewarding instructional experience for you.

References

Boyer, C. (1968), *A History of Mathematics*, New York: J. Wiley & Sons, Inc.

Breidenbach, D., Dubinsky, E., Hawks, J., & Nichols, D. (1992). Development of the process conception of function. *Educational Studies in Mathematics, 23*, 247–285.

Carlson, M., Jacobs, S., Coe, E., Larsen, S., & Hsu, E. (2002). Applying covariational reasoning while modeling dynamic events: A framework and a study. *Journal for Research in Mathematics Education, 33*(5), 352–378.

Carlson, M., Oehrtman, M., & Engelke, N. (submitted). The development of an instrument to assess precalculus students' conceptual understandings: The Precalculus Concept Assessment Instrument.

Carlson, M. P. (1998). A cross-sectional investigation of the development of the function concept. In A. H. Schoenfeld, J. Kaput, & E. Dubinsky (Eds.), *CBMS Issues in Mathematics Education: Research in Collegiate Mathematics Education III, 7*, 114–162.

————— (1999). The mathematical behavior of six successful mathematics graduate students: Influences leading to mathematical success. *Educational Studies in Mathematics, 40*, 237–258.

Carlson, M. P., Smith, N., & Persson, J. (2003). *Developing and connecting calculus students' notions of rate of change and accumulation: The fundamental theorem of calculus.* Paper presented at the 2003 Joint Meeting of PME and PME-NA, Honolulu, HI.

College Entrance Examination Board. (1959). *Program for college preparatory mathematics [Report].* New York: Commission on Mathematics.

Cooney, T., & Wilson, M. (1996). Teachers' thinking about functions: Historical and research perspectives. In T. A. Romberg & E. Fennema (Eds.), *Integrating research on the graphical representation of functions* (pp. 131–158). Hillside, NJ: Lawrence Erlbaum.

Cottrill, J., Dubinsky, E., Nichols, D., Schwingendorf, K., Thomas, K., & Vidakovic, D. (1996). Understanding the limit concept: Beginning with a coordinated process schema. *Journal of Mathematical Behavior, 15*(2), 167–192.

Dubinsky, E. (1991). Reflective abstraction in advanced mathematical thinking. In D. Tall (Ed.), *Advanced mathematical thinking* (pp. 95–126). Boston: Kluwer.

Dubinsky, E., & Harel, G. (1992). The nature of the process conception of function. In G Harel & E. Dubinsky (Eds.), The concept of function: Aspects of epistemology and pedagogy. *MAA Notes, 25*, 85–106.

Dubinsky, E., Schwingendorf, K., & Mathews, D. M. (1994). *Calculus, concepts & computers.* New York: McGraw Hill.

Hamley, H. R. (1934). Relational and functional thinking in mathematics. In *The National Council of Teachers of Mathematics yearbook* (pp. 48-84). New York: Teachers College, Columbia University.

Hedrick, E. R. (1922). Functionality in the mathematical instruction in schools and colleges. *The Mathematics Teacher, 15*, 191–207.

Hughes-Hallet, D., & Gleason, A. M. (1994). *Calculus.* New York: Wiley.

Kaput, J. (1992). Patterns in students' formalization of quantitative patterns. In G. Harel & E. Dubinsky (Eds.), The concept of function: Aspects of epistemology and pedagogy. *MAA Notes, 25.*

————— (1994). Democratizing access to calculus: New routes to old roots. In A. H. Schoenfeld (Ed.), *Mathematics and Cognitive Science.* Washington, D. C.: Mathematical Association of America, 77–156.

Klein, F. (1883). Ueber den allgemeinen Functionbegriff und dessen Darstellung durch eine willknerliche Curve. *Mathematischen Annalen, XXII*, 249.

Kleiner, I. (1989). Evolution of the function concept: A brief survey. *The College Mathematics Journal, 20* (4), 282–300.

Lakatos (1976). *Proofs and Refutations.* Cambridge: Cambridge University Press.

Luzin, N. (1998a). The Evolution of... Function: Part I. *The American Mathematical Monthly, 105*(1), 59-67.

————— (1998b). The Evolution of... Function: Part II. *The American Mathematical Monthly, 105*(3), 263-270.

Monk, S. (1987). *Students' understanding of functions in calculus courses.* (Unpublished manuscript, University of Washington, Seattle, WA).

Monk, S. (1992). Students' understanding of a function given by a physical model. In G. Harel & E. Dubinsky (Eds.), The concept of function: Aspects of epistemology and pedagogy, *MAA Notes, 25*, 175–193).

Monk, S., & Nemirovsky, R. (1994). The case of Dan: Student construction of a functional situation through visual attributes. *CBMS Issues in Mathematics Education: Research in Collegiate Mathematics Education 1994*(4), 139–168.

National Council of Teachers of Mathematics. (1934). *Relational and Function Thinking in Mathematics, 9*. Reston, VA: National Council of Teachers of Mathematics.

———— (1989). *Curriculum and evaluation standards for school mathematics*. Reston: National Council of Teachers of Mathematics.

———— (2000). *Principles and standards for school mathematics*. Reston: National Council of Teachers of Mathematics.

Nemirovsky, R. (1996). A functional approach to algebra: Two issues that emerge. In N. Dedrarg, C. Kieran & L. Lee (Eds.), *Approaches to algebra: Perspectives for research and teaching* (pp. 295-313). Boston: Kluwer Academic Publishers.

Oehrtman, M. (2002). *Collapsing dimensions, physical limitation, and other student metaphors for limit concepts: An instrumentalist investigation into calculus students' spontaneous reasoning* (Doctoral dissertation, University of Texas, Austin, TX).

Ostebee, A., & Zorn, P. (1997). *Calculus from graphical, numerical, and symbolic points of view*. New York: Houghton-Mifflin.

Rasmussen, C. L. (2000). New directions in differential equations: A framework for interpreting students' understandings and difficulties. *Journal of Mathematical Behavior, 20*, 55–87.

Schwarz, B. B., & Hershkowitz, R. (1999). Prototypes: Brakes or levers in learning the function concept? The role of computer tools. *Journal for Research in Mathematics Education, 30*(4), 362–389.

Sierpinska, A. (1992). On Understanding the Notion of Function, In G. Harel & E. Dubinsky (Eds.), *The Concept of Function: Aspects of Epistemology and Pedagogy*. MAA Notes, Volume 25. Washington, DC: Mathematical Association of America.

Sfard, A. (1991). On the dual nature of mathematical conceptions: Reflections on process and objects as different sides of the same coin. *Educational Studies in Mathematics, 22*, 1–36.

Tall, D. (1992). The transition to advanced mathematical thinking: Functions, limits, infinity and proof. In D. A. Grouws (Ed.), *Handbook of research on mathematics teaching and learning* (pp. 495–511). New York: MacMillan.

Thompson, P. W. (1994a). Images of rate and operational understanding of the fundamental theorem of calculus. *Educational Studies in Mathematics, 26*, 229–274.

———— (1994b). Students, functions, and the undergraduate curriculum. In A. H. Schoenfeld & J. J. Kaput (Eds.), *CBMS Issues in Mathematics Education: Research in Collegiate Mathematics Education, 4*, 21–44.

Zandieh, M. J. (2000). A theoretical framework for analyzing student understanding of the concept of derivative. In E. Dubinsky, A. Schoenfeld, & J. Kaput (Eds.), *CBMS Issues in Mathematics: Research in Collegiate Mathematics Education, IV*(8), 103–127.

Acknowledgement Research reported in this paper was supported by National Science Foundation Grants No. EHR-0412537 and EHR-0353470. Any conclusions or recommendations stated here are those of the authors and do not necessarily reflect official positions of NSF.

4

The Concept of Accumulation in Calculus

Patrick W. Thompson, *Arizona State University*
Jason Silverman, *Drexel University*

The concept of accumulation is central to the idea of integration, and therefore is at the core of understanding many ideas and applications in calculus. On one hand, the idea of accumulation is trivial. You accumulate a quantity by getting more of it. We accumulate injuries as we exercise. We accumulate junk as we grow older. We accumulate wealth by gaining more of it. There are some details to consider, such as whether it makes sense to think of accumulating a negative amount of a quantity, but the main idea is straightforward.

On the other hand, the idea of accumulation is anything but straightforward. First, students find it hard to think of something accumulating when they cannot conceptualize the "bits" that accumulate. To understand the idea of accomplished work, for example, as accruing incrementally means that one must think of each momentary total amount of work as the sum of past increments, and of every additional incremental bit of work as being composed of a force applied through a distance. Second, the mathematical idea of an accumulation function, represented as $F(x) = \int_a^x f(t)dt$, involves so many moving parts that it is understandable that students have difficulty understanding and employing it.

Readers already sophisticated in reasoning about accumulations may find it surprising that many students are challenged to think mathematically about them. The ways in which it is difficult, though, are instructive for a larger set of issues in calculus. As such, our intention in this chapter is to:

(1) Explicate the complex composition of a well-structured understanding of accumulation functions,
(2) Illustrate students' difficulties in understanding accumulation mathematically,
(3) Point out promising approaches in helping students conceptualize accumulation functions, and
(4) Place students' understandings of accumulation functions within the calculus as a whole.

Composition of a Well-Structured Understanding of Accumulation Functions

Accumulation functions can be represented generally by $\int_a^x f(t)dt$.[1] It is worthwhile to unpack the meanings behind this formula in order to see all that it entails. We will do this in two passes, first without addressing ideas of Riemann sums, then addressing them.

Accumulation Functions

For sake of illustration, let $f(x) = 2e^{-\cos(x)} - 2.5$. To understand an accumulation function involving f, students must have a *process conception* of the formula $2e^{-\cos(x)} - 2.5$.[2] This means that they must hold the perspective that though

---

[1] For purposes of this chapter, we will speak only of Riemann integrals over an interval.

[2] Briedenbach et al., Dubinsky and Harel, and Carlson speak of a *process conception of function*. We also speak of a *process conception of formulae*. To us, for students to have a process conception of "$f(x) = \ldots$" requires that they have a process conception of the right-hand side. A process

it might require actual effort to calculate any particular value of this formula, in the end it represents a number, and the number it represents depends only on the value of x (Breidenbach, Dubinsky, Hawks, & Nichols, 1992; Dubinsky & Harel, 1992; see also Oehrtman, Carlson, & Thompson, this volume). To have a process conception of a function's defining formula implies that one has what Gray and Tall (1994) call a *proceptual* understanding of what the formula represents. One has in mind a well-structured set of procedures for evaluating the formula together with ability and inclination to see the formula as "self-evaluating" (P. W. Thompson, 1994b), meaning that one sees it as evaluating itself instantaneously for any number.[3]

To understand an accumulation function, students also need a covariational understanding of the relationship between x and f (Carlson, Jacobs, Coe, Larsen, & Hsu, 2002; Saldanha & Thompson, 1998; P. W. Thompson, 1994b, 1994c). In the case of the current example, this means understanding that as the value of x varies, the value of $2e^{-\cos(x)} - 2.5$ varies accordingly. It also entails creating an image of *how* the value of $2e^{-\cos(x)} - 2.5$ varies as the value of x varies, thus generating the relationship expressed by the graph in Figure 1.

Figure 1. Graph that depicts values of x and $2e^{-\cos(x)} - 2.5$ varying simultaneously.

Students who have mastered the process conceptions of formulae and covariational conceptions of function must then coordinate a third aspect with them—imagining accumulation and its quantification. Students must coordinate the value of x as it varies from some starting point, the value of $2e^{-\cos(x)} - 2.5$ as it varies accordingly, and, in addition, imagine the bounded area accumulating (Figure 2) as x and $2e^{-\cos(x)} - 2.5$ vary. Moreover, the student must attend to *how* these values are varying in tandem.[4]

Figure 2. Area accumulates as x varies.

To conceive of an *accumulation function* defined in x is to imagine a total accumulated area for each value of x. This introduces a third dimension into the conceptualization of accumulation functions—students must coordinate three values simultaneously:

$$x, \quad 2e^{-\cos(x)} - 2.5, \quad \text{and} \quad \int_a^x \left(2e^{-\cos(t)} - 2.5\right) dt \,.$$

Figure 3 illustrates this coordination graphically. Points on the space curve are ordered triplets

$$\left(x, 2e^{-\cos(x)} - 2.5, \int_a^x \left(2e^{-\cos(t)} - 2.5\right) dt \right).$$

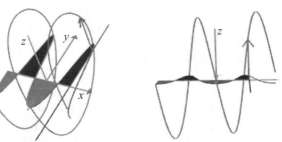

While we do not claim that students must conceptualize the space curve in Figure 3 in order to understand the mathematical idea of accumulation, we do claim that expecting them to understand accumulations *as functions* is tantamount to expecting them to understand a space curve.[5]

Figure 3. Coordination of x, $2e^{-\cos(x)} - 2.5$, and $\int_a^x \left(2e^{-\cos(t)} - 2.5\right) dt$, and its projection into the x-z plane.

Before covering these issues again from the perspective of Riemann sums, we would like to point out a notational issue. The role of t in the expression $\int_a^x f(t)dt$ often is a mystery to students. When textbooks address it at all they treat it as a "dummy variable," or a variable that will disappear when the expression is evaluated (Weisstein, 2006). We propose that t actually serves a conceptual role in making sense of the expression $\int_a^x f(t)dt$. In Figure 2, f cannot be thought of as having the same argument as does $\int_a^x f(t)dt$. In a sense, the graph of f must "pre-exist" when imagining

---

conception of a function entails more than does a process conception of a formula. Our intent is to avoid adopting an "all or none" stance toward what it means to understand a function.

[3] We will give specific examples later in this article of students having and not having a process conception of an integral.

[4] A simpler example involves coordinating changes in x with changes in $3x^2$ while simultaneously imaging how the bounded area under the graph of $3x^2$ is accumulating.

[5] A colleague thought we are implying that accumulation functions should therefore not be taught, since they are so sophisticated. On the contrary, we argue that they *should* be taught, but they should be taught with full awareness of what it means to understand them.

the accumulation of area between it and an axis. Thinking of t as already having varied through f's domain prior to x varying through a subset of f's domain then allows one to think of $\int_a^x f(t)dt$ as representing the accumulation of area within an already bounded region.

Riemann Sums

Calculus texts typically offer Riemann sums as a way to approximate areas bounded by a curve. The question of how a bounded area itself can represent a quantity other than area requires us to examine ways to understand Riemann sums and how they arise.

If f is a function whose values provide measures of a quantity, and x also is a measure of a quantity, then $f(c)\Delta x$, where $c \in [x, x + \Delta x]$, is a measure of a derived quantity. The simplest case is when $f(x)$ and x are both measures of length. Then $f(c)\Delta x$ is a measure of area. If $f(x)$ is a measure of speed and x is a measure of time, then $f(c)\Delta x$ is a measure of distance. If $f(x)$ is a measure of force and x is a measure of distance, then $f(c)\Delta x$ is a measure of work. If $f(x)$ is a measure of cross-sectional area and x is a measure of height, then $f(c)\Delta x$ is a measure of volume. If $f(x)$ is a measure of electric current and x is a measure of time, then $f(c)\Delta x$ is a measure of electric charge. A Riemann sum, then, made by a sum of incremental bits each of which is made multiplicatively of two quantities, represents a total amount of the derived quantity whose bits are defined by $f(c)\Delta x$. Therefore, for students to see "area under a curve" as representing a quantity other than area, it is imperative that they conceive of the quantities being accumulated as being created by accruing incremental bits that are formed multiplicatively.

Our account of how "area under a curve" comes to represent quantities other than area clearly holds an undertone of thinking with infinitesimals. Though a large portion of 19th-century activity in the foundations of mathematics was motivated by the desire to eliminate infinitesimals, we see no way around explicitly supporting students' reasoning about them as part of their path to understanding accumulation functions (and functions in general). Much of this support should be given in middle school and high school, but given that this does not happen in the United States, it must be addressed in introductory calculus courses.[6]

Students' Difficulties in Understanding Accumulation Mathematically

While our analysis of the ideas entailed in understanding accumulation functions also points to ways that students have difficulty with them, we must also note that the major source of students' problems with the idea of accumulation functions is that it is rarely taught with the intention that students actually understand it. We anticipate the objection that definite integrals already receive clear and explicit attention in every calculus textbook. Our reply is that a definite integral is to an accumulation function as 4 is to x^2. No one would claim to teach the idea of function by having students calculate values of a specific function. Similarly, we should not think that we are teaching the idea of accumulation function by having students calculate specific definite integrals.

We say this without hubris. In a teaching experiment (P. W. Thompson, 1994a) conducted with the intention of investigating advanced undergraduates' difficulties forming the ideas of accumulation and rate of change of accumulation, the first author failed to anticipate both the difficulty they would have conceptualizing accumulation functions and the importance that they actually do so for understanding the Fundamental Theorem of Calculus (FTC).[7]

Lastly, the idea of limit and the use of notation are two of the most subtle and complex aspects of understanding accumulation functions. Research on students' understanding of limit (Cornu, 1991; Davis & Vinner, 1986; Ferrini-Mundy & Graham, 1994; Tall, 1992; Tall & Vinner, 1981; Williams, 1991) shows consistently that high school and undergraduate students understand limits poorly, even after explicit instruction on them. We located only two empirical studies that addressed students' reasoning about limit in the context of integration (Oehrtman, 2002; P. W. Thompson, 1994a). Thompson studied advanced undergraduate and graduate students' understanding of the FTC, and in that context found students concluding, for example, that the rate of change of volume with respect to height in a cone was equal to the cross-sectional area at that height because *as you make an increment in height smaller,*

---

[6] The legacy of infinitesimal reasoning in calculus is reflected in the continued use of the integral notation that Leibniz introduced in 1675, when he used "*dt*" in $\int f(t)dt$ to represent the difference between successive values of t (O'Connor & Robertson, 2005).

[7] This was a classic case of an outcome being harder than someone expected even though he anticipated it would be harder than he expected.

the incremental cylinder of volume gets closer and closer to an area (P. W. Thompson, 1994a, p. 34). Oehrtman (2002) named this way of thinking "the collapsing metaphor," meaning that students reasoned that the object being considered (e.g., a cone, a secant, etc.) approached another object having one less dimension. He found one-third of his subjects (first-year calculus students after instruction) employing this metaphor in one setting or another.

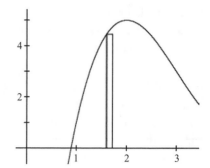

Figure 4. As Δx approaches 0, $f(c)\Delta x$ approaches $f(c)$.

Oehrtman points out that though the collapsing metaphor is mathematically incorrect, it sometimes enables students to educe mathematically correct results from incorrect reasoning. Students sometimes justify the FTC by the incorrect reasoning that as the interval width decreases, the rectangle collapses to its height (Figure 4). Put another way, students reason that $\Delta x \to 0$ implies that $f(c)\Delta x \to f(c)$. They were thinking of an image (e.g., a rectangle) instead of the quantity (e.g., electrical charge) and the value of its measure. We do note that although the collapsing metaphor enables students' intuitive, albeit incorrect, "justification" of the FTC, it also divorces their understanding of the fundamental theorem from any idea of rate of change.

Finally, we address the issue of what one might take as evidence as to whether students understand the representation of an accumulation function. The goal is to avoid accepting as evidence of understanding what Vinner (1997) calls pseudo-analytic and pseudo-conceptual behavior. Students exhibit conceptual behavior when their words and symbols refer to ideas and relationships. They exhibit pseudo-conceptual behavior when their words and notations refer to other words, to notations, or to iconic images. They exhibit pseudo-analytic behavior by applying pseudo-conceptual thinking in the course of their reasoning. The following student's response to the prompt, "Explain what $\int_a^x f(t)dt$ means," illustrates the subtleness in distinguishing between conceptual and pseudo-conceptual behavior.

$\int_a^b f(x)dx$ gives the area bounded by the graph of $f(x)$ and the lines $y=0$, $x=a$, and $x=b$. Therefore, $\int_a^x f(t)dt$ gives the area bounded by $f(t)$, $y=0$, $t=a$, and $t=x$. As x varies, the bounded area varies.

This answer, on the surface, appears quite acceptable. The problem is that we cannot tell which of several possible meanings this student gives to the integral notation. We highlight this by making a notational substitution in this student's answer so that it responds to the question, "Explain what $A_a^x\left(f(t)\right)$ means when $A_a^b\left(f(x)\right)$ stands for the area bounded by the graph of $f(x)$ and the lines $y=0$, $x=a$, and $x=b$."

$A_a^b\left(f(x)\right)$ gives the area bounded by the graph of $f(x)$, $y=0$, $x=a$, and $x=b$. Therefore, $A_a^x\left(f(t)\right)$ gives the area bounded by $f(t)$, $y=0$, $t=a$, and $t=x$. As x varies, the bounded area varies.

In other words, students could be imagining no more than a concrete image of a region "filling up" with paint as one moves one of its vertical edges (Figure 5), and at the same time could be using integral notation referentially (Figure 6) to describe that image. While this understanding of $\int_a^x f(t)dt$ would be a process conception of the notation, and indeed is useful as a shorthand for anyone having an in-depth understanding of the function, the process that students conceive when they have *only* the shorthand has nothing to do with the meaning of integration as the limit of Riemann sums.

We mentioned earlier a teaching experiment that investigated difficulties inherent in coming to understand the FTC (P. W. Thompson, 1994a). It involved 19 advanced undergraduate mathematics and masters mathematics education students. One aspect of the teaching experiment emphasized students' development of a process conception of integrals that entailed ideas of accumulation, variation, and Riemann sums as the root ideas of integration. We joined these ideas by defining Riemann sums as one would for fixed intervals, but modifying the definition so that Δx was a parameter and x was a variable. That is, we held Δx constant and let x vary instead of holding x constant and

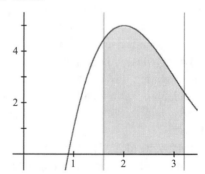

Figure 5. A student's image of a region filling up with paint as she moves one edge.

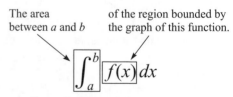

The area between a and b of the region bounded by the graph of this function.

Figure 6. Referential understanding of integral notation.

letting Δx vary.[8] This can be expressed generally as

$$F_{\Delta x,a}(x) = \sum_{i=0}^{\left\lceil \frac{x-a}{\Delta x} \right\rceil} f(i\Delta x + a)\Delta x, \quad a \le x \le b.$$

However, we did not provide this general representation of the Riemann accumulation function. Instead, we worked with students to model the accumulation of a quantity that accrued in bits created by joining two of its constituent attributes (like work from power and time) and to develop a representation of the total accumulation.

At the end of the teaching experiment we used Item 6, among others, to assess the extent to which students had developed a process conception of Riemann accumulation functions.

Item 6. Let $q(x)$ be defined by $q(t) = \sum_{i=1}^{t/\Delta t} f(i\Delta t)\Delta t$, $0 \le t \le b$. Explain the process by which the expression

$$q(t) = \sum_{i=1}^{\frac{t}{\Delta t}} f(i\Delta t)\Delta t$$

assigns a value to q(t) for each value of t in the domain of f.

Responses to this item showed that understanding Riemann sums as functions was a complex act for students. Student 1 wrote:

First the value of a certain chunk is measured by $i\Delta t$. This is then multiplied by the change which is Δt. This is repeated for every value of t and then added up. Each value of t is cut up into $t/\Delta t$ intervals, and added. $t/\Delta t$ is the number of intervals the piece is to be divided up into.

Student 1 had a number of problems, one being that he was imagining a "chunk" of a quantity independently of an analytic expression that established its measure—$i\Delta t$ does not "measure" the chunk, it just puts you at the right place to make it. The student also failed to note the role played by the function f in creating a "chunk"— it is $f(i\Delta t)\Delta t$ that gives the chunk's measure. Also, the student was unclear about what was being summed: "Each value of t is cut up into $t/\Delta t$ intervals, and then added." However, the subintervals are not summed. A more serious problem, though, is that this student appeared to be imagining t and i varying simultaneously instead of, first, varying i from 1 to $\lfloor t/\Delta t \rfloor$ for a fixed value of t, and then varying t.

Student 2 wrote:

- Here Δt represents the size of each interval that f is being broken up into.
- So $t/\Delta t$ equals the number of intervals the graph of f is broken up into.
- So our i starts out at 1 and then goes to $t/\Delta t$.
- The expression first finds f and then it finds the ith interval of f that we are dealing with. Then it finds the value of the function f at that interval and then multiplies by Δt. This finds the area of that particular rectangle. Then we add it to the previous areas found and plot that point. You then connect all the points to get your curve.

The process that Student 2 understood is much more well-structured than Student 1's. While some of her phrasing is imprecise ("… the size of each interval that *f is being broken up into*") and suggests that she is reasoning about a graph, she does seem to be imagining the process being played out for each value of t. One missing element in this student's explanation is that the value of t does not vary. Rather, she seems to imagine that she "samples" values of t and then connects the points that get plotted for each value. This suggests that she was imagining a Riemann sum over a fixed interval, which would normally correspond to an approximation of a definite integral $\int_a^b f(t)dt$ instead of the indefinite integral $\int_a^x f(t)dt$. When students do not see the upper limit as varying, it is difficult, if not impossible, for them to conceive that the accumulation function has a rate of change for every value at which it is defined (Smith, in preparation).

---

[8] Our justification for this approach, fixing Δx and letting x vary, was our intention to have students understand integral accumulation functions as being rooted in Riemann accumulation functions. In earlier attempts to probe students' understandings of accumulation functions, we got only the "paint filling" metaphor alluded to in the discussion of Figure 5. We will say more about the benefits of this approach in the last section.

Only seven of the 19 students expressed an appropriate order of variation for the index variable of the Riemann sum and the argument of the function. Five students appeared to have mixed images of definite and indefinite integrals (e.g., i varied, but t did not). The remaining seven students had confounded the two variations so that everything was happening at once. For us, these results imply that the idea of accumulation function is far more complex than is commonly assumed and that it is still unclear what instructional trajectories will best support students learning them.

We note in closing that Item 6, above, was useful in seeing the extent to which students had developed a process conception of Riemann sums and Riemann accumulation functions. In subsequent years, we have found that Item 6′, below, is a better task for determining that students have developed a process conception of a Riemann accumulation function.

Item 6′. Suppose f is a continuous real-valued function on (a,b) and $\Delta x > 0$. Let g be defined as

$$g(x) = \sum_{i=0}^{\left\lfloor \frac{x-a}{\Delta x} \right\rfloor} f(i\Delta x + a)\Delta x, \quad a \le x \le b.$$

Explain why g is a step function.

To see why g is a step function, let x_0 be $2\Delta x + a$. Then $\left\lfloor \frac{(x-a)}{\Delta x} \right\rfloor$ is 2 for every value of x in $[x_0, x_0 + \Delta x)$, and thus $g(x)$ is $f(0\Delta x + a)\Delta x + f(1\Delta x + a)\Delta x + f(2\Delta x + a)\Delta x$ for every value of x in $[x_0, x_0 + \Delta x)$. Therefore, $g(x)$ is constant as x varies within each interval $[i\Delta x + a, (I + 1)\Delta x + a)$.

Promising Approaches In Helping Students Conceptualize Accumulation Functions

Carlson, Persson, and Smith (2003), building upon Thompson's (1994a) work, conducted a teaching experiment with first-semester calculus students using instruction that addressed the conceptual difficulties experienced by Thompson's students in learning accumulation functions and their rate of change. Their approach to teaching accumulation functions was embedded in a larger effort to have students conceptualize the FTC as the course's culmination. The course began with a review of functions that emphasized covariation of quantities and then leveraged that reasoning to develop rates of change, limits, derivatives, and accumulations in terms of covarying quantities. Carlson et al. spent six sessions over two weeks developing notions of accumulating quantities and accumulation functions and another five sessions over 10 days on the FTC. Their coverage of the FTC had students examine in detail the incremental accumulation of various quantities, tying the idea of accumulation to the notation by which it is represented. They then had students examine the rates at which total accumulations changed by looking at average rates of change of the total accumulation over the interval of incremental accrual. Carlson et al. reported that their approach led to a high success rate, both in terms of students' conceptions of accumulation functions and their ability to use and explain the FTC.[9]

At the same time that Carlson et al.'s (2003) findings point to the promise of building students' understanding and skill with calculus on a strong conceptual foundation of covariation, function, rate of change, and accumulation, their post-instruction interviews suggest that students had not clarified some important issues. For example, one student, Chad, was shown a graph of a piecewise-linear function f whose values gave the rate in gallons per hour at which water filled a container. He was asked to explain the meaning of $g(x) = \int_0^x f(t)dt$ and how to evaluate $g(9)$. Carlson et al. reported that Chad gave an acceptable explanation of the meaning of $g(x)$ and also provided Chad's explanation of how to evaluate $g(9)$.

1. Interviewer: So, how do you think about evaluating $g(9)$?

2. Chad: I see that as finding the time that passes from 0 to 9 and thinking about how much area gets added under the curve as I move along. I see that water is coming into the tank, first at an increasing rate, then at a decreasing rate. Then after 4½ hours, water starts to go out of the tank. As you add up the area under the curve you see that the same amount of water comes in between 0 and 4 ½ that goes out between time 4½ and 9 …. so, the result is that there is no water in the tank after 9 hours have passed.

---

[9] We do not compare performances of Thompson's and Carlson's students because the two studies used different assessment tasks and therefore the results are not directly comparable.

3. Interviewer: How are *g* and *f* related?

4. Chad: The derivative of *g* gives the graph of *f*. What I don't get is why *t* is the variable that is used in *f*. I never really understood this on some of the other problems we did either.   (Carlson et al., 2003, p. 270)

Paragraph 2 suggests that Chad could think about accumulation functions and rate of change to support his evaluation of *g*(9). He also was thinking about the quantities that *x*, *f*(*x*), and *g*(*x*) represented (*viz.*, number of hours, the rate at which water filled the container, and the amount of water in the container). He also appeared to attend to how *g* changes while imagining changes in *x* and *f*. However, we observe that Chad's statements in paragraph 2 also are reminiscent of the "paint filling" notion of accumulation discussed in conjunction with Figure 5. As a result, without querying Chad further, we have no way of knowing if he is using the "paint metaphor" in a pseudo-conceptual way—i.e., does he understand that infinitesimal amounts of multiplicative bits are being accrued as *x* varies. In addition, paragraph 4 suggests that Chad had not worked through the conceptual issues behind the use of *t* in the accumulation function's definition.

It is unclear to us the extent to which Chad's understanding was rooted in Riemann sums as opposed to being rooted in the paint-filling metaphor. We see this as once again pointing to the need for further analysis of what it means for students to understand accumulation functions and how to assess their levels of understanding. It also points to the need for further investigation into the implications that various ways of understanding accumulation have for learning related ideas in the calculus, and the kinds of instruction that will support students in developing those understandings.

Students' Understandings of Accumulation Functions within the Calculus as a Whole

Accumulation functions would not be important if understanding them well did not pay off elsewhere. In this section we argue that the kind of understanding we have depicted as well-structured not only pays off in other areas, they are part of understanding many related ideas and they are essential for understanding many advanced ideas in the calculus. But even beyond the connections with other ideas that we will outline here, we feel that the precise thinking and thoughtful use of notation required to understand accumulation functions well is in itself valuable mathematical activity.

Connections with Other Ideas

Rate of change. The idea of accumulation both grows out of and contributes to a coherent understanding of rate of change (Carlson et al., 2003; P. W. Thompson, 1994a). When something changes, something accumulates. When something accumulates, it accumulates at some rate. To understand rate of change well, then, means that one sees accumulation and its rate of change as two sides of a coin. Thus, students' success in the integral calculus can begin in middle school if rate of change is taught substantively (A. G. Thompson & Thompson, 1996; P. W. Thompson, 1994a, 1994c; P. W. Thompson & Thompson, 1994).

Function. The obvious connection between the ideas of accumulation functions and function is that an accumulation function is precisely that, a function. It is nontrivial for students to understand this. There are three additional important connections that can be exploited in a calculus curriculum.

- Accumulation functions are, in all likelihood, the first functions students meet that are defined in terms of a complex process instead of in terms of an algebraic, trigonometric, or exponential expression. The challenge is to avoid leading them to pseudo-conceptual understandings of accumulation (see discussion of Figure 5) and pseudo-analytic interpretations of the notation (see Figure 6).

- Accumulation is the root of accumulation functions, and hence covariational conceptions of functions must be a key connection.

- Riemann accumulation functions that are specified by the formula

$$g(x) = \sum_{i=0}^{\left\lfloor \frac{x-a}{\Delta x} \right\rfloor} f(i\Delta x + a)\Delta x, \quad a \leq x \leq b$$

are step functions. Computer programs that allow Riemann sums as defined here will also support students' explorations of convergence.[10] The issue of convergence, however, expresses itself differently in this context than it does in typical treatments of Riemann sums. In the typical case, the issue is whether there is a *number* that is the limit of a Riemann sum as $\Delta x \to 0$. The accumulation function $\int_a^x f(t)dt$ is then defined so that each value of the function is a pointwise limit. In the case of a Riemann accumulation function, the issue is whether there is a *function* that is the limit of the family of Riemann accumulation functions that is generated as Δx approaches 0. That is to say, the idea of function is always at the forefront in this approach, even when working with approximations. We anticipate the objection that the issue of limits of function sequences is beyond first-term calculus students' conceptual capacity. Our experience is quite the contrary. Students find it visually compelling when they see a sequence of graphs approaching what appears to be the graph of *some* function, and willingly entertain the question, "What is the function that this sequence appears to approach? Does it approach it pointwise or uniformly?[11] How can we determine it analytically?" For example, the graphs of $\cos(x)$ and

$$g(x) = \sum_{i=0}^{\left|\frac{x-a}{0.01}\right|} \cos\left(0.01i + {}^{-}7\right)(0.01), \quad {}^{-}7 \le x < \infty$$

are given in Figure 7. The accumulation function's graph appears to be of a trigonometric function, but which one? What are its coefficients? Is something added? How is it related to $\cos(x)$? We feel that these questions can be mined fruitfully to develop students' understandings of a web of related ideas—approximation, limits, functions, convergence, and antiderivate, to name a few. How these connections might be developed instructionally, though, requires further investigation.

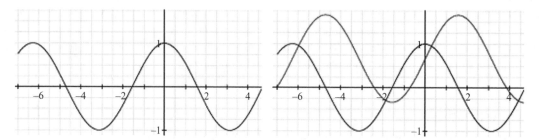

Figure 7. Graphs of $\cos(x)$ and its Riemann accumulation function.

Functions of two variables. Carlson, Oehrtman, and Thompson (this volume) argue that ideas of function-as-covariation in the case of two variables can be extended naturally to functions as covariation in the case of three variables. Were one to take that approach, then accumulation functions defined over lines and regions in a plane and over surfaces would be natural extensions from accumulation functions defined over intervals. We suspect that one needn't fall back to Riemann accumulation functions to make these cases meaningful if students have understood them thoroughly when they were first taught. We stress that ideas of covariation must remain at the forefront even with multiple integrals, and the importance of attending to issues of scope of variation is even greater than what we saw earlier in students' understanding of Riemann accumulation functions over intervals.

Conclusion

As we mentioned in the beginning of this chapter, the concept of accumulation is almost trivial yet, at the same time, quite complex. One aspect of the complexity described in this chapter was the focus on accumulation functions as opposed to the traditional focus on the calculation of a number representing the area bound by the curve over a specific

[10] We used Graphing Calculator from PacificTech to generate the graphs of accumulation functions contained in Figures 2–5. The functions graphed in those figures were specified as Riemann accumulation functions.

[11] The distinction between pointwise and uniform convergence arises quite naturally in classroom discussions when looking at the behavior of Riemann accumulation functions for functions with unbounded derivates.

interval. The emphasis of this chapter, though, was not on the differences between these two related notions of integral calculus; it was on the underlying images students bring to bear on such problems and the implications of those images. The first image involved covering a region, where the result was a number equivalent to "the amount of paint needed" to cover the area between the x-axis and the function on the interval $[a, b]$. The second image involved measuring the accumulation of a quantity that is created from bits that themselves are made from measures of two quantities, one whose measure is a function of the other on the interval $[a, b]$, by summing values of $f(c)\Delta x$, $c \in [i\Delta x,(i+1)\Delta x)$. The connection between the second image and area is simply that if $f(c)$ and Δx are represented by lengths, then $f(c)\Delta x$ gives the area of a rectangle made from those lengths. The former image is difficult to apply to quantities other than area, while the second necessitates understandings that both $f(c)$ and Δx can be measures of quantities (for example force and distance) and $f(c)\Delta x$ is a measure of a derived quantity (work).

It could appear that these images (painted area and accumulated quantities) are the same. We note in reply that they are only the same when one has constructed a scheme of understandings within which area can represent a quantity other than area. Further, the intricacies of understanding accumulation are often reduced to calculating products and limits without understanding the significance of either. Without additional focus on constructing, representing, and understanding Riemann Sums, there is little reason to believe that students will understand accumulation functions as playing a central role in the FTC.

This paper presents a call for increased emphasis on the FTC as explicating an inherent relationship between accumulation of quantities in bits and the rates at which an incremental bit accumulates. Understanding this relationship entails a clear emphasis on covariation as a foundational idea in calculus instruction. We make this call with awareness of the difficulties involved in developing a well-structured understanding of accumulation functions, and that this difficulty stands in contrast with the efficiency of teaching students to calculate definite integrals as area under a curve. We believe that the benefits make the effort worthwhile. Understanding $\int_a^b f(x)dx$ as an expression that yields the area bound by the x-axis and $f(x)$ is efficient but not generative. It supports a superficial understanding of $\int_a^x f(t)dt$. We believe that understanding accumulation so that $\int_a^b f(x)dx$ is simply $\int_a^x f(t)dt$ evaluated at $x=a$, where $\int_a^x f(t)dt$ itself has a well-developed meaning, can be part of a coherent calculus that focuses on having students see connections among rates of change of quantities, accumulation of quantities, functions as models, limits, antiderivatives, pointwise and uniform convergence, and functions of two (or more) variables. Though more work is needed to flesh out instruction that achieves this, we believe that a focus on accumulation functions as discussed in this chapter will be central to it.

References

Breidenbach, D., Dubinsky, E., Hawks, J., & Nichols, D. (1992). Development of the process conception of function. *Educational Studies in Mathematics, 23*, 247–285.

Carlson, M. P., Jacobs, S., Coe, E., Larsen, S., & Hsu, E. (2002). Applying covariational reasoning while modeling dynamic events: A framework and a study. *Journal for Research in Mathematics Education, 33*(5), 352–378.

Carlson, M. P., Persson, J., & Smith, N. (2003, July). *Developing and connecting calculus students' notions of rate-of-change and accumulation: The fundamental theorem of calculus.* In Proceedings of the 2003 Meeting of the International Group for the Psychology of Mathematics Education - North America, (Vol 2, pp. 165–172). Honolulu, HI: University of Hawaii.

Cornu, B. (1991). Limits. In D. Tall (Ed.), *Advanced mathematical thinking* (pp. 153–166). Dordrecht: Kluwer.

Davis, R. B., & Vinner, S. (1986). The notion of limit: Some seemingly unavoidable misconception stages. *Journal of Mathematical Behavior, 5*(3), 281–303.

Dubinsky, E., & Harel, G. (1992). The nature of the process conception of function. In G. Harel & E. Dubinsky (Eds.), *The concept of function: Aspects of epistemology and pedagogy* (pp. 85–106). Washington, D. C.: Mathematical Association of America.

Ferrini-Mundy, J., & Graham, K. (1994). Research in calculus learning: Understanding of limits, derivatives, and integrals. In J. J. Kaput & E. Dubinsky (Eds.), *Research issues in undergraduate mathematics learning.* Washington, D. C.: Mathematical Association of America.

Gray, E. M., & Tall, D. O. (1994). Duality, ambiguity, and flexibility: A "proceptual" view of simple arithmetic. *Journal for Research in Mathematics Education, 25*(2), 116–140.

O'Connor, J. J., & Robertson, E. F. (2005). Gottfried Wilhelm von Leibniz. *The MacTutor History of Mathematics Archive* Retrieved January 2, 2006, from www-groups.dcs.st-and.ac.uk/~history/Mathematicians/Leibniz.html

Oehrtman, M. C. (2002). *Collapsing dimensions, physical limitation, and other student metaphors for limit concepts: An instrumentalist investigation into calculus students' spontaneous reasoning.* Unpublished Dissertation, University of Texas at Austin, Austin, TX.

Saldanha, L. A., & Thompson, P. W. (1998). *Re-thinking co-variation from a quantitative perspective: Simultaneous continuous variation.* In Proceedings of the Annual Meeting of the Psychology of Mathematics Education - North America. Raleigh, NC: North Carolina State University.

Smith, N. (in preparation). *Students' emergent conceptions of the fundamental theorem of calculus.* Unpublished Doctoral Dissertation, Arizona State University, Tempe, AZ.

Tall, D. (1992). The transition to advanced mathematical thinking: Functions, limits, infinity and proof. In D. Grouws (Ed.), *Handbook for research on mathematics teaching and learning* (pp. 495–514). New York: Macmillan.

Tall, D., & Vinner, S. (1981). Concept images and concept definitions in mathematics with particular reference to limits and continuity. *Educational Studies in Mathematics, 12*, 151–169.

Thompson, A. G., & Thompson, P. W. (1996). Talking about rates conceptually, Part II: Mathematical knowledge for teaching. *Journal for Research in Mathematics Education, 27*(1), 2–24.

Thompson, P. W. (1994a). Images of rate and operational understanding of the Fundamental Theorem of Calculus. *Educational Studies in Mathematics, 26*(2-3), 229–274.

Thompson, P. W. (1994b). Students, functions, and the undergraduate mathematics curriculum. In E. Dubinsky, A. H. Schoenfeld & J. J. Kaput (Eds.), *Research in Collegiate Mathematics Education, 1* (Vol. 4, pp. 21–44). Providence, RI: American Mathematical Society.

Thompson, P. W. (1994c). The development of the concept of speed and its relationship to concepts of rate. In G. Harel & J. Confrey (Eds.), *The development of multiplicative reasoning in the learning of mathematics* (pp. 179–234). Albany, NY: SUNY Press.

Thompson, P. W., & Thompson, A. G. (1994). Talking about rates conceptually, Part I: A teacher's struggle. *Journal for Research in Mathematics Education, 25*(3), 279–303.

Vinner, S. (1997). Pseudo-conceptual and pseudo-analytic thought processes in mathematics learning. *Educational Studies in Mathematics, 34*, 97–129.

Weisstein, E. W. (2006). Dummy variable. *Mathworld--A Wolfram web resource.* Retrieved January 2, 2006, from mathworld.wolfram.com/DummyVariable.html

Williams, S. R. (1991). Models of limit held by college calculus students. *Journal for Research in Mathematics Education, 22*(3), 219–236.

Part I
Student Thinking

b. Infinity, Limit, and Divisibility

5

Developing Notions of Infinity

Michael A. McDonald, *Occidental College*
Anne Brown, *Indiana University South Bend*

The various notions of infinity are among the most intriguing and challenging concepts in mathematics. Despite their important role in the undergraduate mathematics curriculum, concepts related to infinity typically receive little direct instructional attention. While a number of mathematics education researchers have examined students' understanding of topics related to infinity, this is still an area rich with unanswered questions. In this chapter, we do not attempt to summarize all of the research in this area and its possible implications for instruction. Rather we look at three specific examples — comparing infinite sets, determining limits of sequences, and constructing infinite iterative processes — and use them to illustrate the types of research being done, and the ways in which research may provide insights for classroom instruction. We begin by introducing the examples and follow up by discussing the results of the related research.

Situation 1. How might students who have not been exposed to Cantor's theory respond to the following question:

Which infinite set, *A* or *B*, below has more elements, or do they have the same number of elements?

$$A = \{1, 2, 3, 4, 5, 6,...\} \qquad\qquad B = \{4, 8, 12, 16, 20, 24,...\}[1]$$

Would their responses change if sets *A* and *B* were listed one above the other? What if they were given in a geometric presentation where set *A* is given as the set of line segments of length 1 cm, 2 cm, 3 cm, 4 cm, and so on, and set *B* is given as the set of perimeters of squares with sides of length 1 cm, 2 cm, 3 cm, 4 cm, and so on?

Would students say that sets *A* and *B* have the same number of elements because they are both infinite sets and there is only one infinity? Or would they say that *A* clearly has more elements than *B* because *B* is a proper subset of *A*? Would some argue that the two sets have the same number of elements because they can set up a one-to-one correspondence between the sets, the answer consistent with Cantorian set theory? And do different presentations suggest different answers to students?

Situation 2. What response will the following problem elicit from the typical Calculus student?

There is a stairway consisting of two steps, with height and width of each step one foot. From this stairway we construct another with twice the number of steps, by halving the height and width.

Staircase 1   Staircase 2   Staircase 3

---

[1] Adapted from Tsamir and Tirosh (1999).

Following this process inductively, we construct a whole family of staircases. What can we say about the perimeter of the staircases? What is the final result of the inductive process?[2]

Will students say that the final result of the geometric constructions is a staircase with an infinite number of stairs that are infinitesimally small, or will they say that it is a ramp with slope 1? And what will they say is the perimeter of this "final" result? Finally, is this a good task to give students who are still developing their notions of limit?

Situation 3. Can the typical mathematics major in a transition course on proofs determine whether the indicated statement concerning the union of the power sets of finite initial segments of the set of natural numbers **N** is true or false?

Prove or disprove: $\displaystyle\bigcup_{k=1}^{\infty} P(\{1,2,\ldots,k\}) = P(\mathbb{N})$.[3]

Do they use the notion of set inclusion to prove or disprove the equality of the two sets? Or do they focus on the nth partial union and note that $\bigcup_{k=1}^{n} P(\{1,2,\ldots,k\}) = P(\{1,2,\ldots,n\})$? Does this knowledge lead them to claim that the infinite union must equal $P(\mathbb{N})$ so that the equality holds? Or do they observe that the final result of the infinite union has only finite sets as elements while $P(\mathbb{N})$ contains infinite sets as elements so the equality does not hold?

We now examine each problem situation more thoroughly and summarize the results of the related research.

Comparing Infinite Sets

Over the past two and a half decades, one group of researchers has carefully examined students' intuitions regarding infinity, particularly the comparison of infinite sets. Primary intuitions that are most often invoked by students in those settings include:

- a single infinity intuition — there is only one infinity, thus all infinite sets have the same size;

- an inclusion or part-whole intuition — the smaller infinite set is the one that is a subset of the other given set; and

- an incomparability intuition — you cannot compare infinite sets, therefore there are no right or wrong answers.

Obviously there is also the conventional mathematical interpretation based on Cantorian set theory that two sets are of the same size if and only if they can be put in one-to-one correspondence with each other. This view is often not very intuitive to students.

These researchers have embarked upon a long-term research agenda where each study reveals more about how students' intuitions impact (and often inhibit) their mathematical understanding of infinity, and subsequent studies build upon what is learned from the previous ones. In this section, we briefly highlight several of the recent studies by Dina Tirosh and Pessia Tsamir and discuss the implications of these studies. Although many of their studies focus on middle and high school students in Israel, some of their studies have been conducted with college students preparing to be teachers. Additionally, our own research related to infinity as well as our experience teaching college and university students convinces us that the ideas discussed below are very relevant to college-aged students.

In Tirosh and Tsamir (1996), the researchers tried to determine whether the representation of the infinite sets to be compared affected the answers and justification students gave. In this study, 189 tenth through twelfth graders in Israel were asked to compare the sizes of two infinite sets presented in one of four ways:

numerical-horizontal, such as: $C = \{1, 2, 3, 4, 5, 6, \ldots\}$ and $D = \{\ldots, -3, -2, -1, 0, 1, 2, \ldots\}$

numerical-vertical, such as:
$$E = \{1, 2, 3, 4, 5, \ldots\}$$
$$F = \{\tfrac{1}{2}, 1, 1\tfrac{1}{2}, 2, 2\tfrac{1}{2}, \ldots\}$$

numerical-explicit, such as: $G = \{1, 2, 3, 4, 5, \ldots\}$ and $H = \{1^2, 2^2, 3^2, 4^2, 5^2, \ldots\}$

---

[2] Adapted from Mamona-Downs (2001).

[3] Adapted from Brown, McDonald and Weller (in press).

geometric, such as
$$G = \text{the set of sides of squares of length } 1, 2, 3, 4, 5, \ldots$$
$$H = \text{the set of areas of squares of length } 1, 2, 3, 4, 5, \ldots$$

They were also asked to justify their answers. The researchers found that equivalent set arguments using one-to-one correspondence justifications were given most often when sets were in numerical-explicit or geometric representations. However, when arguing that the sets are equivalent in terms of number of elements, students still used single infinity arguments more often than one-to-one correspondence.

In Tsamir and Tirosh (1999), the researchers used a variation of Situation 1 presented earlier to examine the thinking of 32 thirteen to eighteen year old Israeli public school children who were advanced in mathematics. Based on previous research findings, the researchers encouraged part-whole (inclusion) thinking as follows. They first showed the student a card with the natural numbers listed, $\{1, 2, 3, 4, 5, \ldots\}$. Students were asked to circle all multiples of four and then list those numbers in a new set, $\{4, 8, 12, 16, 20, \ldots\}$, and asked to compare the size of this set to the previous one. The researchers then encouraged thinking about one-to-one correspondences. They showed them a card with line segments of increasing length and asked students to write down the lengths in a set, resulting once again in $\{1, 2, 3, 4, 5, \ldots\}$. Students were then asked to create a sequence of squares using each of the line segments as the side of one of the squares, to write down the perimeters of the squares in a set, resulting in $\{4, 8, 12, 16, 20, \ldots\}$, and again asked to compare the size of these two sets. The four sets were then shown to the students and they were presented with their own, often contradictory, comparison results in the two tasks.

The researchers found that 12 of the 32 students argued that all the sets had the same number of elements using either single infinity or incomparability arguments. The remaining 20 students gave contradictory answers in the two comparison tasks. Some students noted that their answers were incompatible, but most needed some prompting to realize this. Ultimately, a few students decided that contradictions were acceptable in mathematics, while the others resolved their contradictions using one of the reasoning patterns indicated at the beginning of the section, with only two choosing the one-to-one correspondence criterion.

This study has two implications for instruction on the comparison of infinite sets. First, they confirm that students' responses are not independent of the way the sets are represented. Second, the fact that not all students see the need to avoid contradictions suggests that simply encouraging the use of the Cantorian approach, while pointing out that inconsistencies occur when other comparison criteria are used, will not guarantee that students will dismiss their primary intuitions and other criteria in favor of the Cantorian approach.

In Tsamir (2001), the researcher followed up on the previous study with a new problem situation that she proposes may help students reject the part-whole criteria by adding a numerical-explicit representation to the geometric representation in the problem. Students were first given the two sets $\{1, 2, 3, \ldots\}$ and $\{3, 4, 5, \ldots\}$ in a numerical-horizontal representation, usually evoking part-whole thinking on the part of the students. The second representation of these same sets given in an attempt to evoke one-to-one correspondence was a geometric representation with a sequence of trapezoids where the top base had measure equal to a natural number n and the bottom base was always two units larger than the top. Finally, students were given a numerical-explicit representation given vertically consisting of the two sets $\{1, 2, 3, \ldots\}$ and $\{1+2, 2+2, 3+2, \ldots\}$.

Tsamir found that this sequence of presentation of sets helped students provide consistent responses and to use one-to-one correspondence as the sole criteria of comparison. However, in follow-up interviews, she found that this was the case only if students could specify the rule of correspondence (in the previous case "+2"). When asked to compare the set of natural numbers and the set $\{0, 1, 3, 6, 10, 15, 21, \ldots\}$, where the correspondence may not be obvious, none of the students participating in the follow-up used one-to-one correspondence. In fact, about half used a single infinity argument and about half reverted to a part-whole criterion. Tsamir notes that while this teaching intervention seems to encourage the use of one-to-one correspondence, what it does not do is convince students that one-to-one correspondence should be the only criterion one uses to compare infinite sets.

In all the previous studies, students were asked to give their own criteria for each comparison task. In Tsamir (2003), the researcher presented students with possible approaches to a comparison task and asked whether the given criteria (either one-to-one correspondence, inclusion, or single infinity) were acceptable. Participants in this study were 110 prospective secondary school teachers who had taken a yearlong Cantorian set theory class and 71 prospective teachers who had not taken the course. They were given a questionnaire with illustrations of a student explaining their belief in the use of one of the three criteria, and participants were asked to discuss whether these were suitable.

They were then given a second questionnaire that illustrated a student who claimed one should only use one-to-one correspondence as the criteria for comparing infinite sets along with four specific comparison tasks along with his explanations of how he used one-to-one correspondence. Again, participants were asked if each explanation was a suitable argument for showing the equivalency of infinite sets. The first questionnaire was then re-administered. Finally, 10 participants from each of the two cohorts who gave insufficient justifications were interviewed.

In the initial questionnaire, over half the students who had not taken the set theory class, and just over three-quarters of those who had, said that one-to-one was an acceptable criterion. However a substantial number of both cohorts said the other two criteria were also acceptable. After the intervention of the second questionnaire, there was an increase in both groups' acceptance of one-to-one correspondence as a valid criterion when retaking the first questionnaire. But there was no change in the rate of acceptance of the inclusion criteria in either group and an increase in the rate of acceptance of the single infinity criteria in both groups! Thus, while reminding and familiarizing students with one-to-one correspondence as a criterion strengthens their tendency to accept it, it does not weaken their tendency to accept other criteria. As Tsamir concluded, "Even after studying set theory, participants still failed to grasp one of its key aspects, that is, that the use of more than one ... criteria for comparing infinite sets will eventually lead to contradiction" (Tsamir, 2003, p. 90).

In teaching about comparison of sets, Tsamir recommends using cognitive conflict teaching, a method in which tasks are sequenced with the aim of making students aware of the contradictions in their thinking. An example of a task that might spark cognitive conflict is asking students to compare the number of elements in $A = \{1, 2, 3, 4,...\}$ and $B = \{1, 4, 9, 16,...\}$, and then to compare the number of elements in $A = \{1, 2, 3, 4,...\}$ and $C = \{1^2, 2^2, 3^2, 4^2,...\}$. By providing this task, where students often use the inclusion criterion to say that A has more elements than B but use a one-to-one comparison criterion to say that A and C have the same number of elements, one might raise awareness with students' of inconsistencies in their answers because they can easily see that B and C contain the same elements. Tsamir (2003) presents a possible cognitive conflict based teaching sequence in the appendix of the paper. In part 1.a, she gives students various finite sets, such as

$$A = \{5, 10, 15, 20, 25, 30\} \text{ and } B = \{10, 20, 30\},$$

asks if the two sets have an equal number of elements, and asks students to explain how they reached their conclusion. After reflecting on the different criteria they used in part 1.a, in part 1.b, students revisit each example from 1.a and, for each of the three criteria, decide if it is an applicable criteria to use for the given sets. After they complete this for all pairs of sets with all three criteria, they are asked to reflect on whether different conclusions are reached using different criteria. Tasks 2.a and 2.b are essentially identical to 1.a and 1.b, but involve infinite sets such as

$$F = \{5, 10, 15, 20, 25, 30,...\} \text{ and } K = \{10, 20, 30,...\}.$$

At the end of the teaching sequence, students are asked to reflect on what happens when they move from dealing with finite to infinite sets.

In addition to the specific results just presented regarding infinity, this discussion highlights the role of intuition in the development of mathematical knowledge. As Tirosh (1991) reports, contradictory intuition has been shown to be an obstacle to acquiring formal mathematical knowledge. In addition, intuition, particularly in this case with regard to the concept of infinity, is highly resistant to change from direct instruction. The result of instruction, such as in the sequence of studies considered here, may only help students develop an awareness of their intuitions, recognize the need to control their primary intuitions, and encourage them to use explicit deductively developed means of solving problem situations (Tirosh, 1991). These are important issues for teachers of mathematics to understand as we interact with students in our classrooms on difficult concepts for which students bring their varied, and often strongly embedded intuitions.

The brief survey of the sequence of studies presented in this section also shows how research and teaching practice intertwine and can inform each other. A researcher conducts a careful empirical study, reflects on what might be learned from the study about student learning of a given topic, and designs instructional content and pedagogical approaches based upon theory and empirical results. The implementation of this instruction and pedagogy then serves as the next stage in the research program, whereby the researcher conducts further empirical studies to find out more about the impact of the instruction on student learning and mathematical performance. Next, we look at an example of a theoretical research report and its potential impact on practice.

Limits of Sequences

While cardinal infinity is theoretically grounded in set theory, as dealt with in the previous section, other aspects of infinity play a critical role in a variety of conceptual domains. For example, aspects of infinity are ubiquitous in calculus. From the use of limits to examine the behavior of functions or to define both the derivative and the integral, students' ability to reason about infinite processes is essential. In Mamona-Downs (2001), the researcher presents a didactical approach that may help students better understand the concept of the limit of a sequence. [We also refer readers to the chapter by Oehrtman in this volume for a more thorough review of studies related to the limit concept.] Mamona-Downs' report differs from those in the previous section as it takes a purely theoretical approach, based on existing research results and the author's experience and knowledge, rather than an empirical study of students' reasoning or mathematical performance.

Mamona-Downs notes that one of the problems with the limit concept is that students possess a "surprisingly rich intuitive base" (p. 259) for the concept of infinity, in agreement with the results of Tirosh and Tsamir in the previous section. A second problem the author identifies in students' development of their understanding is "the clash between the 'dynamic approaching' intuitive image and the 'static' image evinced by the definition of limit" (p. 261). A central issue in much of the literature concerning limits, this conflict is between the mental image of a sequence in motion and the formal expressions that uniquely definite the limit of a sequence. These observations lead Mamona-Downs to present a didactical approach that uses classroom discussion to raise awareness of and further develop intuition, followed by an analysis of the formal definition of the limit using a particular representation, and concluding with a comparison of these ideas with students' intuitions. Her approach is intended to help students move from intuitions of mathematical ideas to more formal approaches and it may be adaptable to other situations. [Another framework for the process of formalization is described in the chapter by Rasmussen and Ruan in this volume.]

Mamona-Downs begins by presenting tasks that are intended to help raise students' intuitive understandings explicitly so that they can be considered, and then accepted or rejected. The first task she presents is that of a ping-pong ball that is dropped from a given height and which continues to bounce, each time coming back exactly half the prior height. Students are asked to discuss how many times the ball bounces (and if they say "infinity", asking them what this means in this context), and how far the ball travels in total. Issues raised may include the tension between the physical perception (that the ball will stop) and the theoretical possibility (that the ball bounces infinitely often). By considering the process of counting the number of bounces, the students may also develop a belief that "infinity is the number that you get if you count forever" which might in turn lead them "to consider an infinite sequence (a_n) to possess a final term a_∞" (p. 268). In considering how far the ball travels, students may at first assume that if the ball bounces an infinite number of times, it must travel an infinite distance. But as they begin discussing the fact that the lengths decrease each time, this assumption comes into question. The author presents a simulated classroom debate on this issue that is consistent with the results of the research literature, Sierpinska (1987) in particular.

In the ping-pong ball problem, there is a natural candidate for the limit when summing to get the distance the ball travels. However, there are problems, like Situation 2 involving the sequence of staircases given at the beginning of the paper, where the actual limit is not the natural candidate. In the staircase problem, we start out with two stairs, each with height and width one foot. For the purpose of this discussion, we will ignore the base and back of the staircase, and define the perimeter of the staircase to be the sum of the lengths of the rise and tread of the steps. Thus the total perimeter of the first staircase is four feet. If we double the number of steps and halve their heights and widths, we still have perimeter four feet. One might expect the limiting object of this sequence of staircases also to have perimeter four feet. If we allow for non-standard analysis, students may argue that the result is a staircase with an infinite number of steps that are infinitesimally small. But in standard analysis, the limiting object is a ramp with slope one, and perimeter $2\sqrt{2}$ not 4! While the problem is intriguing, Mamona-Downs cautions that tasks with paradoxical results might be better left until students have debated the more basic issues and have developed some confidence in their thinking. Thus, the ping-pong ball problem, or a problem where one approximates the circumference of a circle using a sequence of inscribed polygons, might be more appropriate in these initial stages of students' development of the limit concept.

The overall purpose of the class discussions that Mamona-Downs advocates is to allow students' intuitions and ideas to be brought up rather than to deny them and advocate only the formal ideas. But her study also highlights that the choice of task can influence what issues are raised and whether it leads to productive discourse or a "non-useful impasse" (p. 273). Students also may focus on the infinite process and then wonder how this process is completed, or they may focus on the completion itself. This already raises for the students the "clash" Mamona-Downs noted between the dynamic and the static.

Having raised students' awareness of their intuitions, a teacher then needs to help them grapple with the definition and to reconcile their understanding of the definition with their intuitions. While Mamona-Downs provides a long discussion of the various issues and difficulties embedded in building a solid and coherent image of the typical epsilon-N definition of limit, and how to connect students' intuitive understanding with this image, we present only a few selected issues here. She also recommends using an illustrating graph to help students understand the details of the definition.

This is likely one of the first times in students' mathematical experience where they must deal with such a formal definition and potentially confusing symbolism. Some examples of how students must digest the meanings of these symbols follow. The focal point of the standard definition is the inequality $\left| a_n - L \right| < \varepsilon$. But students often will have difficulty interpreting this. It is also often the last statement in the definition, requiring students to suspend all the information introduced about these various symbols earlier in the definition until they reach the end. The individual symbols require a great deal of care to understand, as L should have the characteristic of a constant, a_n the characteristic of a variable (while n varies through the natural numbers), and ε the characteristic of a parameter. Further, the definition introduces a new symbol, N, that seems completely independent of the focal point of the definition, the inequality. Finally, the inequality must be interpreted not as a statement of truth, but rather as forming the basis for a decision process. [A more thorough examination of the issues related to student understanding of literal symbols appears in the chapter by Trigueros and Jacobs, this volume.]

Mamona-Downs discusses how an illustrating graph (see Figure 1) can assist students in developing the understandings just described, and ultimately may help the student distinguish between the sequence and the limit of that sequence.

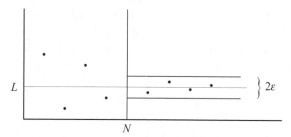

Figure 1. Distinguishing between a sequence and its limit.

As noted earlier, a student might believe that the sequence has an ultimate term, a_∞, which then must be equal to the limit, L. But visually the limit on the graph is a horizontal line, $y = L$, whereas the terms of the sequence are identified with discrete points obtained using the functional relationship $y(n) = a_n$. This may help the students "identify L as a number and a construct related ... to the sequence but not as an integral part of it" (p. 286). We return to the importance of distinguishing between the sequence and its limit in the next section where we talk more generally about infinite processes.

As we see from the example just presented, some research reports are theoretical in nature, and can provide an informed perspective on what might happen when implementing a didactical intervention. Reports that are more theoretical in nature are usually based on other empirical studies, personal experience, existing theories, and deep understanding of the mathematics involved. Careful reflection on theoretical reports like this can help teachers improve their own practice. As Mamona-Downs herself states, a research report like hers is often written to "inform teachers what to expect, or what to include in their class plan" (p. 261).

Infinite Iterative Processes

While concepts related to infinity are ubiquitous in calculus, infinity is often formally examined for the first time in an introductory set theory or introduction to proofs course. The next research study we consider (Brown, McDonald,

& Weller, in press) was conducted with students who had recently completed such a course. The problem situation examined in the research was presented in Situation 3: to determine whether or not $\bigcup_{k=1}^{\infty} P(\{1,2,...,k\}) = P(\mathbb{N})$. Given the content covered in the course, it was expected that the students would observe that the infinite union contains only finite sets as elements whereas $P(\mathbb{N})$ also contains infinite sets as elements. In fact, although they demonstrated knowledge of the definitions of the objects involved, all of the students tried to make sense of the infinite union by constructing one or more infinite processes. Thus, the primary focus of the study was to examine, using a particular theory of learning called APOS theory (Asiala et al., 1996), how an individual might cognitively construct an understanding of an infinite iterative process in a context involving actual infinity.

In Brown, et al., the term infinite iterative process refers to the repeated application of a transformation of mental or physical objects, involving one or more parameters that change with each repetition. For example, perhaps the most simple infinite iterative process is enumerating through the natural numbers where one simply "adds 1" at each step. In the problem considered in this study, all of the students attempted to mentally construct an infinite iterative process that produced the sequence of partial unions $\bigcup_{k=1}^{n} P(\{1,2,...,k\})$. Most of the students quickly noticed that, since the power sets are nested, the partial unions "collapse" in the sense that

$$\bigcup_{k=1}^{n} P(\{1,2,...,k\}) = P(\{1,2,...,n\}) \quad \text{for each} \quad n \geq 1.$$

Thus, for these students, the key to solving this problem was to imagine the ultimate result of continuing this process indefinitely, and to consider whether this result is indeed equal to $P(\mathbb{N})$. We will now discuss their attempts more specifically. In our discussion, we will introduce and use the language of APOS theory, as it applies to this research.

All of the students began to think about the infinite union by looking at the first few partial unions, writing: $P(\{1\}) \cup P(\{1,2\}) = P(\{1,2\})$, $P(\{1\}) \cup P(\{1,2\}) \cup P(\{1,2,3\}) = P(\{1,2,3\})$, and sometimes one or two steps further. In APOS theory, this is called performing an *action*, transforming mathematical or mental objects using external cues such as the formula for the infinite union and explicit values for the index $k=1$, $k=2$, etc. As students realized that they could do this over and over again, and could imagine doing it mentally without necessarily having to write down each step, they had mentally constructed what is called a *process* in APOS theory. So, for instance, many of the students in the study were able to describe the nth partial union and understood that this finite union consisted of all subsets of the set $\{1,2,3,...,n\}$. They could describe the relationship between successive partial unions or explain how to obtain one from the previous one. To continue, they had to find a way to extend their understanding of the finite processes to construct a corresponding infinite process.

One of the key results of Brown, et al. (in press) is the observation that students who made significant progress in solving the problem were the ones who could conceive of the process of taking the union of an infinite number of sets of the form $P(\{1,2,3,...,n\})$ as *complete*. Unlike a finite process where one can imagine completing the process by reflecting on the last step, in an infinite process one must imagine completing all the steps even though there is no last step. This was not easy for most of the students interviewed. For example, one student could imagine a process of taking the union through the nth partial union and obtaining $P(\{1,2,3,...,n\})$, but from that point she simply imagined the index n increasing continually and obtained $P(\mathbb{N})$ as the result. Thus, she did not focus on all steps of the process being completed and missed the significance of the fact that all of the power sets produced by the process of forming partial unions are actually finite.

Only two students in the study could see the infinite process as complete, and could hold in mind the properties of the result at each step in this process even as they talked about the process being "done." As one student so clearly articulated about the infinite process, "it doesn't ever get to infinity, 'cause it's always just one more than it was previous. But you just keep going and going and going. So this is the union of an infinite number of finite sets." She was able to see that in the infinite process no infinite power sets are computed, which allowed her to move forward in solving the problem.

Beyond seeing the process as complete, students must also be able to reason about the infinite union as a single entity in order to compare the result with the set $P(\mathbb{N})$. Being able to move from a dynamic process of generating partial unions term by term to thinking of the "unioning" process as a single operation is an example of what is referred to in APOS theory as seeing the process as a *totality*. In general, seeing a process as a totality may be followed by *encapsulating* the process into an *object*. Often one needs to encapsulate a process into an object in order to perform an action. Since an action can only be performed on a static object, one has to encapsulate the dynamic process. For the problem in this study, an appropriate action might be asking "What is the final result of the infinite union?" Only the one student quoted in the previous paragraph was able to view the completed process as a totality, and she clearly articulated that what is obtained is an infinite collection of finite sets.

Brown, et al. (in press) introduce the term *transcendent object* for the state at infinity that is the encapsulation of an infinite process. The name was chosen to indicate that this object must be understood as not being produced by the process itself, but instead as transcending the process. So while the process of computing partial unions produces only finite collections of finite sets, what we get as a final result of the infinite union, the transcendent object of this process, is an infinite collection of finite sets.

Similar to the previous section on limits of sequences where we saw that many students consider an infinite sequence (a_n) to possess a final term a_∞, or to the first section where we saw that students carry over their intuitions of comparing finite sets over to comparing infinite sets, many students in this study carried over ideas that worked for them in constructing finite processes to their construction of infinite ones. For example, one student who noted that the kth partial union $P(\{1\}) \cup P(\{1,2\}) \cup \cdots \cup P(\{1,2,\ldots,k\})$ equals $P(\{1,2,\ldots,k\})$ observed that it reduced to the set "farthest to the right", and therefore concluded that the final result of the infinite union would be $P(\mathbb{N})$. One fallacy in this reasoning is that while the result of every finite partial union is a power set, the result of the infinite union is not the power set of any set, so there is a "discontinuity at infinity." Returning to Situation 2, the staircase problem, recall that Mamona-Downs cautions against using such counter-intuitive situations too early in students' development of their understanding of a limit of a sequence. In comparison, the results by Brown, et al. suggest that we *should* use a problem like the staircase problem eventually, as it foreshadows in a simpler situation a subtle phenomenon that students apparently find quite difficult: an infinite process with a discontinuity at infinity.

The interview data in Brown, et al. (in press) suggests that students construct infinite iterative processes when faced with certain situations involving actual infinity. More research is needed in order to describe the variety of situations in which this might apply. However, it is clear that when that phenomenon occurs, it is important to help students conceive of infinite processes as completed totalities, and help them identify the appropriate action, such as asking "What is the final result?", which may spur them to encapsulate the process into an object. Once the infinite process is encapsulated, the student may be ready to reconsider the situation in terms of formal definitions. That is, construction of an infinite iterative process might be seen as an exercise that builds understanding of the objects involved; this understanding then plays a role when the formal definitions are used.

Since this research is in its early stages in that no instructional strategies have been tested yet, we hesitate to make strong suggestions concerning pedagogy. It may be useful, though, to consider one simple example that raises the idea of a completed infinite process and its transcendent object. Imagine the indicated sequence of regular polygons, one created at each step:

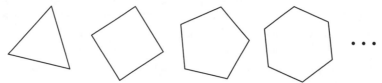

One might ask students to describe what results from each of the steps of the process, and to propose an ultimate result. This may raise some interesting ideas (as in Situation 2), but one helpful idea that seems likely to arise is that a circle is an appropriate transcendent object. Since it is not a polygon, it cannot be directly produced by any step. Yet, it can be imagined as a natural transcendent object for this sequence.

A more complex example is to consider the set of all finite strings of two letters a and b. This infinite set can be imagined to be generated recursively, using ordering by size. That is, an iterative process could be constructed in this way: take a string S of length n at step n, form Sa and Sb, and adjoin all such strings of length $n+1$ to the existing collection. Students might be asked to compare the size of the collection at consecutive steps, to describe all objects produced at the steps, and then to suggest an ultimate result. As in Situation 3, one of the key ideas is to understand that the ultimate result is an infinite set in which each element is finitely represented.

Other APOS-based empirical and theoretical studies of concepts involving infinity are ongoing (see Weller, Brown, Dubinsky, McDonald, & Stenger, 2004). One research group is examining students' understanding of iteration through natural numbers and an infinite iterative process that arises from a very different type of problem situation.

Another is examining how individuals might construct a deep understanding of $P(\mathbb{N})$, an uncountable set. Yet another group (Dubinsky, Weller, McDonald, & Brown, 2005a, 2005b) is using cognitive explanations from APOS Theory to describe the thinking about infinity in the historical and philosophical literature and to give new cognitive resolutions of various historical paradoxes of the infinite. Further initiatives in research on conceptions of infinity will no doubt arise from these projects. The potential for instructional implications based on a careful examination of the mental constructions students may or may not make while engaged in mathematical problem solving situations related to concepts of infinity is just beginning to be realized.

Conclusion

This report has laid out three specific research agendas that address issues related to concepts of infinity. There are a number of other studies that have been conducted on issues related to infinity, yet there are many issues that have not yet been researched. Concepts of infinity span a wide range of the mathematical curriculum and as researchers continue to examine various aspects of students' thinking about infinity, there is the potential for broad impact on educational practice.

In addition to giving a brief summary of the specific findings related to infinity, we also have presented studies that highlight different ways that empirical and theoretical research reports may be used to implement curricular and pedagogical improvements in the college classroom. While mathematics education research, like research in many other areas in academia, is often basically "theoretical" in nature, its application to our teaching and student learning is the "applied" aspect. We have presented here, in the context of research on concepts of infinity, three different models of how mathematics education research might be "applied" to classroom teaching and student learning.

References

Asiala, M., Brown, A., DeVries, D., Dubinsky, E., Mathews, D., & Thomas, K. (1996). A framework for research and curriculum development in undergraduate mathematics education. In J. Kaput, A. H. Schoenfeld, & E. Dubinsky (Eds.), *Research in collegiate mathematics education. II* (pp. 1–32). Providence, RI: American Mathematical Society.

Brown, A., McDonald, M.A. & Weller, K. (in press). Step by step: infinite iterative processes and actual infinity. In F. Hitt, D. Holton & P. Thompson (Eds.) *Research in collegiate mathematics education. VII*. Providence, RI: American Mathematical Society

Dubinsky, E., K. Weller, M.A. McDonald & A. Brown. (2005a). Some historical issues and paradoxes regarding the concept of infinity: An APOS-based analysis: Part 1. *Educational studies in mathematics*, 58(3), 335–359.

—— (2005b) Some historical issues and paradoxes regarding the concept of infinity: an APOS-based analysis: Part 2. *Educational studies in mathematics,* 60(2), 253–266.

Mamona-Downs, J. (2001). Letting the intuitive bear on the formal: A didactical approach for the understanding of the limit of a sequence. *Educational studies in mathematics*, 48, 259–288.

Sierpinska, A. (1987). Humanities students and epistemological obstacles related to limits. *Educational studies in mathematics*, 18(4), 371–397.

Tirosh, D. (1991). The role of students' intuitions of infinity in teaching the Cantorian theory. In D. Tall (Ed.), *Advanced mathematical thinking* (pp. 199–214). Dordrecht: Kluwer.

Tirosh, D. & Tsamir, P. (1996). The role of representations in students' intuitive thinking about infinity. *International journal of mathematical education in science and technology*, 27(1), 33–40.

Tsamir, P. & Tirosh, D. (1999). Consistency and representations: the case of actual infinity. *Journal for research in mathematics education*, 30(2), 213–219.

Tsamir, P. (2001). When the same is not perceived as such: The case of infinite sets. *Educational studies in mathematics*, 48, 289–307.

—— (2003). Primary Intuitions and Instruction: the Case of Actual Infinity. In A Selden, E. Dubinsky, G. Harel, & F. Hitt (Eds.), *Research in collegiate mathematics education. V* (pp. 79-96). Providence, RI: American Mathematical Society.

Weller, K., Brown, A., Dubinsky, E., McDonald, M., & Stenger, C. (2004). Intimations of infinity. *Notices of the American Mathematical Society*, 51(7), 741–750.

6

Layers of Abstraction: Theory and Design for the Instruction of Limit Concepts

Michael Oehrtman
Arizona State University

Imagine asking a first-semester calculus student to explain the definition of the derivative using the epsilon-delta definition of a limit. Given the difficulty of each of these concepts for students in such a course, you might not be surprised at the array of confused responses generated by a question requiring understanding of both. Since the central ideas in calculus are defined in terms of limits, research on students' understanding of limits and the ways in which they can develop more powerful ways of reasoning about them has significant implications for instructional design. Throughout this paper we will focus on calculus courses intended as an appropriate introduction for students who have never seen limits or derivatives and that are not intended to be a rigorous treatment of analysis. The following typical response to the question relating the definitions of limit and the derivative illustrates the confusion that students exhibit when trying to make such connections. This response was offered by an A-student, who we will call Bob, during a clinical interview late in a first-semester course:

> Your epsilon — this — the slope of this tangent line. You want to pick a set of *x*'s, and that's here [*points at graph*]. This *x*, it's barely changing such that it's equal to or less than this tangent line. That would be your delta. The slope — oh, OK. The slope of this tangent line [*points at tangent*] — that's epsilon. The slope of this line [*points at secant*] that you're making is your delta at 2. Take a delta — a slope of this line [*points at secant*] less — such that it is less than the slope of this tangent line.

Bob's language is confused, but it seems he was identifying epsilon as the slope of the tangent line and possibly both *x* and delta as the slope of a secant line and indicates that he wants the latter to be smaller. It is quite likely that this question was beyond Bob's conceptual resources and that he was simply trying to make any connection possible to appease the interviewer. Had he been present, Bob's professor might have been rather discouraged by this response given the efforts he made in class, during special study sessions, and through creatively designed homework to help his students understand the formal definitions of both limits and derivatives.

Only a few seconds later, however, the interviewer asked Bob to explain the same idea in terms of approximation and received this response:

> There will be — there could be a difference in the slopes of these lines. You could say that the slope of this line [*points at secant*] is approximately equal to this [*points at tangent*] with a margin of error of such and such, and that margin of error can be less than that [*points at the word "bound"*]. You can choose a slope that's less than the margin of error — less than whatever you need it to be.

When asked to explain his use of language about "error" in more detail, Bob explained

> Your margin of error is here [*holds up hands facing each other to indicate a distance*] and here's your limit [*waves one hand*] and you have to be at least in so far closer to it [*waves other hand across the space in*

between]. You can always get closer to it, you know? That's the way I was looking at bounding. You can always get closer to it.

Bob's characterization of the limit is noticeably different in this excerpt. He described approximating the slope of a tangent line using the slopes of secant lines with an error that can be made smaller than some predetermined bound. Much of the logic involved in this statement is identical to the logic of the epsilon-delta statement that he completely failed to interpret only moments earlier.

What did Bob understand about limits? What about derivatives? What bearing did his understanding of limits have on his understanding of derivative? In this chapter, we will explore the pedagogical implications of this sort of discrepancy in students' ability to articulate mathematical concepts involving limits in a wide variety of situations. This will lead into a design perspective on how we might better help students learn and use limit concepts. Before engaging in that task, however, we will address how abstract concepts develop in general and identify various goals for teaching limits.

The Nature and Process of Abstraction

One of Jean Piaget's most forcefully repeated conclusions from careful observations of the nature of abstraction relates to the source of abstract concepts. Specifically, he argues that the source of conceptual structure such as that found in mathematics is an individual's *actions* or *coordinations of actions* on physical or mental objects (Piaget, 1970a, 1970b, 1975, 1980, 1985, 1997). As illustrated in Figure 1, the significance of this statement is that it emphasizes actions rather than other potential sources, such as objects, their properties, or even relationships among objects. To serve as the source of an abstracted concept, such actions must be engaged repeatedly while receiving and incorporating feedback under the specific constraints of a system that is being explored. Piaget (1970b) used the structure of the algebraic group as a prototype for all conceptual structure to emphasize the way in which actions embody structure that can be abstracted by the individual. He emphasized the role of operations (such as physically or mentally performing a symmetry transformation of a figure) and ways in which they can be coordinated (such as realizing an associative property or inverse condition) in addition to the elements with their properties and relations. Conceptual structure, such as that representing a dihedral group, is formed as a whole from the inseparable interplay between these elements and operations.

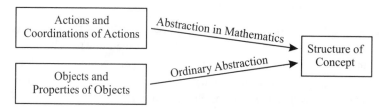

Figure 1. The source of complex abstract concepts are actions and coordinations of actions.

Piaget's characterization of the process of abstraction applied to limit concepts suggests three important features of instructional activity. First, the structure of understanding we hope our students will achieve should be systematically reflected in the actions we ask them to perform. Since these activities form the basis of conceptual understanding and thus also precede such understanding, they must be stated in terms more accessible to students than formal definitions. If a formal understanding (such as an ability to interpret and apply epsilon-delta definitions in a variety of contexts) is to eventually develop, it will be built on conceptual structures that already make sense to students because of their previously internalized activity. On the other hand, if students will never be expected to develop such formal understandings, then their conceptual structures and abilities can still reflect rigorous and appropriate mathematics. For example an engineering student, such as Bob who was introduced at the beginning of this chapter, might be able to develop a general understanding of techniques by which the error for an approximation may be made smaller than some required bound without this being formalized as finding a $\delta > 0$ such that $|f(x) - L| < \varepsilon$ whenever $0 < |x - a| < \delta$.

The second feature of instructional activity we infer from Piaget's theory of abstraction is that students' actions should be repeated and coordinated in ways that help them attend to feedback obtained from the inherent constraints of the system being explored. In the preceding example, the dependence of delta upon epsilon is a key feature of the structure. Students struggle with this dependence, however, since it moves in the opposite direction from the action of the function (i.e., it moves from a condition in the range to a condition in the domain). A student such as Bob may only attend to the appropriate dependence due to encountering a difficulty that otherwise arises, for example, through a real need to find an approximation with sufficient accuracy for some purpose.

The third implication we draw from Piaget's theory is that instruction on the limit concept should not be isolated, but extend throughout its many applications in the calculus curriculum in ways that foster mutual growth. The concepts defined in terms of limits provide fertile ground for continued exploration into important issues related to limits. Reciprocally, an emerging understanding of the depth of limit structures can help guide students' explorations into these other concepts.

Goals for Teaching Limits

Instructional decisions regarding the teaching of limits will ideally follow from specific objectives for students' learning. We will briefly outline a small number of possible goals and trace research related to the implications of pursuing each one.

Exposure to formal definitions and proofs. A possible objective for the instruction of limits is for students to develop facility with formal epsilon-delta and epsilon-N definitions and arguments. In fact, the *Principles and Standards,* published by National Council of Teachers of Mathematics (NCTM) argue that throughout their mathematical careers, students should continually engage in proof and argumentation. Construction of simple epsilon-delta proofs provides opportunities to interpret and then use definitions in a rigorous fashion (cf., Edwards and Ward, this volume, for a discussion of the role of definitions in formal proofs). Other objectives that would lead to a similar approach include fostering students' understanding of limit concepts in terms of epsilon-delta (or epsilon-N) arguments preparing students for more advanced mathematical study and establishing a rigorous foundation for the entire calculus curriculum. Such goals were especially pursued during the 1960s and early 1970s in efforts to increase the rigor of mathematics curricula to support a growing demand for scientists, engineers, and mathematicians.

Most calculus textbooks include a section on the formal definitions of limits, providing basic epsilon-delta and epsilon-N definitions, some pictures and intuitive explanations using graphs of functions and sequences, simple existence proofs for specific limits, and proofs of basic properties of limits (such as linearity). These ideas are typically presented in the text shortly after limits are introduced, but since most introductory calculus courses are not intended to provide a rigorous treatment of analysis, they are rarely raised afterwards. Several education researchers and curriculum committees have concluded that carrying formal limit proofs forward throughout an introductory calculus sequence might be successful in preparing a small number of the most talented students for further studies in advanced mathematics, but it leaves the vast majority of students with little more than a procedural understanding and an impression of mathematics as personally incomprehensible (Davis, 1986; Tall, 1992; Tucker & Leitzel, 1995). It is unclear whether introducing formal definitions even conveys to students a sense that there is a rigorous foundation for the mathematics. Consequently, these definitions and proofs are often de-emphasized in current curricula and courses as explicitly recommended in the report of the content workshop for the MAA publication *Toward a Lean and Lively Calculus* (Tucker, 1986).

A more modest goal for introducing limit proofs than providing a rigorous treatment of the entire calculus curriculum is to engage students in a limited amount of formal mathematical argumentation. Unfortunately, many instructors find little time to devote to this goal under the pressures of an expansive curriculum. As a result, most students only learn the basic patterns to complete simple algebraic proofs or learn the rules and peculiarities of a particular representation (e.g., games where you "keep the graph in the box" on a calculator or player 1 challenges with an ε and player 2 finds a δ) without understanding the connections to other representations, potential applications, or other content in the course (Jacobs, Larsen, & Oehrtman, 2003).

From Piaget's characterization of the process of abstraction, we can understand some of the difficulties students have with formal limit concepts. Instruction that begins with formal definitions attempts to move in the opposite direction from which abstraction naturally occurs. When students are first exposed to the concepts in calculus, there is no conceptual structure through which they can meaningfully interpret key features of formal limit structures. Based on Piaget's theory of abstraction and refined through a series of clinical interviews with students, Cottrill et al. (1996)

have proposed a progression of actions that students must abstract, generalize, and relate to one another in order to construct such a conceptual structure. For the limit $\lim_{x \to a} f(x) = L$, they suggest that students must first abstract the actions of evaluating f at points near a, then develop and coordinate domain and range processes of x approaching a and $f(x)$ approaching L. Then this coordinated structure must be reinforced by performing actions on limits, such as by considering limits of combinations of functions. Only at this stage in Cottrill et al.'s framework are students able to reconstruct these coordinated processes in terms of inequalities, apply a consistent understanding of the universal and existential quantifiers, and develop a complete epsilon-delta conception for a specific situation.

Attempts to support students' understanding of a formal definition with an intuitive rephrasing such as "You can make $f(x)$ arbitrarily close to L by making x sufficiently close to a" also provide neither appropriately structured activity nor underlying meaning, so are likely to fail as well. Instead, under the burden of making some sense out of what is being said, students attach simpler meanings to these phrases. In interviews with students throughout three semesters of calculus being exposed to such language, nearly all interpreted the modifiers "arbitrarily" and "sufficiently" in the simplest way possible: as indicators of degree (Oehrtman, 2002). To them, "sufficiently small" meant "very small" and "arbitrarily small" meant "very very small." These students did not have any experiences from which the intended logical entailments of these phrases could be generated.

If students are not expected to use epsilon-delta definitions and arguments throughout a course, the corresponding conceptual structure is neither continually reinforced nor developed for use as a powerful tool. Consequently, it is unclear how formal limit definitions and proofs could guide students' exploration into subsequent topics without offering a nearly complete analysis course.

Intuitive understanding. Most secondary and introductory undergraduate calculus courses and textbooks take an approach to limits that focuses on intuitive ideas and phrasings, such as "when x gets close to a, $f(x)$ gets close to L." Even if formal definitions are introduced and used to prove some basic properties of limits, they are de-emphasized or abandoned when advancing to subsequent concepts, even those that are defined in terms of limits. The definition of derivative is rarely treated in terms of epsilons and deltas in an introductory calculus course, for example. One purpose of treating limits with an intuitive approach is to provide a common and accessible introduction to other concepts throughout the course (cf., Speiser & Walter, this volume). Derivatives are discussed in terms of secant lines where the points are made closer and closer to each other, and definite integrals are defined as summing products over intervals that get smaller but increase in number. Another objective leading to similar approaches is the need to teach a number of techniques for algebraic computations that will be used later in the course, such as finding certain derivative formulas, determining specific values for improper integrals, or applying convergence tests for infinite series. Since these skills do not require students to understand epsilon-delta and epsilon-N structures, formal definitions are often de-emphasized and intuitive descriptions of limits viewed as sufficient.

When subsequent topics are introduced through an informal understanding of limits, the role of limits is typically suppressed. Operationally, the limit concept is often concealed conceptually by definitions of derivatives in terms of slope, definite integrals in terms of area, Taylor series as actual sums, etc. Corresponding to this conceptual shift, limits may also be de-emphasized notationally. For example, $f'(a)$ may be described as the slope of the tangent to the graph of f at $(a, f(a))$, the definite integral $\int_a^b f(x)\,dx$ may be described as the area under the graph of f on the interval $[a,b]$, or the Taylor series

$$\sum_{n=0}^{\infty} \frac{f^n(a)}{n!}(x-a)^n$$

as being equal to $f(x)$ for most functions students will see. Limit notation is absent from all of these descriptions and limit structures (as encapsulated by epsilon-delta or epsilon-N definitions) are even further in the background. Intuitive images such as tangent lines, areas, and infinite sums are often used as a proxy for limits since they are conceptually accessible to students and can be extremely powerful for intuitive reasoning (Monk, 1987, 1992; Rodi, 1986; Tall, 1992; Thompson, 1994).

Informal language and reasoning about limits can also lead to misconceptions for even advanced students who are supposedly equipped with the formal tools to avoid such errors. Twenty-two students in their final year of university mathematics and who had dealt with the formal epsilon-delta definition of limits for two years were asked the following question:

> True or false: Suppose as $x \to a$ then $f(x) \to b$, and as $y \to b$ then $g(y) \to c$. Then it follows that as $x \to a$ then $g(f(x)) \to c$.

All but 1 of these 22 students responded "true" and refused to change their answer even when pressed (Tall & Vinner, 1981). Whether considered explicitly or subconsciously, the logic of this statement as typically verbalized establishes a false syllogism by stating "If x approaches a then $f(x)$ approaches b, and if y approaches b then $g(y)$ approaches c." If the first premise holds, then $f(x)$ satisfies the hypothesis of the second premise, i.e., it qualifies as a y that approaches b, thus leading to the incorrect conclusion that $g(f(x))$ approaches c. Additionally, the arrows and their verbalizations as "approaches" or "goes to" are represented in the same way for both dependent and independent variables in this problem. The suppression of different technical meanings by using the same notation for both contributes to students' misconceptions.

In another study on the use of intuitive dynamic language, when faced with challenging problems involving limiting situations, students did not rely on images of motion to reason about the problems (Oehrtman, 2003). This is particularly surprising given the predominance of motion language used when talking about limits and abundant proclamations that intuitive, dynamic views of functions should help students understand limits. While students frequently used words such as "approaching" or "tends to," these utterances were not accompanied by any description of something actually moving. When asked specifically about their use of such phrases, students denied thinking of motion and gave an alternate explanation for their words.

In terms of abstraction, the informal approaches to limits described in this section are susceptible to a similar problem as purely formal approaches. Without other supports, they do not provide students with a structure that can guide their investigation of the relevant mathematics of subsequent topics. Instead of providing an incomprehensible structure, they provide little to no structure, but the result is still that students are left to construct an understanding based on disjoint and possibly unguided connections and images. The work of understanding subsequent topics is then shifted to representations specific to each one (a lack of gaps for continuity, steepness for derivatives, area for definite integrals, etc.). Each of these understandings, then are bound to a specific representation (typically graphical), and for students to reason conceptually requires a translation back and forth between that representation and the problem context. It is difficult for students to see and work with the commonalities between these images as required, for example, to understand the fundamental theorem of calculus. Further, since the central concepts in calculus are defined in terms of limits, important aspects of these structures are also lost to the same degree that limits are de-emphasized.

De-emphasis of limits and alternative foundations of calculus. Due to the well-documented difficulty of limit concepts, several researchers and reformers have suggested providing a more intuitive starting point for calculus. The most common is based on using infinitesimals, rounding, and local linearity. This approach mirrors some aspects of Newton's reasoning with fluxions and fluents and Leibniz's notational encapsulation of infinitesimal quantities. Concerns about lack of rigor are addressed by referring to Robinson's set-theoretic work in the 1960s establishing nonstandard analysis as a logical foundation for an infinitesimal approach. Promising aspects of infinitesimal instruction are that the foundational concepts are accessible and that, in some cases, students used those ideas as integral parts of their reasoning.

Citing historical and cultural difficulties related to concepts of function, limit, infinity, and proof, Tall (1986, 1990, 1992) suggests that a better cognitive starting point for calculus might be "local straightness." Students are introduced to tangents via magnification of the graph of a function at a point. This approach, he suggests, allows for the investigation of a rich source of concepts: different left and right gradients, functions that are locally straight nowhere, etc. Students taught with this approach were much better at recognizing, drawing, and reasoning about graphical information for derivatives than students in a control group. On the other hand, they tended to describe a tangent as passing through two or more very close points on the graph. At least part of these students' difficulties seems to be conflation of the tangent line and the actual graph caused by the appearance of the graph as a straight line after sufficient magnification.

In the infinitesimal approach, computations are performed using an infinitesimal element, ε, and standard algebra extended to the infinitesimals. This process is followed by "rounding off" infinitesimal terms, so that an expression like $2x + \varepsilon$ is replaced by $2x$. Tangents are then treated by magnification as described above with the addition that the graph is magnified to an infinitesimal scale. Frid (1994) found that although students who were given instruction with infinitesimals did not perform significantly better on standard computations, they did use the language and notation of rounding as an integral part of their explanations. Whether the students' use of everyday language was of help or a hindrance depended on the extent to which they integrated that informal language with technical language or symbols in ways congruent with the corresponding concepts.

Michèle Artigue (1991) conducted a study with 85 third-year university students enrolled in multivariable calculus and physics courses to investigate their understanding and use of differential elements. In their courses, students were provided with a tangent linear approximation definition, which dominated their descriptions of differentials. At the procedural level, however, they reverted to treating differentials algebraically in algorithms involving partial derivatives and Jacobian matrices. Students were not able to identify conditions in specific contexts necessitating the use of differentials and gave incorrect justifications about convergence of approximations based on convergence of geometric "slices" in a diagram. Such arguments also may be common for students who have not received infinitesimal instruction, however. Thompson (1994) observed advanced mathematics students incorrectly justify the fundamental theorem of calculus by arguing geometrically that the shape of a three-dimensional object with thickness Δx converges to a two-dimensional object as $\Delta x \rightarrow 0$. Oehrtman (2002, 2003) classified a ubiquitous category of such reasoning in terms of "collapsing dimensions" among freshmen calculus students and secondary mathematics teachers in a wide variety of problem contexts.

Ideas of local linearity contain conceptual pitfalls when used to supplement a standard treatment of calculus. In a class regularly exposed to descriptions of zooming in on graphs, students did not ever develop these concepts for use in any of their own explanations about limits (Oehrtman, 2002). When directly probed about what they would see when zooming in on the graph of a function, only 10 out of 77 gave a response that was relevant to the mathematics, and these were all incorrect, suggesting that one would see a straight line because the vertical change is reduced to a very small amount (although this argument seems to imply that one would see a horizontal line). All of the other students appealed to non-mathematical interpretations such as images of the line becoming thicker or blurrier under magnification or that you would see individual calculator pixels or atoms of paper. This indicates that images of zooming did not provide these students with sufficient structure to guide their reasoning and the related instructional process lacked the necessary feedback to prevent major misconceptions. Additionally, subsequent analysis-based mathematics courses are rarely taught in terms of nonstandard analysis and science and engineering rarely use mathematical models that incorporate infinitesimals.

A Design Approach to Limit Instruction

The main thrust of this chapter is to frame a set of objectives related to the learning of limits, taking into account many of the goals introduced above, and to outline an instructional approach based on research and refined through several teaching experiment cycles. One of our main objectives is to base the instruction on activities that are conceptually accessible to students. As discussed above, this has been achieved by others, notably through infinitesimal approaches, and we have drawn from their successes. A second goal is to structure students' understanding in ways that reflect formal definitions. The purpose of this is to lay a conceptual groundwork from which formal understandings may later emerge but not necessarily to provide those formalizations themselves. Such an approach could, of course, leave open the option for an instructor to develop these definitions at an appropriate time. Third, we strive to establish an instructional approach for limits that serves as a guide for the investigation of all other concepts defined in terms of limits in ways that enhance exploration of their underlying structures. Finally, the approach should allow and encourage flexible application in all representations (algebraic, graphical, numerical, contextual/descriptive, etc.). The diversity of these goals leads to the consideration of an important additional constraint: we require an approach that is coherent. That is, the treatment should be mutually reinforcing across the entire calculus curriculum, and the process of achieving each goal outlined above should support the attainment of the others.

A design process. From a design perspective we have sought to achieve these goals via the following process:

1. Identify the mathematical structures (elements, operations, relations that result from coordinating operations, etc.) that must be reflected in the instructional activities.

2. Identify a structurally equivalent conceptual system and language base that is accessible to students. This is achieved by documenting students' natural reasoning, developing possible frameworks of mathematical

expressions for this reasoning, then evaluating the effectiveness of structuring students' activities around these mathematical versions of their natural reasoning.

3. Develop, test, and refine instructional activities in which students apply the framework to particular applications. Students work in groups on structurally similar problems in a variety of contexts and then present results to each other, reinforcing the structure across novel contexts and problems. Design whole-class discussion to elicit the common features across all applications. Initial activities should focus on familiarizing students with the language, notation, and procedures of the framework and assisting them in choosing and applying its tools (e.g., focusing on types of questions generally asked, common procedures that may be used, and relevant representations of the results). Later activities should encourage students to reason through solving problems on their own.

4. Repeat Step 3 for a variety of applications of the concept. This establishes a second level of activities in which students are encouraged to see similarities across different uses of the concept and develop a more general and robust abstraction of the concepts.

5. Design tasks to foster formalization as an end result. This includes naming or symbolizing a structure that has already been abstracted and can lead to discussion and use of formal definitions and proofs.

The overarching principle is that students should engage in multiple activities that reveal and encourage the abstraction of a common structure, and the results of many such abstractions should share common features to allow for further levels of abstraction. At each level, students should participate in experientially real activities designed to engage them in the relevant structures of the underlying mathematics (although not necessarily the formal representations) and in seeing common structures across multiple experiences. This allows an abstract understanding to emerge over a long period of time with significant reinforcement at a variety of conceptual levels. The concept may be formalized near the end of this process as a way to concisely capture a well-understood structure.

In the case of limits, a particular application developed in an iteration of Step 3 would stem from the limit structure involved in the definition of the derivative (see Figure 2). The activities are designed to reflect the predetermined

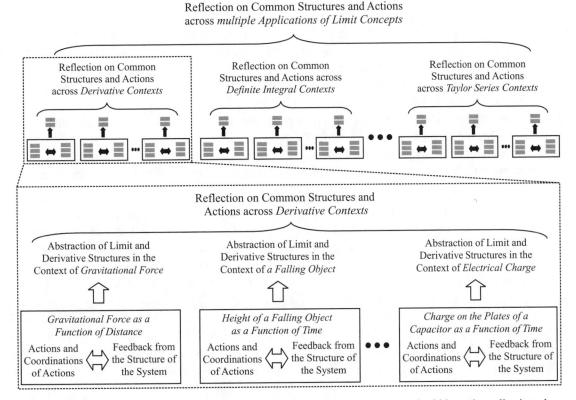

Figure 2. Layers of Abstraction: A common structure for limit concepts is repeated within each application, then across multiple applications to provide coherence throughout the calculus curriculum.

limit structures to provide an appropriate source for later abstraction. Different groups of students present their work detailing the operations and relations involved in applying the limit framework to different rate of change contexts. Although the contexts are different, the underlying structure is the same, and classroom discussion is focused on drawing out the common features. As this is repeated for the limit of a function at a point, the definite integral, the fundamental theorem of calculus, Taylor series, etc., there is variation in the structures of these different topics but certain consistencies in the underlying limit structures.

We have identified in our goals, the formal limit definitions as capturing the underlying structure to be modeled through instruction. This means students should develop to use conceptual tools corresponding to the algebraic entities and expressions in the definitions, guided by the types of operations possible within the underlying logical connections between these expressions. We have also identified the need to apply this structure to solve problems in the contexts of other concepts within the calculus curriculum.

Students' spontaneous reasoning with approximation concepts. Formal limit definitions and structures are often considered beyond the reach of most introductory calculus students. Students, however, often naturally reason about limit concepts in terms of approximations in ways that are structurally equivalent to aspects of formal epsilon-delta and epsilon-N definitions (Oehrtman, 2004). For example, students may be able to construct an idea such as "the slope of the tangent line is approximated by slopes of secant lines, and the errors (differences between the two slopes) can be made smaller than any predetermined bound" even though merely interpreting an abstract statement such as "for every $\varepsilon > 0$, there is a $\delta > 0$ such that whenever $0 < |x - x_0| < \delta$ then $\left| \frac{f(x) - f(x_0)}{x - x_0} - m \right| < \varepsilon$" may be entirely beyond their reach. Furthermore, instruction can foster the development and application of appropriate versions of such reasoning so that it may become a basis for understanding the formal statements and incorporating aspects into their reasoning with approximations (Oehrtman, 2004). For this reason we take the stance that epsilon-delta and epsilon-N definitions should only be introduced after multiple rounds of instruction that reinforce the conceptual structure of limits in different settings as depicted in Figure 2.

In a study to characterize calculus students' spontaneous reasoning patterns while working with limits, Oehrtman (2002) collected responses to short writing assignments from an entire class of 120 students and more in-depth descriptions of students reasoning from 25–35 students from regular online writing assignments. Nine students participated in initial clinical interviews during which the interviewer prompted for detailed explanations of their reasoning about the meaning of limits through standard problems, and follow-up interviews were conducted with an additional 11 students. Approximation ideas emerged as the strongest and most frequently applied metaphor for limits in this study, and students' reasoning while thinking about approximations were more likely to reflect the correct mathematical structures than any of the other contexts that emerged. These results may not be surprising since much of calculus is historically motivated by needs for numerical estimation techniques, and these ideas continue to influence our classroom and textbook presentations. Consider the following quote from a typical second-semester calculus student as she explains her understanding of the equality $\sin x = x - \frac{1}{3!}x^3 + \frac{1}{5!}x^5 - \frac{1}{7!}x^7 + \cdots$. Attend to her use of the words "approximation", "error", and "accuracy" and how their usage matches the structures in epsilon-N convergence arguments.

> When calculating a Taylor polynomial, the accuracy of the approximation becomes greater with each successive term. This can be illustrated by graphing a function such as $\sin(x)$ and its various polynomial approximations. If one such polynomial with a finite number of terms is centered around some origin, the difference in y-values between the points along the polynomial and the points along the original curve ($\sin x$) will be greater the further the x-values are from the origin. If more terms are added to the polynomial, it will hug the curves of the sin function more closely, and this error will decrease. As one continues to add more and more terms, the polynomial becomes a very good approximation of the curve. Locally, at the origin, it will be very difficult to tell the difference between $\sin(x)$ and its polynomial approximation. If you were to travel out away from the origin, however, you would find that the polynomial becomes more and more loosely fitted around the curve, until at some point it goes off in it's own direction and you would have to deal once again with a substantial error the further you went in that direction. Adding more terms to the polynomial in this case increases the distance that you have to travel before it veers away from the approximated function, and decreases the error at any one x-value. Eventually, if an infinite number of terms could be calculated, the error

would decrease to zero, the distance you would have to travel to see the polynomial veer away would become infinite, and the two functions would become equal. This is a very important and useful characteristic, as it allows for the approximation of complicated functions. By using polynomials with an appropriate number of terms, one can find approximations with reasonable accuracy.

This student received no special instruction related to ideas about approximation, yet the language of approximation, errors, and accuracy figured prominently and systematically in her reasoning. Furthermore, the structure of these ideas for this student reflect a sophisticated understanding of limits and is integrated with her understanding of various aspects of Taylor series, such as the relationship between graphs of a function and several Taylor polynomials, pointwise convergence, and the radius of convergence. These types of statements were common among students trying to make sense out of limit concepts in their own language (Oehrtman, 2002).

The main components of students' spontaneous use of approximation ideas to reason about limits consisted of an unknown actual quantity and approximations that are believed to be close in value to the unknown quantity. For each approximation, there is an associated error,

$$\text{error} = |\text{ unknown quantity} - \text{approximation }|.$$

Consequently, a bound on the error allows one to use an approximation to restrict the range of possibilities for the actual value as in the inequality

$$\text{approximation} - \text{bound} < \text{unknown quantity} < \text{approximation} + \text{bound}.$$

An approximation is contextually judged to be accurate if the error is small, and a good approximation method allows one to improve the accuracy of the approximation so that the error is as small as desired. An approximation method is precise if there is not a significant difference among the approximations after a certain point of improving accuracy.

The structure of this schema parallels the logic of epsilon-N and epsilon-delta definitions of limits. For the latter, bounding the error corresponds to the statement "then $|f(x) - L| < \varepsilon$". The need to obtain *any* predetermined degree of accuracy evokes the requirement that the condition hold "for any $\varepsilon > 0$." A mechanism to generate better approximations corresponds to the phrase "there exists a δ such that whenever $0 < |x - a| < \delta$." Linking these structures together gives the practical statement of being able to find a suitable approximation for any degree of accuracy on the one hand and the formal epsilon-delta definition on the other. Students' intuitive descriptions of precision such as "There will not be a significant difference among the approximations after a certain point." reflect the structure of Cauchy convergence, if $n > N$ then $|a_m - a_n| < \varepsilon$. These structures are consistent even with a generalized definition of definite integral as a net, with partitions partially ordered by refinement. In terms of approximations such a description may be even more intuitively accessible than a simple limit definition and its restrictions on the types of approximations considered.

Instructional Activities. The types of tasks discussed in this section have been tested and refined in use with three different student populations: a standard introductory calculus class, a supplemental calculus workshop, and a summer professional development workshop for high school teachers. Research teams videotaped all instructional settings and analyzed data to discern which aspects of the activities reinforced the desired conceptual structures and provided students with powerful reasoning tools and where potential difficulties might arise. In each round, the instructional activities were refined accordingly.

An initial task that must be accomplished by the instructional activities is to systematize students' spontaneous understandings related to approximations so that a relevant and standard set of ideas and language can be developed by the class to use in further explorations. Williams (1991) found students' exhibited strongly held sets of beliefs typically surrounding the contexts in which they were first exposed to limits and that their viewpoints were extremely resistant to change, even in response to explicit discussions about contradictory examples. Students viewed counterexamples as minor exceptions rather than reasons to abandon an incomplete concept and evaluated the appropriateness of any particular conceptualization based on its usefulness in a given setting rather than on its rigor, consistency, or correctness. These are hallmarks of spontaneous reasoning which is not volitional or structured (Vygotsky, 1987). The development of students' scientific concepts alongside their spontaneous concepts can be slow and difficult, but in any instructional process related to limits, something similar will be necessary. The key is to have a strong set of spontaneous concepts (as is the case with students' approximation ideas) to enable and mediate this process. To accomplish this, we have developed a variety of heavily "scaffolded" tasks (tasks with significant initial instructional support designed to be gradually removed throughout subsequent activities as students develop proficiency). Examples are shown in Figures 3 and 4.

These highly scaffolded activities introduce the students to limit structures using applications that are fully developed later in the course, such as the derivative or continuity. Other initial tasks engage students in similar activities using different applications of limits such as the definite integral structure presented as approximating how far a wind-up car would travel and infinite series presented as "mystery" sums. Although there are opportunities for some discussions about these other topics at this stage, best results were obtained by focusing group and class discussions on the limit structures within each application in order to reinforce the use of ideas about approximations, errors, and bounding errors. In each case, students were able to reason within the given context to determine whether specific approximations were overestimates or underestimates, then to find an actual bound on the size of the error. They were then asked to reverse this process and find several approximations with errors smaller than pre-determined bounds. In students' work and presentations, continual emphasis that these two processes are the reverse of one another was necessary to help students understand the distinction and when one way of reasoning would be required over the other.

Subsequent activities provide fewer step-by-step instructions for the students with the expectation that they will begin to remember or develop appropriate strategies to solve increasingly more sophisticated problems. Through the

---

In the following problem, you will approximate the slope of the tangent line to a curve at a point. There are several important ideas about approximation that are embedded in these exercises that have a close relationship to the limit concept. You will need a graphing calculator or a graphing program on a computer.

Graph $y = 2^x$ on a calculator or computer over the interval $[-3,3]$ and take careful note of the general shape of the curve. Now zoom in on the graph at $x = -1$. That is, change the window to show the graph over a smaller interval around $x = -1$, like $[-2,0]$. Notice that the graph appears less curved and more like a straight line. If you keep zooming in around $x = -1$, the graph will appear more and more like a straight line. This is called the tangent line to the graph of $y = 2^x$ at $x = -1$. The details of tangent lines will be developed more fully later in this course. For now, you will approximate the slope of the tangent line.

1. Look at a region of the curve where it appears fairly straight but still has a slight, noticeable curvature, e.g., on $[-2,0]$. Take a point on the curve to the *right* of the point at $x = -1$, and find the slope between these two points. (Make sure to keep as many decimal places in your calculation as possible since this exercise will require precision.)

2. Take a point on the curve to the *left* of the point at $x = -1$. Find the slope between these two points.

3. Are the two slopes from parts a and b both underestimates, both overestimates, or one of each? Explain how you know. (Hint: Use the fact that the graph of $y = 2^x$ is concave up, i.e., it curves upwards.)

4. Using your work from above, give a range of possible values for the slope of the tangent line. Using the center of this range as an approximation, what is a bound on the size of the error?

5. Explain why your bound is just an *upper bound* for the error and not *exactly* the error.

6. Zoom in and use points to the left and right of $x = -1$ to find an approximation of the slope of the tangent line with error less than 0.0001. Record your work for each computation you do.

7. Explain why any points between $x = -1$ and the points you used in Part f would result in an approximation with error less than 0.0001.

8. What other x-values can you use for the second point and have the error be less than 0.0001.

Be prepared to answer:

1. What unknown value were you approximating?

2. What were your approximations?

3. Describe what the error for each approximation was. Why is the exact value of the error impossible for you to determine?

4. How did you bound the error?

5. Explain a procedure for getting an approximation with error smaller than any pre-determined bound.

Figure 3. A scaffolded activity on the slope of a tangent line designed to reinforce approximation structures relevant to limit concepts. This task is used before limits or derivatives are formally introduced to lay a foundation for the conceptual structure.

The graph of

$$f(x) = \frac{\sqrt[3]{x+7} - 2}{x-1}$$

has a hole. Your task is to determine the location of this hole using the approximation techniques you have learned.

1. Identify what unknown numerical value you will need to approximate. Give it an appropriate shorthand name.

2. Determine what you will use for approximations. Write out your answer algebraically.

3. Draw the graph using your entire whiteboard. Depict your answers to #1 and #2 on the graph with labels for each.

4. What is an algebraic representation for the error in your approximations? Add a graphical representation to your picture.

5. List three fairly decent approximations. For each one, give a bound for the error and use this to determine a range of possible values for the actual value. Add one of these values to your picture and depict both the error bound and the range of possible values. Don't forget to label everything!

Approximation	Error Bound	Range of Possible Values

6. Find an approximation with error smaller than 0.0001. Then describe **all** of the approximations that would have an error smaller than 0.0001. Add this to your picture.

7. For any pre-determined bound, can you find an approximation with error smaller than that bound? Explain in detail how you know.

Figure 4. A scaffolded activity on the limit of a function designed to reinforce approximation structures relevant to limit concepts.

teaching experiments, we have determined that once students are able to complete introductory activities such as the ones above, they are ready to begin group work on less scaffolded tasks as shown in Figure 5. With some preliminary discussion about average rate of change and intuitive interpretations of instantaneous rate of change, these activities prepare students for the introduction of the definition of the derivative.

Eventually, students are given problems with very few prompts regarding approximation structures. Consider, for example, the problems posed in Figure 6. Typically, such contexts would be presented to students as tasks to construct a definite integral and evaluate using the fundamental theorem of calculus. The slight change in this formulation requires students to coordinate the product, sum, and limit structures of the definite integral across multiple representations. Table 1 provides brief descriptions of typical responses expected of and provided by students in previous teaching experiments for Context 2 of Figure 6.

Algebraic and numerical representations of the actual value, such as shown in the shaded cells of Table 1, are usually found last by students. Even when students attempt to begin their work by computing an integral, they are typically unable to determine the correct integrand until they have found and represented several particular approximations. Other than the shaded cells, students typically proceed sequentially down the columns by finding and representing the elements of each row in the table in turn (Sealey & Oehrtman, 2005). The need to express answers using all representations arose naturally as students needed to accomplish various sub-goals. For example, to figure out how to compute specific numerical values, students needed to carefully express their ideas on their picture of the dam, going through several revisions of their diagram and labeling. For larger computations that required a calculator or computer, students were forced to express their work algebraically in order to determine an appropriate command.

A typical group of Sealey's & Oehrtman's students wrestled with an incorrect definite integral $\int_0^{50} 62.5x \, dx$, then multiplied the pressure at the middle depth of the dam by its area to get what they believed was the actual answer. They then quickly determined how to find under and overestimates for partitions with ten subintervals then for 50 subintervals, and thought momentarily that they could not subdivide the partitions any further since they were now one foot each. Once they realized how to represent finer partitions algebraically and how to enter them into their calculators, they

Instructions: You will approximate the instantaneous rate of change for one of the situations below by answering each of the following questions algebraically, numerically, and by representing each answer in your diagram:

    0. Draw a large picture of the physical situation for the value of the variable given.

    1. Imagine how things are changing in this situation. List all of the quantities that you think are changing. Describe how they are changing.

    2. On the same picture, draw several "snapshots" of the situation.

    3. Label the changing and constant quantities in your drawing.

    4. Describe in more detail what you have been asked to approximate.

    5. What can you use for approximations?

    6. What are the errors?

    7. Find an approximation and a bound for the error. What is the resulting range of possible values for your instantaneous rate?

    8. How can you find an approximation with error smaller than a predetermined bound?

Context 1: An object is falling according to the equation $h(t) = 100 - 16t^2$ feet (with t measured in seconds). Approximate the speed when $t = 2$ seconds.

Context 2: Approximate the instantaneous rate of change of the area of a circle with respect to its radius when the radius is 3 cm.

Context 3: The force of gravity between two objects is inversely proportional to the square of the distance separating them. Approximate the instantaneous rate of change of the gravitational force with respect to distance when two objects are 230 km apart. (Note that all of your answers will involve the constant of proportionality.)

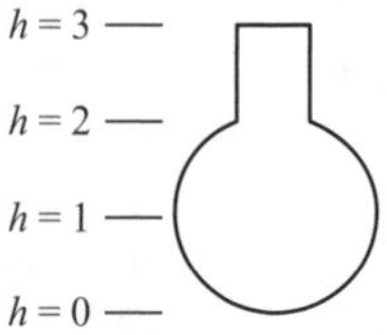

Context 4: Approximate the rate of change of the height of water in this bottle with respect to the volume of water when the height is 1.5. (Note that your answers will involve the size of the spherical portion of the bottle.)

Context 5: The half-life of Iodine-123, used in some medical radiation treatments, is about 13.2 hours. Thus a sample that originally has 6.4 μg of Iodine-123 will decay so that the amount left after t hours will be roughly $I(t) = 6.4\left(\frac{1}{2}\right)^{t/13.2}$ μg. Approximate the instantaneous rate at which the Iodine-123 is decaying after 5 hours.

Figure 5. Typical partially scaffolded activities developing limit and derivative structures through approximation ideas.

Instructions: Draw a picture of the situation, labeling everything possible. Determine a way to approximate the quantity requested. Be prepared to explain exactly how you obtained your approximations, what your errors are, how you can bound the errors, and how you can find an approximation with an error smaller than any predetermined bound. Express your answers algebraically and numerically, labeling appropriate quantities in your diagram.

Context 1: For a constant force F to move an object a distance d requires an amount of energy equal to $E = Fd$. Hooke's Law says that the force exerted by a spring displaced by a distance x from its resting length is equal to $F = kx$, where k is a constant that depends on the particular spring. If the spring constant is $k = .155$ N/cm, approximate to within 1000 ergs the energy required to stretch the spring from a position 5 cm beyond its natural length to 10 cm beyond its natural length. (Note that 1 erg = 10^{-5} N·cm.)

Context 2: A uniform pressure P applied across a surface area A creates a total force of $F = PA$. The density of water is 62.5 lb per cubic foot, so that under water the pressure varies according to depth, d, as $P = 62.5d$. Approximate to within 1000 pounds the total force of the water exerted on a dam 100 feet wide and extending 50 feet under water.

Context 3: The mass M of an object with constant density d and volume v is $M = dv$. A 10-meter long, 10-cm diameter pole is constructed of varying metal composition so that its density increases at a constant rate from 3 grams per cubic centimeter at one end to 20 grams per cubic centimeter at the other. Approximate the mass of the pole with an accuracy of 100 grams.

Context 4: Fluid traveling at a velocity v across a surface area A produces a flow rate of $F = vA$. Poiseuille's law says that in a pipe of radius R, the viscosity of a fluid causes the velocity to decrease from a maximum at the center $(r = 0)$ to zero at the sides $(r = R)$ according to the function $v = v_{max}\left(1 - r^2/R^2\right)$. Find an approximation of the rate that water flows in a 1-inch diameter pipe if $v_{max} = 2$ ft/s with an accuracy of 0.01 cubic feet per second (cfs).

Context 5: The volume V of an object with constant cross-sectional surface area, A, and height, h, is $V = Ah$. A large spherical bottle of radius 1 foot is filled to a height of 16 inches. Approximate the volume of water in the bottle to within 0.01 cubic feet.

Figure 6. Typical non-scaffolded definite integral questions in terms of approximation.

found under and overestimates for 100 and 800 subintervals (800 was the largest number of terms allowed by their calculators for computing a sum). At each of these steps, they noted that the actual force was somewhere between their values, that the error was bound by the difference, and that it was much larger than the desired 2000 pounds thus requiring further work. At this point, they proceeded to determine that achieving an error less than 2000 pounds would require 7750 subintervals. They became eager to actually try this and broke the problem into ten sub-problems with 800 or fewer terms each. All students in the group then agreed to find the sum for the first 800 terms to check their work, to work on different sums, and finally combine their results at the end. The students worked in a highly collaborative and engaged manner with these activities, and there was constant talk throughout that reflected the structure of both limits (finding

Table 1. Descriptions of typical responses in multiple representations for the approximation questions applied to the question about the force of water on a dam. Students typically produced responses in the order of descending rows with the exception of the two shaded cells which were often produced last.

	Contextual	Graphical	Algebraic	Numerical
Unknown Value: The force of water against the dam			$F = \int_a^b p(x) \cdot w(x) \ dx$ $= \int_0^{50} 62.5 \cdot 100 \cdot x \ dx$	$F = 7{,}812{,}500$ pounds
Approximation: Assume constant pressure across strips and use $F = P \cdot A$. Using pressure from the bottom of strips yields an overestimate. Pressure from top yields underestimate			$A = \sum_{i=1}^{10} 62.5 x_i \cdot 100 \cdot \Delta x$ where $x_i = 5i$ and $\Delta x = \dfrac{50-0}{10} 5$	$F \approx 156{,}250 + 312{,}500$ $+468{,}750 + 625{,}000$ $+781{,}250 + 937{,}500$ $+1{,}093{,}750 + 1{,}250{,}000$ $+1{,}406{,}250 + 1{,}562{,}500$ $= 8{,}593{,}750$ pounds
Error: The difference between the actual force and approximation			$E = \|F - A\|$ $= \left\| \int_0^{50} 62.5 \cdot 100 \cdot x \ dx - \sum_{i=1}^{10} 62.5 x_i \cdot 100 \cdot \Delta x \right\|$	Error $= E =$ $\|7{,}812{,}500 - 8{,}593{,}750\|$ $= 781{,}250$ pounds
Bound on the Error: The difference between an overestimate and underestimate, equal to the estimated force on the deepest strip.			$E <$ overestimate $-$ underestimate $= \sum_{i=1}^{10} 62.5 x_i \cdot 100 \cdot \Delta x$ $- \sum_{i=0}^{9} 62.5 x_i \cdot 100 \cdot \Delta x$ $= 62.5 \cdot 50 \cdot 100 \cdot \Delta x$	$E < 1{,}562{,}500$ pounds
Method to Achieve Desired Accuracy: Use thin strips: estimated force on the deepest < allowable error			For an error bound of ε, choose $n > \dfrac{62.5 \cdot 50^2 \cdot 100}{\varepsilon}$	For an error less than 2000 pounds, choose $n > 7812.5$

approximations, determining bounds for how far off they were, and determining how to achieve the desired accuracy) and definite integrals (breaking the problem into sections where pressure is nearly constant, computing forces using products, summing the results, and developing a general Riemann sum).

Results from these studies also indicate that the activities are effective in helping students systematize their reasoning around approximation ideas. Several of these students were able to make sense out of the epsilon-delta definition in terms of their approximation language, at which point they began interchanging language and symbols related to approximation and the formal definition and referred to them as being the same thing. This is an indicator of structured, scientific reasoning since it is only possible if the student is able to recognize the underlying structure despite different sets of terminology.

In subsequent activities, students are given progressively fewer prompts for techniques such as finding under and overestimates and are expected to apply these techniques appropriately on their own. For each task, they are asked to identify contextual, graphical, numerical, and algebraic referents for each of the following questions:

1) What are you approximating?

2) What are the approximations?

3) What are the errors?

4) What are bounds on the size of errors? and

5) How can the error be made smaller than any predetermined bound?

Again, questions four and five are emphasized as reciprocal processes so that students see and remember the purpose of each. Throughout all of these materials, the structure of the underlying limit concepts determines the nature of the instructional activities. Further, answering these questions encourages students' explorations into the relevant structures of the concept defined in terms of limits. In activities about the derivative, the need to approximate a rate of change to a given quantity results in the exploration of average rates of change over small intervals and the analysis of underestimates and overestimates based on arguments of increasing or decreasing rate derived from the context. In the previous activity on the definite integral and the examples shown in Figure 6, the structure of refinements to Riemann sums emerges as a result of engaging in the need to approximate to a given accuracy. Bob's quote at the beginning of this chapter in which he interpreted the definition of the derivative in terms of approximation is illustrative of the type of reasoning that has emerged consistently in the teaching experiments. Exploration, presentations, and discussion of multiple contexts exhibiting a common structure encourage the abstraction of the limit concept within the particular conceptual strand of calculus being covered. Figure 6 shows an example of such an activity for students to explore definite integral structures.

Summary

We have outlined several approaches to instruction related to limit concepts discussed in the mathematics education research literature. A typical class is often not represented by any one of these approaches but reflects a mixture of them. Regardless of the approach, however, the literature indicates students have major difficulties understanding limit concepts, which in turn impedes their understanding of other fundamental ideas in the calculus. In the first part of this chapter, we applied Piaget's theory of abstraction to characterize potential sources of student difficulties for various approaches. By highlighting these difficulties, we hope to assist individuals responsible for calculus instruction to address typical pitfalls. For example, an initial step in this direction might be to directly address common misinterpretations of imagery such as zooming in on a graph or viewing the fundamental theorem of calculus as being true as a result of an area collapsing in dimension to a line.

The second half of the chapter is intended to provide an example of designing an approach to calculus instruction that is coherent with respect to its treatment of limit concepts. The example provided uses common notions about approximations, is based on Piaget's theory of abstraction, and builds a structural understanding through repeated engagement in activities that reflect that structure. Certainly many other approaches could be designed to accomplish similar results. Our research shows that this design-based approach to instruction on limits develops a rich cognitive structure that reflects the standard mathematical definitions and applications and is powerful in supporting instruction on the other major concepts in calculus defined in terms of limits. This approach provides a facility with these major concepts grounded in ideas of approximation and bounding error which are the basis for many applied applications of mathematics (e.g., in physics and engineering) and for a rich understanding of the mathematical formulas, theorems, and tools used in computational techniques. Finally, students are encouraged to develop an intuitive facility with the structures that can form a foundation for later abstraction to epsilon-delta and epsilon-N constructions, the basis of formalization and proof in upper-division and graduate analysis courses and of computational techniques in many applied mathematics and differential equations courses.

References

Artigue, M. (1991). Analysis. In D. Tall (Ed.) *Advanced mathematical thinking*. Boston: Kluwer, 167–198.

Cottrill, J., Dubinsky, E., Nichols, D., Schwingendorf, K., Thomas, K., & Vidakovic, D. (1996). Understanding the limit concept: Beginning with a coordinated process schema. *Journal of Mathematical Behavior, 15,* 167–192.

Davis, R. (1986). Calculus at university high school. In Douglas, R.G. (Ed.). *Toward a lean and lively calculus*. Washington, DC: Mathematical Association of America.

Frid, S. (1994). Three approaches to undergraduate calculus instruction: Their nature and potential impact on students' language use and sources of conviction. *Research in College Mathematics Education I,* Conference Board of the Mathematical Society, Issues in Mathematics Education, Volume 4, American Mathematical Society, 69–100.

Jacobs, S., Larsen, S., & Oehrtman, M. (2003). *Taking command of the limit concept,* Presentation at the 29th Annual Conference of the American Mathematical Association of Two-Year Colleges, Salt Lake City, UT.

Monk, G. (1987). *Students' understanding of functions in calculus courses.* Unpublished manuscript, University of Washington, Seattle.

——— (1992). Students' understanding of a function given by a physical model. In G. Harel & E. Dubinsky (Eds.), *The Concept of Function: Aspects of Epistemology and Pedagogy.* MAA Notes, Volume 25. Washington, DC: Mathematical Association of America.

Oehrtman, M. (2002). *Collapsing dimensions, physical limitation, and other student metaphors for limit concepts: An instrumentalist investigation into calculus students' spontaneous reasoning,* Doctoral dissertation, The University of Texas, Austin, TX.

——— (2003) Strong and weak metaphors for limits. In N. Pateman, B. Dougherty & J. Zilliox (Eds.), *Proceedings of the Twenty-seventh Annual Meeting of the International Group for the Psychology of Mathematics Education.* Honolulu, HI: University of Hawaii, (3) 397–404.

——— (2004) Approximation as a foundation for understanding limit concepts. In D. McDougall, & J. Ross (Eds.), *Proceedings of the twenty-six annual meeting of the North American chapter of the International Group for the Psychology of Mathematics Education.* Toronto: University of Toronto, (1) 95–102.

Piaget, J. (1970a). *Genetic epistemology.* (E. Duckworth, Trans.). New York: Columbia University Press.

——— (1970b). *Structuralism.* (C. Maschler, Trans.). New York: Basic Books, Inc.

——— (1975). Piaget's theory. In P. B. Neubauer (Ed.), *The process of child development.*

——— (1980). *Adaptation and intelligence.* (S. Eanes, Trans.). Chicago: University of Chicago Press.

——— (1985). *The equilibration of cognitive structures.* (T. Brown & K. J. Thampy, Trans.). Chicago: The University of Chicago Press.

——— (1997). *The principles of genetic epistemology.* (W. Mayes, Trans.). New York: Routledge.

Rodi, S. (1986). Some systemic weaknesses and the place of intuition and applications in calculus instruction. In R.G. Douglas (Ed.). *Toward a lean and lively calculus.* Washington, DC: Mathematical Association of America.

Sealey , V. & Oehrtman, M. (2005). Student understanding of accumulation and Riemann sums. In G. M. Lloyd, M. R. Wilson, J. L. M. Wilkins, & S. L. Behm (Eds.), *Proceedings of the 27th annual meeting of the North American Chapter of the International Group for the Psychology of Mathematics Education* [CD-ROM]. Eugene, OR: All Academic.

Tall, D. (1986). *Graphic calculus.* London: Glentop Press.

—— (1990). Inconsistencies in the learning of calculus and analysis. *Focus on learning problems in mathematics, 12,* 49–62.

—— (1992). The transition to advanced mathematical thinking: Functions, limits, infinity, and proof. In Grouws, D. (Ed.), *Handbook of research on mathematics teaching and learning* (pp. 495–511). New York: Macmillan.

Tall, D. & Vinner, S. (1981). Concept image and concept definition in mathematics with particular reference to limits and continuity. *Educational Studies in Mathematics, 12,* 151–169.

Thompson, P. (1994). Images of rate and operational understanding of the fundamental theorem of calculus. *Educational Studies in Mathematics, 26,* 229–274.

Tucker, T. (Chair), (1986). Calculus syllabi: Report of the content workshop. In R. G. Douglas (Ed.), *Toward a lean and lively calculus* (pp. vii–xiv). Washington, DC: The Mathematical Association of America.

Tucker, A.C. & Leitzel, J.R. (Eds.). (1995). *Assessing calculus reform efforts: A report to the community.* Washington, DC: Mathematical Association of America.

Vygotsky, L. (1987). The development of scientific concepts in childhood. In R. W. Rieber and A. S. Carton (Eds.), *The collected works of L. S. Vygotsky.* Vol. 1, 167–241.

Williams, S. (1991). Models of limit held by college calculus students. *Journal for Research in Mathematics Education, 22,* 219–236.

7

Divisibility and Transparency of Number Representations

Rina Zazkis
Simon Fraser University

It is easy for colleagues to agree that students' understanding is one of the main goals of instruction. It is considerably more difficult to agree on what good understanding of a specific concept entails and how it is possible to achieve it or to assess it. I believe that understanding of any mathematical concept includes the ability to deal with various representations of this concept. As suggested by the title of this article, I focus here on the concept of divisibility and how it may be understood by considering various representations of natural numbers.

Divisibility is one of the main concepts in elementary number theory in that it underlies the multiplicative structure of natural numbers. However, it is impossible to discuss divisibility without addressing the concepts of factors, multiples, prime and composite numbers and prime factorization. In what follows I present snapshots from recent research on college students' understanding of these concepts. I then point out some common themes emerging in these examples and present a theoretical perspective as a lens for considering various approaches. I conclude with a discussion of one particular pedagogical method, *consideration of "big" numbers*, which I used in an attempt to improve students' understanding. I describe the benefits of this method and exemplify how it can be applied to concepts beyond elementary number theory.

Snapshots from Research

As stated above, I start with examples from recent research on learning topics in elementary number theory. To situate these examples it is important to note that, unless otherwise stated, the participants in these research studies were preservice elementary school teachers. The participants were enrolled in mathematics content courses designed specifically for this population as a part of the requirements for teaching certification at the elementary level. The topic of introductory number theory — factors, multiples, prime and composite numbers, divisibility and divisibility rules, prime decomposition and the fundamental theorem of arithmetic — was discussed in these courses before the research data were collected.

Research in mathematics education has repeatedly shown that elementary school teachers' understanding of even elementary mathematical concepts is weak, incomplete, and fragile. Concepts of elementary number theory are not exceptions. However, I believe that rather than making general claims about weak understanding, it is important for both teachers and researchers to get a better insight into the nature of these possible weaknesses.

Example 1

Zazkis (2000) investigated preservice elementary school teachers' understanding of the concepts of factors, divisors and multiples, focusing on the mathematical connections among these concepts. In one of the interview questions used in this study participants were asked to identify prime factors of $117 = 3^2 \times 13$. Darlene systematically built a factor tree for 117, summarizing the results as $3 \times 3 \times 13$. Then the following conversation took place:

Interviewer:	And you have found 3, 3 and 13 and you have written it down as $3^2 \times 13$. Now I'm asking you, what I wrote here before, when I gave you the number 117, I wrote it here, it is $3^2 \times 13$.
Darlene:	Um hm.
Interviewer:	Could you use this?
Darlene:	As an explanation for this? [Darlene pointed to her factor tree]
Interviewer:	Not an explanation, maybe a hint.
Darlene:	Oh no, I, like just looking at that, like I didn't, I wouldn't understand that.

Example 2

Zazkis and Campbell (1996) studied preservice teachers' understanding of the notion of divisibility and its relation to multiplication and division. In one of the interview questions participants in their study were asked to consider the number M, where $M = 3^3 \times 5^2 \times 7$, and determine divisibility of M by several numbers, such as 7, 5, 2, 11, 9, 63, 15. Below is the excerpt from the interview with Armin in which part of this task is attempted.

Interviewer:	I'm asking you to look at the number which is $3^3 \times 5^2 \times 7$, do you think this number is divisible by 7?
Armin:	Okay, first I'm just multiplying $27 \times 25 \times 7$ and I get 4,725 and now I need to divide them all by 7.
Interviewer:	Okay.
Armin:	So (I) get 675, so you have it divisible.
Interviewer:	So this number is divisible by 7. Could you know this without using the calculator and without finding out the product of all the numbers?
Armin:	Could I know it? Um, well, I know we discussed something in class about if, if one number is divisible by 7, then another number is divisible, or what was it, this number is divided by 7, and this number is divided by 7, then the sum of those numbers should divide by 7.
Interviewer:	If I asked you whether this number was divisible by 5, what would you do?
Armin:	I'd do the same thing.

Though the responses presented in examples 1 and 2 do not represent the majority of students, they are not unique occurrences either. For example, 8 out of 21 participants in the Zazkis and Campbell (1996) study interpreted divisibility exclusively through carrying out division.

Example 3

Similar approaches of preservice elementary school teachers were identified by Brown, Thomas and Tolias (2002). In one of the tasks, interviewees in their study were asked to identify multiples of 24 in the list of numbers that included 2400, 2401 and 2412. Adam's solution included the following:

Adam:	So take 24 hundred divided by 24 [carries this out on his calculator]. I find it does go evenly and it's 100, so that tells me that 24 hundred is a multiple, um, 24 hundred and one doesn't work because I don't think that goes evenly [divides with calculator] and it doesn't, so I know 2401's not a multiple and then I do the same thing for 2412 and it doesn't go evenly so it's not a multiple.

Brown et. al. commented that Adam didn't find the expression "24 hundred" meaningful and believed it was necessary to carry out the division. They further pointed out that even though Adam predicted that 2401 "doesn't work", he still found it necessary to perform the calculation in order to confirm.

Example 4

Participants in Ferrari's (2002) study were first year undergraduate computer science students. In the tasks presented to these students, Ferrari defined the number M to be $M = 3^4 \times 5^3 \times 7^6 \times 19^8$ and asked whether M was divisible by 63. Though the numbers were carefully chosen to avoid the use of a calculator to perform the computation, Roberto, as well as some of his classmates, carried out the task by performing division $(M \div 63)$ considering the prime factored form of the numbers; that is, representing 63 as $3^2 \times 7$, "canceling" the terms, and finding the result of division to be

$3^2 \times 5^3 \times 7^5 \times 19^8$ in order to conclude divisibility. Roberto's approach demonstrates this student's ability to manipulate the factored form of the number, but not the ability to consider the features it entails. In other words, his ability to consider numbers in their prime factored form appears to be limited to manipulation of symbols.

Example 5

Zazkis & Liljedahl (2004) collected part of their data through written questionnaires rather than interviews. This tool made it possible to assess a larger number of participants. The following question was presented to a group of 116 preservice elementary school teachers.

Consider $F = 151 \times 157$. Is F a prime number?
Circle YES/NO and explain your decision.

Of the 74 students who claimed correctly that F is composite only 52 justified this by its structure as a product, focusing the explanation on the definition of either prime or composite numbers. A popular solution was to calculate F and apply the learned algorithm of checking for primality. Responses from two students using this method are presented below:

Andy circled: *NO* (F is not prime.)
$\sqrt{23707} = 153.9$. *Now we check if any of the prime numbers lower than* 153 *divide* 23707. 23707 *is divisible by* 151 *and* 157 *so it is not prime.*

Terry circled: *YES* (F is prime.)
$151 \times 157 = 23707$, $\sqrt{23707} = 154$. *Check all prime numbers lower than* 154 *to see if the number is prime and if none of them can divide* 23707 *then the number is prime.*

Terry's response was accompanied by the list of primes up to 29. The strategy described in these responses is not incorrect, but rather unnecessary and inelegant in the given case; it is the incomplete implementation of the strategy that led Terry to a wrong conclusion. As shown here, the strategy of applying the algorithm resulted in both correct and incorrect conclusions, as some students checked divisibility of F only by a few "small" primes.

Reflecting On Examples

There are several common themes that emerge in students' responses presented in these examples. One such theme is a procedural disposition and search for algorithms, what is referred to in terms of the APOS theoretical lens (Dubinsky, 1991) as an "action conception" of divisibility or of primality of numbers. Another theme that may explain the procedural disposition is a lack of connectedness of knowledge, that is, a lack of, or insufficient links in students' understanding between factors, divisors and multiples or, more generally, between multiplication and division. However, the theme I wish to focus on in this article is that of representation and structure, that is, the lack of attention to the multiplicative structure of natural numbers as represented in their prime factored form.

Many of my mathematical colleagues have been surprised not only with the students' responses, but also with the questions themselves. They often consider the questions as "trivial" and not worth asking. However, this perception of simplicity is what guided the choice of many of the problems used in prior research: they are simple only for those who are attending to, or who "are seeing," the embedded structure. It is the variety of approaches that students choose for seemingly simple problems that provide researchers with a "window" on their understanding.

Formalization of "Seeing"

As mentioned earlier, one of the unifying themes in the above examples is that students do not "see," or do not attend to, the properties of numbers that are easily recognizable in the numbers' representations. We say that students do not attend to the *transparent* features of representation.

In general, a representation is *transparent* with respect to a certain property if this property "can be seen" considering the representation. Otherwise the representation is *opaque* with respect to the property in question. For example, representing the number 784 as 28^2 emphasizes—or makes transparent—that this number is a perfect square, but de-emphasizes — or leaves opaque — the divisibility of this number by 98. Representing the same number as $13 \times 60 + 4$ makes it transparent that the remainder of 784 in division by 13 is 4, but leaves opaque this number's property of being a perfect square. In this sense, all the representations of natural numbers, including the canonical decimal representation, are opaque, however, each one has transparent features.

To exemplify further transparency in representation of numbers, Zazkis and Gadowsky (2001) invited the reader to consider the following five numbers:

(a) 216^2,   (b) 36^2,   (c) 3×15552,   (d) $5 \times 7 \times 31 \times 43 + 1$,   (e) $12 \times 3000 + 12 \times 888$

From representation (a) it is transparent that the number is a perfect square; representation (b) shows that the number is a perfect cube; from (c) we conclude that the number is a multiple of 3 and of 15552. Of course it is possible to derive that the number is a multiple of 3 from (a) and (b) as well, but (a) and (b) do not give us a clue regarding 15552. From (d) we conclude that the number leaves a remainder of 1 in division by 5, 7, 31 and 43, a conclusion that is not apparent — or properties that are not transparent — in representations (a), (b) and (c). From representation (e) we see that the number is a multiple of 12 and, acknowledging distributivity, that it is a multiple of 3888. It is not apparent, however, that all these expressions represent the same number, 46656. However, following Mason (1998), we say that each representation shifts our attention to different properties of the number.

More Examples from Research

The above examples from research concerned numbers and their representations in prime factored form. However, the idea of transparency in representation presented a broader view on what can be derived from number structure.

Example 6

Campbell (2002) discussed in detail the division theorem (also referred to as the division algorithm), that is, the fact that for any two whole numbers A and D, there are unique whole numbers Q and R, $D > R \geq 0$, such that $A = QD + R$. A common utilization of this theorem for elementary school is in division of whole numbers, so-called division with remainder, where the division of A by D results in a whole number quotient Q with a remainder R and is denoted symbolically as $A \div D = Q$ remainder R. A very illustrative example of participants' lack of attentiveness to the mathematical meanings represented in the structure is presented in responses to a request to determine the quotient and the reminder in division by 6 of the number A, where $A = 147 \times 6 + 1$. Fifteen out of 21 participants in Campbell's study calculated the dividend A and used a long division algorithm in order to answer this question. Further, even those who could derive the information considering the given structure regressed to calculating A when, solving a similar task, they were asked to find the quotient and remainder in division of A by 2. Campbell described this as a lack of connection between division with remainder and multiplication with addition, which extends the previous findings of a weak connection between multiplication and division. This is illustrated in the following excerpt where Anita discusses how she sees the relationship between multiplication, division and division with remainder:

> Anita: Division is an inverse of multiplication. […]. Well, if you just, by definition sense, multiplication is how many groups of something there is, and division is how many groups can go in a number, so you're basically just going backwards. Like um, it's like addition and subtraction, like $3 + 4$ is 7, $7 - 4 = 3$; multiplication 3×4 is 12, 12 divided by 4 is 3.

> Interviewer: Alright, Um, how about division with remainder? Is there any way in which you see that as being an inverse to multiplication?

> Anita: [pause] Well, whenever you're multiplying, you always get a whole number, you never get a …, not, not a whole number, but you always, well you always have a number, you never have a remainder in multiplying. […] an inverse of something is just the opposite. Once you have a remainder it makes it totally different from multiplication.

While Anita recognizes division as the inverse of multiplication and is able to exemplify it by drawing an analogy with addition-subtraction, she comments about division that "once you have a remainder it makes it totally different from multiplication." This lack of connection explains students' difficulty in recognizing quotient and remainder — terms connected to division — in the transparent representation of these quantities in number A, $A = 147 \times 6 + 1$.

Example 7

The multiplicative structure of arithmetic sequences was the theme of a study conducted by Zazkis and Liljedahl (2002). In one of the interview questions participants were asked to consider whether a given number was an element in a given sequence. The results revealed that participants recognized the structure in the sequence of multiples, such as 3, 6, 9 … or 7, 14, 21 …, but experienced difficulty when the given sequence was a sequence of "non-multiples", such as 2, 5, 8 or 8, 15, 22.… In the sequences of multiples, being a multiple of the common difference served as a distinguishing criteria in determining whether the given numbers were elements in the sequence. However, in most cases, the participants were not able to adjust this strategy when considering sequences of non-multiples. In the following excerpt Sally considers the arithmetic sequence 8, 15, 22… and the number 704.

Interviewer:	So 704 is not divisible by 7, none of these elements in this sequence you believe will be divisible by 7, so can you draw conclusions from what you have now?
Sally:	It's, it's um very possibly in this set.
Interviewer:	Um hm. What, what will convince you?
Sally:	(laugh) Well just because it's not divisible by 7, doesn't mean it's in the set, right?
Interviewer:	Can you give me an example of a number that you know for sure that is not in this arithmetic sequence?
Sally:	Um hm, um 700…
Interviewer:	Another one …
Sally:	Um, 77.
Interviewer:	Okay. And how about 78?
Sally:	It may be in the set, but it's not divisible by 7…
Interviewer:	(laugh) So 77 you're sure is not, 78 you're not sure.
Sally:	Right.
Interviewer:	79?
Sally:	Could be …
Interviewer:	Could be. 80?
Sally:	Could be …

Zazkis and Liljedahl (2002) noted that a number's property of "being a multiple" or "being divisible by" gives a clear indication of its belonging to a sequence of multiples or not belonging to a sequence of non-multiples; however, the property of "being a non-multiple" identifies that a number does not belong to a sequence of multiples, but gives no explicit hint with respect to the number's membership in a given sequence of non-multiples. As the above excerpt illustrates, Sally clearly claims that any given multiple of 7 is not an element in a sequence of "non-multiples." Nevertheless, she is not able to draw a definite conclusion when testing the membership of a number that is not a multiple of 7. Her expressions "very possible" or "could be" suggest that she identified the dichotomy between multiples and non-multiples. She is aware of the multiplicative structure in the sequence of multiples, however, she is not attending to the inherent multiplicative structure of the arithmetic sequence of non-multiples, that is, she is not recognizing the sequence as "multiples of 7, plus 1."

Pedagogical Considerations

Is it possible (and if so, how is it possible) to help students recognize the structure embedded in the representation of numbers? This is a pedagogical question that emerges from the aforementioned snapshots from research. Having introduced the terminology, this pedagogical question can be refined as follows: Is it possible (and if so, how is it possible) to make it transparent to the eyes of students what is transparent to the eye of a mathematician?

I have struggled with this question in my teaching, with varying degrees of success. My students' desire to follow algorithms and verify everything by computation was strongly embedded in their approaches. One of the strategies suggested by a colleague was to disallow the use of calculators. It was believed that this restriction could help students focus on structure, rather than engage in long efforts with paper and pencil. Surprisingly, students rarely attempted to avoid even messy and long computations. While the temporary ban on calculators had the desired effect for some students, others expressed frustration with the regulation, made mistakes in computation, or used the "forbidden tool" under the desk.

What if — I thought — calculators were not banned, but simply made obsolete? In what follows I introduce one possible pedagogical approach based on creating problem situations in which the algorithmic approach is unworkable.

Introducing Big Numbers

Before I introduce the pedagogical strategy that emerged, consider the approach of Armin in Example 2 and the approach of Roberto in Example 4. The similarity is that both participants performed division in order to conclude divisibility. The difference is that Roberto performed division using prime decomposition, while Armin found the decimal representation of the number and performed division using her calculator. It is reasonable to assume that Roberto, a computing science major, had a stronger mathematical background than Armin, a preservice elementary school teacher. However, it is unclear whether the method preferred by Roberto was due to his stronger mathematical background or the choice of numbers in the question, numbers that made utilizing decimal representation and the calculator simply impossible. What would Armin do if she were presented with a "big" number? The following conversation with Jenny presents a possible scenario.

In the beginning of the interview, Jenny was presented with number M ($M = 3^3 \times 5^2 \times 7$) and performed division in order to decide whether M was divisible by 7 and 15. When prompted about the possibility of dealing with the task without the use of a calculator, she considered pencil and paper calculations. Then she was presented with a variation on the original question, that invited her to consider another number, $3^{30} \times 5^{20} \times 7$.

Interviewer:	Could you draw these conclusions without the calculator?
Jenny	I guess, It will just take me forever to do the division.
Interviewer:	Please consider another number, let's call it B, where $B = 3^{30} \times 5^{20} \times 7$ Is B divisible by 15? [Jenny attempted to calculate B.]
Jenny:	The calculator isn't much help here.
Interviewer:	And why's that?
Jenny:	It gives all kinds of digits that don't help here
Interviewer:	Do you think it is possible to draw a conclusion without the calculator?
Jenny:	I could spend a day to make this number [pause] Wait. I think 15 will go into it.
Interviewer:	And why is that?
Jenny:	Because 15 is 5 times 3 and we have 3 here and 5 here, so when put together, yeah, when put together, 15 is there, in this number.
Interviewer:	And how about 63?
Jenny:	[pause] Yes, it's also there. It's made of two 3's and a 7, and we have here a 7 and 20 3's. So it goes, I mean it's divisible.
Interviewer:	Let's go back for a moment to a number we considered before, $M = 3^3 \times 5^2 \times 7$. Is M divisible by 63 ?
Jenny:	Yes, you can make exactly the same claim.
Interviewer:	What's exactly the same?
Jenny:	Like 63 is made of 9 times 7, you have 7 here, you don't have 9 here, but 9 is made of two 3's and you have more than two 3's here, so that's why it's the same, I mean 63 can go into this M.

In this excerpt, Jenny appeared to realize that the calculation was out of reach and therefore turned her attention to the structure of the number, in this case to its prime decomposition. When asked why she preferred calculation as her initial choice of strategy, Jenny replied, "It was easy to do so I didn't have to think about it." A move from small number M to a big number B, using yet again the terminology from Mason (1998), has shifted Jenny's attention from "what the number is" to "what the structure of the number is", a shift which is necessary in acquiring skills in algebra.

Algebra? Yes, algebra, accepting the idea that generalization and consideration of structure are among the components of algebraic thinking (Mason, 1996). Though Jenny is considering specific numbers, her reasoning can be seen as algebraic; her reasoning is about form, structure and relationships. Jenny's starting point is in arithmetic calculation. The instructor's goal is to help Jenny understand that $p^x \times q^y \times t^z$ is divisible by $p^a \times q^b \times t^c$ if and only if $a \leq x$, $b \leq y$ and $c \leq z$. In fact, Jenny is rather close to this generalization. She has definitely recognized the factors appearing in prime decomposition and by saying "9 is made of two 3's and you have more than two 3's here" she is considering the exponents of 3. A handful of additional investigations may be necessary before the general conclusion can be drawn and expressed in "proper" mathematical terms, but Jenny is definitely on the right track.

Working with Big Numbers

Following the pathway of generalization, the strategy exemplified above can be described simply as "working with big numbers." By "working with" I mean considering big numbers, rather than computing them.

A natural question that arises is "What is big?" or "How big should a big number be"? This depends on the problem and on the pedagogical situation. In most cases the number is "big enough" if it does not invite the learner to engage in computation or if it is beyond the computational abilities of a hand-held calculator. An example of the latter is presented in the previous section. An example of the former is presented below.

In their consideration of student-generated examples, Hazzan and Zazkis (1999) asked participants to provide an example of an object that satisfied certain properties. They observed that students attempted to choose an object at random and check whether the requested property was satisfied, rather than to construct the object such that it would have the desired specifications. When asked for an example of a 6-digit number that is divisible by 9, students picked 6 digits and checked to determine if their sum was divisible by 9. Similarly, when asked to give an example of a 6 digit number that is divisible by 17, some students preferred to guess and check, or to adjust, that is, if the first choice of a number resulted in a remainder of 5, they added 12 to the choice to obtain a number divisible by 17.

In a classroom activity with a group of preservice elementary school teachers, I attempted to increase gradually the numbers in the tasks. I asked for examples of 5-or-more-digit numbers that left (a) a reminder of 1 in division by 3, (b) a reminder of 3 in division by 5, (c) a reminder of 7 in division by 17, (d) a reminder of 56 in division by 73, (e) a reminder of 123 in division by 247, etc. Students who initially preferred the approach of checking and adjusting gradually abandoned this approach in favor of consideration of structure.

As one example of this movement toward consideration of structure, Jill was proud to present 38,655 as a number that leaves a reminder of 123 in division by 247. She explained, upon request, that she obtained this number by multiplying 247 by 146 and then adding 123. She further explained that "146 doesn't have to be there. It can be anything you wish. Just multiply 247 by something big and then add 123."

Though presented differently than standard algebraic notation, Jill has generated an algebraic form $247 \times n + 123$, where n is a whole number, as a general form for a number that leaves the remainder of 123 in division by 247. I suggest that this migration would not have occurred when she was using 'small' numbers because her computation was too fast and the numbers were too familiar. The examples in (a) and (b) were immediate to generate, so, for Jill, there was no need to explicitly acknowledge the structure. Examples in (c), (d) and (e) invited Jill to search for a strategy. Only after attending to the structure of large numbers could she recognize the structure of small numbers.

Modifying Tasks

Most mathematics textbooks—texts for elementary and middle school students as well as texts for preservice elementary school teachers—have not yet taken into account the availability of a calculator "in every pocket." The examples and the exercises in most textbooks are still utilizing "small" numbers. Standard tasks to derive divisibility can be carried out and verified using division, conventional tasks to find a prime decomposition of a number usually end up with the list of "small" prime numbers. I argue for modification of these tasks to include consideration of "big" numbers.

A variety of examples can be designed to show how "big numbers" can be utilized so that conclusions about divisibility, factors, and multiples can be drawn from considering the transparent structure. One such example, which was presented above, considered $3^{30} \times 5^{20} \times 7$ rather than $3^3 \times 5^2 \times 7$. Of course, once the possibility of computation is not a consideration or constraint, the options are limitless. When asked whether the number 3^{10} is odd or even, some students may prefer to calculate the number. When the same question is posed about 3^{1000}, calculation is not an option.

Let us turn again to the examples presented in the beginning of this article. Would Darlene (Example 1) have calculated the number and built a factor tree if she had been asked to identify prime factors of $3^{200} \times 13$, rather than of $3^2 \times 13$? Would Adam (Example 3) have attempted to perform division if asked whether $24^{200}+12$ or $24 \times 10^{50}+12$ were divisible by 24, rather than 2412? Would Andy (Example 5) have calculated the number F and checked for its divisibility by all the primes up to the square root of F if F had been defined as 157^{151}, rather than 151×157? Even with easy access to a calculator, it is impossible to calculate the decimal representation of these big numbers. Therefore, these students would have either been totally lost, or forced to consider the transparent features of the structure embedded in the representation of numbers. I suggest that with appropriate exposure to these modified tasks, the later option would eventually prevail. I believe — a belief that was confirmed by research and experience — that consideration of "big numbers" helps students understand "big ideas", that at times remain opaque when calculations with "small numbers" are performed.

Seeing and Looking

"How can you expose the genericity of an example to someone who sees only its specificity? Apart from stressing and ignoring, and repeating the general statement over and over again, how can the necessary act of perception, of seeing the general in the particular, be fostered?" (Mason & Pimm, 1984, p.287). This article suggests that a possible way to help students see the general in the particular is to invite consideration of particular examples that are somehow detached from their concrete experiences. Big numbers serve this purpose as they help in attending to transparent features in a given representation.

The possibility of seeing starts with the action of looking. In this section I offer a word of caution on what there is to look for. In my teaching I emphasized the ideas of prime decomposition and directed students' attention to the ways that conclusions about factors and multiples could be derived by considering the presented structure of numbers. Exposure to big numbers helped my students draw inferences on divisibility and indivisibility without carrying out division. They seemed to apply the criteria correctly and appeared confident in their judgment. Therefore, I believed that transparent features of prime decomposition of numbers had become transparent to my students. I was about to celebrate my pedagogical achievement.

However, in one of the more recent experiments I asked students to look at another variation of the "renowned" number M ($M = 3^3 \times 5^2 \times 7$). The students were asked to consider whether the number $K = 3^3 \times 5^2 \times 7+11$ (yes, this is $+$, not a typo) was divisible by 7. About half of the participants claimed that K was divisible by 7 because 7 appeared, that is, "was seen", in the prime decomposition.

This response echoes and contrasts one of the findings reported by Ferrari (2002). In the same sequence of tasks as presented in Example 4, Ferrari (2002) asked participants in his study to consider whether the number $M+5$ was divisible by 10. (M was given as $M = 3^4 \times 5^3 \times 7^6 \times 19^8$). He reported that almost all the participants claimed that $M+5$ was not divisible by 10 because "there was no factor of 2 within $M+5$".

The analogy is rather ironic. My students looked for prime factors and "saw" 7, which was not a factor. Ferrari's students looked for a prime factor 2 and could not conclude its existence without explicitly seeing it within the representation $3^4 \times 5^3 \times 7^6 \times 19^8 + 5$. It could be the case that some students, when exposed to big numbers, perform the action of "looking," rather than the action of computing. However, unlike the case of computing, there is no possibility to confirm whether the conclusion reached by "looking" is a correct one.

All this suggests a word of caution regarding what we are really looking for. From the above examples it appears that students' attention was focused on looking for *prime numbers*—7 and 2 respectfully—rather than on *prime factors*. However, in order to draw conclusions on divisibility, one should focus on possible factors in the given representation of a number. In $3^4 \times 5^3 \times 7^6 \times 19^8 + 5$ the factor 2 is "opaque" and as such was not noticed by the students. In $3^3 \times 5^2 \times 7+11$ it is transparent that 7 is not a factor, as the number leaves the remainder of 11 in division by 7.

Based on these findings, a seemingly successful instructional practice requires an amendment. The strategy of looking for factors and recognizing the transparency of representation in prime decomposition should supplement the focus on *prime* factors with consideration of prime *factors*.

Utility Beyond Number Theory

The "big numbers" strategy described above could be helpful when considering other mathematical topics. In what follows I present examples from different areas in mathematics. The first two cases are taken from Stewart's (2003) study in which she was interested in "procedural change" and what may cause it. That is, Stewart observed that, in a variety of cases, the algorithms used by preservice elementary school teachers were not incorrect, but rather lengthy, messy or inefficient. She investigated what could help participants in adopting better—that is, more efficient and more elegant — procedures.

For example, one possible way to find the least common multiple (LCM) of two numbers is by listing multiples of each and then looking for the smallest element in the intersection of two sets. This approach is helpful in introducing the LCM concept, as it decodes the meaning of a multiple, then common multiples and finally the least of the common multiples—LCM. However, when the task becomes one of finding the LCM, rather than deciphering the meaning of words, this approach is time consuming and at times unfeasible. Still, it could remain a preferred approach for many students, often serving as a safety net. Stewart (2003) reported that what helped students give up the familiar approach and take advantage of a newly introduced approach—that relied on considering prime decomposition—was "big numbers." That is, students were more successful in adopting the new procedure when the old algorithm appeared impractical.

Even simpler arithmetic

Stewart (2003) also reported that her students, preservice elementary school teachers, automatically converted mixed numbers to improper fractions when asked to perform addition and subtraction. For example, when asked to add $4 + 7\frac{1}{3}$, it was not unusual to demonstrate the following calculation:

$$\frac{4}{1} + \frac{22}{3} = \frac{12}{3} + \frac{22}{3} = \frac{34}{3} = 11\frac{1}{3}$$

Stewart concluded that exposure to a simpler and more elegant method — wholes first — didn't create the expected change in students' approaches because they felt that conversion to improper fractions was their "safety net." Students either did not want to — or simply could not — use the method introduced in their college class and therefore relied on their prior knowledge from grade school. However, what did help initiate procedural change was the strategy of big numbers. When the addends were $324\frac{7}{10}$ and $213\frac{4}{5}$, participants, or at least a majority of them, tended to adopt the 'wholes first' approach.

"Short" Multiplication

In a similar scenario, I was unsuccessful in convincing my students to see the difference of squares in $(2x-3y)^2 - (x+2y)^2$ as there was an almost instinctive desire to remove the parentheses by first computing the squares of the sum and difference. However, I found that students generated much more appreciation for the difference of squares when the chosen numbers were not quite "big", but just bigger, such as $(26x-15y)^2 - (24x+15y)^2$.

Beyond Numbers

The strategy of using big numbers can also be helpful in supporting students' understanding of structure in more advanced mathematical topics. Consider for example the following integral:

$$\int_0^1 4x(2x-1)^4 \, dx.$$

It is possible to find the value by expanding the fifth degree polynomial. However, a much more efficient way to solve this integral is by substituting $u = 2x-1$. The first method is usually learned by students before the second, and many students find it difficult to give up their tested methods — even if those are time consuming and inelegant — simply for the sake of efficiency. How is it possible to initiate students' appreciation of a "better method"? In accord with the above discussion, the method I suggest is "big numbers." For example, consider the following minimal variation in the question:

$$\int_0^1 4x(2x-1)^{40} \, dx.$$

Carrying out the expansion method is practically impossible, but this question is no different from the previous one if a student chooses the substitution method. In fact, some calculus textbooks include similar examples to highlight the benefits of the substitution method.

Conclusion

One general conclusion that can be drawn from looking at a variety of research reports that focus on the teaching of elementary number theory is that many students do not attend to structure, even when this structure is transparent for an expert. I presented an argument that consideration of "big numbers" can be helpful in recognizing structure, a recognition that allows learners to deal with problems related to consideration of divisibility and other concepts in elementary number theory without relying on computation.

Small numbers are perceived by students as concrete, meaningful, and easy to manipulate. Algebraic symbols are often perceived as abstract, meaningless and impossible to reason with. On the one hand, big numbers can also be seen as abstract, as they are not available for immediate arithmetic manipulation and are, at times, beyond the computational abilities of a hand-held calculator. On the other hand, they are concrete by the virtue of being particular numbers. As such, they can serve as a bridge between arithmetic and algebraic reasoning, a bridge between the consideration of specific numerical examples and the consideration of general structure. Particular examples that instantiate the generality could lead students to develop insights into the notion of structure. However, a quest for structure will start for some students only when computation is seen as hard or even impossible given the available tools. If teaching is about creating opportunities, then here is one that is not to be missed. Working with preservice elementary school teachers, consideration of big numbers may not only help participants recognize the multiplicative structure of whole numbers, but may also introduce them to a powerful teaching strategy for their future instructional repertoire.

Unlike mathematics, mathematics education has no trusted routines to prove an argument. All that is possible is to exemplify the argument with a variety of examples that — mathematical logic aside — increase the chance of the argument being "true." I believe I have done just that. Consideration of big numbers is not a general magic solution for every problem — it is just one small idea that proved helpful in a small number of instances. Readers are invited to test whether this idea could be implemented in their own practice.

References

Brown, A., Thomas, K., & Tolias, G. (2002). Conceptions of divisibility: Success and understanding. In Campbell, S. R., & Zazkis, R. (Eds.) *Learning and teaching number theory: Research in cognition and instruction* (pp. 41–82). Journal of Mathematical Behavior Monograph. Westport, CT: Ablex Publishing.

Campbell, S. R. (2002). Coming to terms with division: Preservice teachers' understanding. In Campbell, S. R. & Zazkis, R. (Eds.) *Learning and teaching number theory: Research in cognition and instruction* (pp. 15–40). Journal of Mathematical Behavior Monograph. Westport, CT: Ablex Publishing.

Dubinsky, E. (1991). Reflective Abstraction in Advanced Mathematical thinking. In D. Tall (Ed.), *Advanced mathematical thinking* (pp. 95–123). Boston: Kluwer Academic Publishers.

Ferrari, P. L. (2002). Understanding elementary number theory at the undergraduate level: A semiotic approach. In Campbell, S., & Zazkis, R. (Eds.) *Learning and teaching number theory: Research in cognition and instruction* (pp. 97–116). Journal of Mathematical Behavior Monograph. Westport, CT: Ablex Publishing.

Hazzan, O. & Zazkis, R. (1999). A perspective on "give an example" tasks as opportunities to construct links among mathematical concepts. *Focus on Learning Problems in Mathematics*, 21(4), 1–14.

Mason, J. & Pimm, D. (1984) Generic examples: Seeing the general in the particular. *Educational Studies in Mathematics, 15*, 277–289.

Mason, J. (1996). Expressing generality and roots of algebra. In N. Bednarz, Kieran, C., and Lee, L. (Eds.) *Approaches to Algebra: Perspectives for research and teaching.* (pp. 65–86). Kluwer Academic Publishers.

—— (1998). Enabling Teachers to be Real Teachers: Necessary Levels of Awareness and Structure of Attention. *Journal of Mathematics Teacher Education, 1*, 243–267.

Stewart, C. (2003). *Procedural change in mathematics: Tales of adoption and resistance.* Unpublished Doctoral dissertation. Simon Fraser University, British Columbia, Canada.

Zazkis, R. (2000). Factors, divisors and multiples: Exploring the web of students' connections. *Research in Collegiate Mathematics Education.* Vol. 4. 210–238.

Zazkis, R. & Campbell. S. R. (1996). Prime decomposition: Understanding uniqueness. *Journal of Mathematical Behavior, 15*(2), 207–218.

Zazkis, R. & Gadowsky, K. (2001). Attending to transparent features of opaque representations of natural numbers. In A. Cuoco (Ed.), *NCTM 2001 Yearbook: The roles of representation in school mathematics* (pp. 41–52). Reston, VA: NCTM.

Zazkis, R. & Liljedahl, P. (2002). Arithmetic sequence as a bridge among conceptual fields. *Canadian Journal of Science, Mathematics and Technology Education, 2*(1). 91–118.

—— (2004). Understanding primes: The role of representation. *Journal for Research in Mathematics Education, 35*(3) 164–186.

Part I
Student Thinking

c. Proving Theorems

8

Overcoming Students' Difficulties in Learning to Understand and Construct Proofs

Annie Selden and John Selden
New Mexico State University

When a topologist colleague was asked to teach remedial geometry, he used *Schaum's Outline of Geometry* and also wrote proofs on the blackboard. One day a student, who was familiar with two-column proofs having statements such as $\triangle ABD \cong \triangle BCD$ and reasons such as *SAS*, blurted out in utter surprise, "You mean proofs can have words!"

This geometry student's previous experience had led him to an unfortunate view of proof. Other students experience epiphanies about themselves and proof. Asked what she (personally) got out of a transition-to-proof course, one of our students answered, "I learned that I could wake up at 3 A.M. thinking about a math problem."

What do responses like this tell us? Almost all undergraduate mathematics courses are about the concepts and theorems of mathematics — when a matrix has an inverse, how to find it, and when to use it; when a series converges; the distinction between continuous and uniformly continuous; the meaning of compact. However, students in courses like abstract algebra, real analysis, and topology normally demonstrate their competence by solving problems and proving theorems. And, if students go beyond a few lower-division courses such as calculus or first differential equations, this usually involves constructing original proofs or proof fragments. But, often not much time can be devoted to helping students learn how to construct proofs. This might not lead to difficulties, if only students came to university understanding something about the nature of proof and already had some experience constructing simple proofs. Unfortunately, many students, even high-performing ones, do not. And the resulting difficulties they encounter may be one of the reasons many students do not continue in mathematics.

Transition-to-proof courses, also called bridge courses, are meant to ameliorate this situation. Their main focus is not on the concepts and theorems of mathematics, but on helping students learn to construct proofs.[1] This is perhaps best seen as a complex constellation of content knowledge, beliefs, problem solving ability, and skills. These skills include identifying hypotheses and conclusions, locating relevant definitions and theorems, using them appropriately, isolating the mathematical "problem," coming up with "key" ideas to solve it, and finally, organizing them into a logically coherent deductive argument. Their acquisition seems to be considerably aided by practice, and the process of learning to construct proofs may even involve students coming to know themselves better. Indeed, the above student's comment, about waking up with a math problem, suggests that she has learned to persist until she eventually comes up with a solution, even if that's in the middle of the night. Unfortunately, many students believe that they either know how to solve a problem (prove a theorem) or they don't, and thus, if they don't make progress within a few minutes, they give up and go on to something else.

---

[1] However, the introduction to one transition-to-proof textbook asserts something different. It states that the purpose of such courses, often called something like Introduction to Mathematical Reasoning, is to gather together results concerning basic set notions, equivalence relations, and functions so teachers will not have to cover them repeatedly at the beginning of courses like abstract algebra and real analysis.

Of course, undergraduate students do not learn to construct proofs only in transition-to-proof courses. They tend to improve their ability to construct proofs throughout the entire undergraduate mathematics program. Some departments do not even offer transition-to-proof courses, and some combine them with mathematics content courses such as discrete structures. Occasionally, students are offered an R. L. Moore type course,[2] that is, a course in which the textbook and lectures are replaced by a brief set of notes and in which the students produce all the proofs. To some extent, the emphasis in such courses is on a deep understanding of the mathematical content — however, it has been our experience that once students get started in such courses they often improve their proof making abilities very rapidly. Unfortunately, a few students may have great difficulty getting started.

In whatever setting students are to progress in their proving abilities, one might expect the teaching to be somewhat special. In many university mathematics content courses, teachers can profitably explain mathematical theorems and why they are true, but in teaching the skills and problem solving abilities involved in proving, one should also expect to emphasize guiding students' practice. In developing such teaching, it can be useful to ask: What kinds of difficulties do student have, and how might these difficulties be alleviated?

We will describe some results from the mathematics education research literature that address these questions. However, it is important to note that this research typically makes no claim (as one familiar with other social sciences might expect) that all, or even most, students have the described difficulties. Instead, the main point of this literature is to uncover, understand, and describe features of student learning, in this case difficulties that might otherwise go unnoticed, at least to the degree claimed. Evidence in such work is usually directed towards being sure that the observations and descriptions are accurate about the particular (often, small) group of students studied.

As we proceed, we will mention possible pedagogical reasons for the difficulties described, and when they are known, we will give some pedagogical suggestions. In conclusion, we will offer some additional teaching suggestions, based partly on the research literature and partly on our own teaching experience.

The Curriculum and Students' and Teachers' Conceptions of Proof

That the concept of proof in mathematics is a difficult one for students is not surprising given various everyday uses of the word "proof." To a jury, it can mean "beyond a reasonable doubt" or "the preponderance of the evidence." To a social scientist or statistician, it can mean "occurring with a certain (rather large) probability." And, to a scientist, it can mean the positive results of an empirical investigation. To comprehend the special way that "proof" is used in mathematics can take time and such everyday meanings can get in the way.

Views of High School Geometry Students

A number of studies have documented the finding that the concept of mathematical proof is not quickly or easily grasped. For example, in the middle of a year-long U.S. high school geometry course, after being introduced to deductive proof, students in five classes were given a short instructional unit designed to highlight differences between measurement of examples and deductive proof. Seventeen of the students were interviewed and asked to compare and contrast two arguments (for different theorems) — a deductive proof and an argument containing four examples using differently shaped triangles. Some of these students had a nuanced "evidence is proof" view. They considered empirical evidence to be sufficient proof for a statement about all triangles, provided one took measurements of each type of triangle — acute, obtuse, right, scalene, equilateral, and isosceles. Others had a qualified view of deductive proof, believing that a two-column proof only proved a theorem for the type of triangle depicted in the accompanying figure and would need to be reproved, perhaps using the same steps, for other types of triangles. Most surprising and quite disturbing, especially after an instructional unit designed to help them make the distinction between empirical justification and deductive proof, was the result that some of these geometry students simultaneously believed that empirical evidence is sufficient proof and that proof is just evidence for a claim (Chazan, 1993).

The Influence of the Curriculum

It would seem that partly, perhaps even to a large extent, how students see proof is a consequence of how it is portrayed by their teachers and the overall curriculum. To give a well-documented example, we turn to secondary schools in

---

[2] Such courses are also referred to as modified Moore Method or Texas Method courses. For more information, see Mahavier (1999) or Jones (1977).

England where a new National Curriculum, for students aged 5–16 years, was adopted following the 1982 publication of the influential Cockcroft Report. The new curriculum was partly "in response to evidence of children's poor grasp of formal proof in the 60's and 70's" (Hoyles, 1997). The intent of the new curriculum was for students to test and refine their own conjectures in order to gain personal conviction of their truth, and to present justifications for their validity, that is, deductive arguments.

Somehow, in that new National Curriculum, proof was relegated to only one of several strands, called student Attainment Targets (ATs), namely, to AT1: Using and Applying Mathematics.[3] Unfortunately, as was found several years later, even high-attaining secondary students came to see proof in terms of the "investigations" that were undertaken in this applied strand of the curriculum (Coe & Ruthven, 1994; Healy & Hoyles, 1998, 2000). While such investigations were undertaken with the intention of having students see proof as less of a formal ritual demanded by teachers and more as a natural outgrowth of testing, refining, and verifying their own conjectures, the results were disappointing, even disastrous, for students entering England's universities. Apparently, the verifications, that had been intended to be student constructed deductive arguments, were instead turned into standardized templates and empirical arguments (Coe & Ruthven, 1994).

By 1995, the situation had caused so much concern that the London Mathematical Society issued a report on the problems as mathematicians perceived them. The report stated that recent changes in school mathematics "have greatly disadvantaged those who need to continue their mathematical training beyond school level." In particular, the following problems were cited: "serious lack of essential technical facility — the ability to undertake numerical and algebraic calculation with fluency and accuracy," "a marked decline in analytical powers when faced with simple problems requiring more than one step," and "a changed perception of what mathematics is — in particular of the essential place within it of precision and proof" (London Mathematical Society, 1995, p. 2).

After the public outcry of mathematicians, a large-scale study, called *Justifying and Proving in School Mathematics*, was undertaken. The study surveyed 2,459 high-attaining Year 10 students (14–15 years old, that is, comparable to U.S. high school sophomores) in 94 classes from 90 English and Welsh schools. In a series of papers and reports, it was convincingly documented that it was the new National Curriculum, as implemented by teachers, that was, in large part, responsible for the perceived decline in U.K. students' notions of proof and proving (Hoyles, 1997; Healy & Hoyles, 1998, 2000).

What did this large, mostly quantitative, but partly qualitative, study find? In the Executive Summary of the report (Healy & Hoyles, 1998), one finds the following conclusions, amongst others. (1) Students' performance on constructing proofs was "very disappointing." These better-than-average[4] students were asked to judge whether a number of empirical, narrative, and algebraic arguments were correct and convincing, as well as which they would produce and which would get the best marks from their teachers. They were also asked to prove a familiar result, *the sum of any two odd numbers is even*, and an unfamiliar result, *if p and q are two odd numbers, then $(p+q) \times (p-q)$ is a multiple of 4*. Only 40% of students showed evidence of deductive reasoning for the familiar result, and just 10% did so for the unfamiliar result. (2) Even so, most students (84% in geometry, 62% in algebra) were aware that once a statement is proved "no further work was necessary to check if it applied to a particular range of instances," for example, the sum of two odd numbers that are squares is also even. (3) Students who expected to take the higher level GCSE examination at age 16, rather than the middle-level examination, were better at constructing and identifying correct arguments.[5] In conclusion, the report stated:

> The major finding of the project is that most high-attaining Year 10 students after following the National Curriculum for 6 years are unable to distinguish and describe mathematical properties relevant to a proof and use deductive reasoning in their arguments.... at least some of the poor performance in proof of our highest-attaining students may simply be explained by their lack of familiarity with the process of proving. (Healy & Hoyles, 1998, p. 6)

---

[3] The new National Curriculum has undergone several changes in the number of attainment targets (ATs). In 1995, the other three ATs were AT2: Number and Algebra; AT3: Shape, Space, and Measures; and AT4: Handling Data. Relegating proof to AT1 (Using and Applying Mathematics) had the effect of separating the function of proof away from the other mathematical strands (Hoyles, 1997, p. 8).

[4] They had scored an average of 6.56 on a national test (the Key Stage 3) whose overall average is normally between 5 and 6.

[5] British students can opt for one of three levels of examination: foundation-, middle-, and higher-tier. It has been argued elsewhere that the acceptability (for entrance to university) of good results on the middle-tier has discouraged many students from taking the higher-tier GCSE mathematics examination.

Thus, the way a curriculum conveys proof and proving is clearly crucial, but skilled and knowledgeable teachers are also critical for implementing such a curriculum. The current *NCTM Standards* (2000) advocate reasoning and proof across the K–12 curriculum. For example, in a section describing the Reasoning and Proof Standard for Grades 9–12, one reads, "Students should understand that having many examples consistent with a conjecture may suggest that the conjecture is true but does not prove it, whereas one counterexample demonstrates that a conjecture is false." (NCTM, 2000, p. 345.) Are current U.S. secondary school teachers capable of providing the rich opportunities and experiences with proof that would enable students to come to such an understanding?

Secondary Teachers' Views and Knowledge of Proof

In one recent study (Knuth, 2000a, 2000b), seventeen secondary school mathematics teachers, with from 3 to 20 years teaching experience, some with master's degrees, were interviewed on their conceptions of proof and its place in secondary school mathematics. They were asked such questions as: What does the notion of proof mean to you? Why teach proof in secondary school? When should students encounter proof? They were also presented with arguments for several mathematical statements and asked to evaluate them. Some of these arguments were proofs; others were not.

Although all these teachers professed the view that a proof establishes the truth of a conclusion, several also thought it might be possible to find a counterexample or some other contradictory evidence to refute a proof. The interviews produced no evidence to suggest the teachers saw proof as promoting understanding or insight. Three teachers did talk about the role of proof in explaining why something is true, but by this they meant understanding how one proceeded step-by-step from the premise to the conclusion.

In the context of secondary school, the teachers distinguished formal proofs, less formal proofs, and informal proofs. For some teachers, two-column geometry proofs were the epitome of formal proofs. Less formal proofs were not as mathematically rigorous, and informal proofs were explanations or empirically-based arguments. All the teachers considered proof as appropriate only for those students in advanced mathematics classes and those intending to pursue mathematics-related majors in college. All indicated that they accepted informal proofs from students in lower-level mathematics classes. However, doing only this may have the unfortunate consequence that students develop the belief that checking several examples constitutes proof (Knuth, 2002a).

The teachers were given five sets of statements with 3 to 5 arguments purporting to justify them; in all, there were 13 arguments that were proofs and 8 that were not. The teachers rated each argument on a four-point scale with 1 not a proof and 4 a proof and provided rationales for their ratings. Ratings of 2 or 3 were included to allow teachers to express alternative views of validity. In general, the teachers were successful in recognizing proofs, with 93% of the proofs rated as such. However, the number of nonproofs they also rated as proofs was surprising — a third of the nonproofs were rated as proofs. In fact, every teacher rated at least one of the eight nonproofs as a proof and eleven teachers rated more than one as a proof. Indeed, ten teachers considered an argument demonstrating the converse of the statement, *If $x > 0$, then $x + \frac{1}{x} \geq 2$*, to be a proof of it; these teachers seemed to focus on the correctness of the algebraic manipulations, rather than on the validity of the argument (Knuth, 2000b).

Given this result regarding some better and more committed secondary mathematics teachers, can one expect that beginning U.S. university students would be reasonably skilled at proof and proving? Would they, for example, understand the distinction between proof and empirical argument? Probably not.

University Students' Views of Proof

Undergraduate students sometimes come to see proofs and proving as unrelated to their own ways of thinking. In order to cope, they may employ mimicking strategies with the result that they develop various views of proof that are unusual from a mathematician's viewpoint; Harel and Sowder (1998) have classified some of these "proof schemes." These are *not* techniques of (mathematical) proof, but rather kinds of arguments, sometimes incorrect or incomplete, that some university students find convincing, and may even think of as proofs.[6] An example of preservice elementary teachers' views of proof follows.

In the 10th week of a sophomore-level mathematics course, 101 preservice elementary teachers were asked to judge verifications of a familiar result, *if the sum of the digits of a whole number is divisible by 3, then the number*

---

[6] The taxonomy of "proof schemes" also includes various axiomatic proof schemes, that is, arguments that mathematicians would consider proofs.

is divisible by 3, and an unfamiliar result, *if a divides b and b divides c, then a divides c*. For each of these, students were given, in randomized order, inductive arguments based on examples, patterns, and specific large numbers, and deductive arguments — a general proof, a false "proof," and a particular (or generic[7]) proof. These students, who had met the idea of proof in their high school geometry courses and in the current course, rated these arguments on a four-point scale, where 4 indicated they considered the argument to be a mathematical proof and 1 indicated it was not a proof. The results showed that both inductive and deductive arguments were acceptable to the students. Apparently the current course that had given "extensive and explicit instruction about the nature of proof and verification in mathematics" had not achieved its goal. In particular, each of the inductive arguments was rated high (3 or 4) by more than 50% of the students. For both familiar and unfamiliar contexts, 80% gave a high rating (3 or 4) to at least one inductive argument, and over 50% gave a very high rating (4) to at least one inductive argument. Also, while over 60% accepted a correct deductive argument as a valid mathematical proof, 52% also accepted an incorrect deductive argument (Martin & Harel, 1989).

Nonstandard views of mathematical proof can be seen as obstacles to overcome. While it is not clear precisely how to bring students' views of proof in line with mathematicians' views, it seems plausible that working towards mathematical sense-making, explanation, and justification on the part of students would be one possible route, provided one avoids the pitfall, described above, of allowing mathematical "investigations" to conclude with purely empirical justifications.

Understanding and Using Definitions and Theorems

Not only are there everyday uses of "proof" that might compound students' difficulties in coming to know what a mathematical proof is, students can be confused about the role of definitions in mathematics.

Mathematical Definitions

Everyday descriptive, or dictionary, definitions[8] describe both concrete and abstract things, already existing in the world, such as trees, love, democracy, or epistemology. They can be both redundant and incomplete, and it is never clear whether all aspects of a definition must apply for its proper use. In contrast, mathematical definitions[9] bring concepts into existence; the concept, say of group, means nothing more and nothing less than whatever the definition says. While all parts of a mathematical definition definitely need to be considered when producing examples and nonexamples, other features of prospective examples need not be considered. This point is often missed. When asked whether $F = 151 \times 157$ is prime, a number of preservice elementary teachers correctly, but irrelevantly noted that both 151 and 157 are prime, before going on to conclude that their product is composite (Zazkis & Liljedahl, 2004). Furthermore, in proving theorems, one should consider all parts of a definition.

Students may not be aware of, or may not make, the distinction between everyday definitions and mathematical definitions. One could help them become aware of this distinction by discussing it with them and by engaging them in the act of defining (Edwards & Ward, 2004; Chapter 17 of this volume).

Interpreting and Using Theorems

Undergraduate students often fail to use relevant theorems or they interpret the content of theorems incorrectly [see Rasmussen and Ruan in this volume for a notable exception]. Below, we provide some examples that illustrate students' difficulties in using and interpreting theorems.

The Fundamental Theorem of Arithmetic, guaranteeing a unique prime decomposition of integers, is part of the core mathematics curriculum for preservice elementary teachers, but in practice some of these students appear to deny the uniqueness. Zazkis and Campbell (1996) asked preservice elementary teachers whether 17^3 was a square number or whether $K = 16,199 = 97 \times 167$ could have 13 as a divisor, given both 97 and 167 are primes. These

---

[7] A generic proof is a proof of a particular case that can be generalized in a straightforward way. For example, see Rowland (2002).

[8] Dictionary definitions are also referred to as *descriptive*, *extracted*, or *synthetic* definitions.

[9] Mathematical definitions are also referred to as *stipulated* or *analytic* definitions. Such definitions apply in an "all or nothing" sense, that is, a given set, together with an operation, is a group or is not a group. In contrast, one can say that two countries are democracies, yet that one is more democratic than the other.

students took out their calculators — in the first instance, to multiply out and extract the square root, and in the second instance, to divide by 13. When asked to determine (and explain) whether $M = 3^2 \times 5^2 \times 7$ was divisible by 2, 3, 5, 7, 9, 11, 15, or 63, a majority (29 of 54) stated that 3, 5, 7 were divisors since those were among the factors in the prime decomposition. However, sixteen were unable to apply similar reasoning to 2 and 11, some noting instead that "M is an odd number" so "2 can't go into it" or resorting to calculations (like the above) for 11. In addition, many of these students believed that prime decomposition means decomposition into *small* primes (see also Zazkis & Liljedahl, 2004).

Undergraduate students often ignore relevant hypotheses or apply the converse when it does not hold. A well-known instance is the use, by Calculus II students, of the converse of: *If $\sum a_n$ converges, then* $\lim_{n \to \infty} a_n = 0$, as an easy, but incorrect, test for convergence. Some calculus books go on to point out that this theorem provides a Test for Divergence. But, perhaps it would be better to explicitly state the contrapositive, *If* $\lim_{n \to \infty} a_n \neq 0$, *then* $\sum a_n$ *diverges*.

Sometimes undergraduate students use theorems, especially theorems with names, as vague "slogans" that can be easily retrieved from memory, especially when they are asked to answer questions to which the theorems seemingly apply. For example, Hazzan and Leron (1996) asked twenty-three abstract algebra students: *True or false? Please justify your answer. "In S_7 there is no element of order 8."* It was expected that students would check whether there was a permutation in S_7 having 8 as the least common multiple of the lengths of its cycles. Instead, 12 of the 16 students who gave incorrect answers invoked Lagrange's Theorem[10] or its converse. Seven of them incorrectly invoked Lagrange's Theorem to say the statement was false — there is such an element since 8 divides 5040. Another two students inappropriately invoked a contrapositive form of Lagrange's Theorem to say the statement was true because 8 doesn't divide 7. The authors go on to point out that students often think Lagrange's Theorem is an existence theorem, although its contrapositive shows that it is a non-existence theorem: *If k doesn't divide $o(G)$, then there doesn't exist a subgroup of order k.* Perhaps it would be good to state this version explicitly for students.

The above examples refer to students' misuse of theorems when they are asked to solve specific problems, for example, determine whether a number is prime or a series converges, or decide whether a group has an element of order 8. However, it is not hard to imagine similar difficulties when students attempt to use theorems in constructing their own proofs.

A Positive Result on Improving Students' Mathematical Reasoning

That even young children can make remarkable strides towards proof, when challenged with appropriate tasks and probing questions, can be seen from one noteworthy longitudinal study (Maher & Martino, 1996a, 1996b, 1997). The study consisted of a series of relatively small, but coherent, long-term interventions with one group of children over a number of years. It led to some extraordinary instances of mathematical sense-making, explanation, and justification, including the development by children, on their own, of the idea of proof in a concrete case.

We describe the reasoning progress of Stephanie, one of the children with whom Maher and Martino (1996a, 1996b, 1997) began their long-range, but occasional, interventions commencing in Grade 1. By Grade 3, the children had begun building physical models and justifying their solutions to the following problem: How many different towers of heights 3, 4, or 5 can be made using red and yellow blocks? Stephanie not only justified her solutions, she validated or rejected

> her own ideas and the ideas of others on the basis of whether or not they made sense to her.... She recorded her tower arrangements first by drawing pictures of towers and placing a single letter on each cube to represent its color, and then by inventing a notation of letters to represent the color cubes. (Maher & Speiser, 1997, p. 174)

She used spontaneous heuristics like guess and check, looking for patterns, and thinking of a simpler problem, and developed arguments to support proposed parts of solutions, and extensions thereof, to build more complete solutions. Occasional interventions continued for Stephanie through Grade 7. Then in Grade 8 she moved to another community and another school and her mathematics was a conventional algebra course. The researchers interviewed

---

[10] Lagrange's Theorem states: Let G be a finite group. If H is a subgroup of G, then $o(H)$ divides $o(G)$. Here $o(H)$ stands for the order of the group H, i.e., its cardinality.

her that year about the coefficients of $(a+b)^2$ and $(a+b)^3$. About the latter, she said "So there's a cubed . . . And there's three a squared b and there's three ab squared and there's b cubed.... Isn't that the same thing?" Asked what she meant, she replied, "As the towers." It turned out, upon further questioning, that Stephanie had been visualizing red and yellow towers of height 3 in order to organize the products $a^i b^j$. At home, before the interview, she had written out the coefficients for the first six powers of $a+b$. Also, in a subsequent interview, she could explain how counting the towers related to the binomial coefficients and used Pascal's triangle to predict the terms of $(a+b)^n$. (For a more complete discussion, see Maher & Speiser, 1997.) Her understanding prompted Speiser to remark, "I wish some of my [university] students were able to reason that well." The case of Stephanie illustrates possibilities for developing solid mathematical reasoning early on.

However, most undergraduate students do not come to university with such experiences. Rather, as described above, many come with nonstandard views of proof. In addition, they often need to acquire much of the complex constellation of knowledge and skills used in proving theorems.

Understanding the Structure of a Proof and the Order in which it Might be Written

Transition-to-proof course students often say they don't know what the teacher wants them to do or where to begin. This is especially true of intuitively obvious results such as $A \cup B = B \cup A$, where one "follows one's nose" logically and there is no "trick" or mathematical problem-solving aspect. Rather it is a matter of knowing how to use, and sometimes unpack, the relevant definitions, including using them in the "right" order. For mathematicians, this has become automatic, whereas students often don't know what to do. For example, on the final exam in a transition-to-proof course, students were asked to prove: *Let f and g be functions on A. If $f \circ g$ is one-to-one, then g is one-to-one.* In the course, the definition of f one-to-one was that if $f(x) = f(y)$ then $x = y$. However, all but one student unsuccessfully began with the hypothesis — $f \circ g$ is one-to-one — rather than assuming that $g(x) = g(y)$. (Moore, 1994). They did not appear to know how to use the definition of one-to-one and relate that to the structure of their proofs.[11]

Unpacking the Logical Structure of Statements of Theorems

Another difficulty students have when constructing their own proofs is an inability to unpack the logical structure of informally stated theorems — theorems that depart from a natural language version of predicate calculus. That is, theorems that omit specific mention of some variables or depart from the use of *for all, there exists, and, or, not, if-then,* and *if-and-only-if* in a significant way. For example the statement, *Differentiable functions are continuous*, is informal because a universal quantifier and the associated variable are understood by convention, but not explicitly indicated. Similarly, *A function is continuous whenever it is differentiable* is informal because it departs from the familiar *if-then* expression of the conditional as well as not explicitly specifying the universal quantifier and variable.

Being able to unpack the logical structure of such informally stated theorems is important because the logical structure of a mathematical statement is closely linked to the overall structure of its proof. For example, knowing the logical structure of a statement helps one recognize how one might begin and end a direct proof of it. When asked to unpack the logical structure of four informally worded syntactically correct statements, two true and two false, undergraduate mathematics students, many in their third or fourth year, did so correctly just 8.5% of the time. Especially difficult for them was the correct interpretation of the order of the existential and universal quantifiers in the false statement: *For $a < b$, there is a c so that $f(c) = y$ whenever $f(a) < y$ and $y < f(b)$*[12] (Selden & Selden, 1995).

Furthermore, the ability to unpack the logical structure of the statement of a theorem also allows one to know whether an argument proves that statement, as opposed to some other statement. For example, eight mid-level undergraduate mathematics and mathematics education majors were asked to judge the correctness of student-generated "proofs" of

---

[11] In a rather formally written proof, one might begin something like, "Suppose $f \circ g$ is one-to-one." But (with this definition), the hypothesis is not used until one attempts to prove that g is one-to-one by assuming $g(x) = g(y)$. An alternative definition, $x \neq y$ implies $f(x) \neq f(y)$, might have made this particular theorem easier to prove, but apparently the students did not think of using it.

[12] If f were continuous and if it were stated that $a < c < b$, this would be the Intermediate Value Theorem as stated in most beginning calculus textbooks.

a single theorem.[13] Upon finding a proof of the converse particularly easy to follow, four initially incorrectly stated that it was a proof of the original statement, and two of these maintained this view throughout the interview (Selden & Selden, 2003).

Understanding the Effect of Existential and Universal Quantifiers

One source of students' difficulties in discerning the logical structure of theorems is a lack of understanding of the meaning of quantifiers and that their order matters. Undergraduate students often consider the effect of an interchange of existential and universal quantifiers as a mere rewording. For example, in another study, when given the two statements:

- *For every positive number a there exists a positive number b such that b < a.*
- *There exists a positive number b such that for every positive number a, b < a.*

24 of 54 students in undergraduate mathematics courses, such as linear algebra and multivariable calculus, and 3 of 9 students in a beginning graduate abstract algebra course said they were "the same" or were merely reworded (Dubinsky & Yiparaki, 2000).

While understanding the logical structure of a definition or a theorem is certainly not sufficient for constructing a proof, it is definitely necessary. In other words, if you do not understand what something really says, you certainly cannot prove it.

Knowing How to Read and Check Proofs

An integral part of the proving process is being able to tell whether one's argument is correct and proves the theorem it was intended to prove. For this, one must check one's own proof. How do undergraduates read and check proofs?

An Exploratory Study

We conducted an exploratory study of how eight undergraduates (four secondary education mathematics majors and four mathematics majors) from the beginning of a transition-to-proof course validated, that is, evaluated and judged the correctness of, four student-generated "proofs" of a very elementary number theory theorem (Selden & Selden, 2003). The "proofs" were real student work from a similar transition-to-proof course. The theorem was: *For any positive integer n, if n^2 is a multiple of 3, then n is a multiple of 3.* Unbeknownst to the students, for later reference, we had dubbed the student-generated "proofs": (a) Errors Galore, (b) The Real Thing, (c) The Gap, and (d) The Converse, indicating our view of them. Each of the eight students was interviewed individually for about one hour in a semistructured interview consisting of four phases. In Phase 1, the students were asked to explain the statement of the theorem in their own words, give some examples of it, and try to prove it. Two were successful, and after some time, those who could not complete a proof were asked to proceed with the other portions of the interview. In Phase 2, the students were shown the four "proofs," one after the other, and asked to think out loud as they read each one and decided whether it was, or was not, a proof. In Phase 3, having seen and thought about the "proofs" one after the other, they were given an opportunity to reread them all together and rethink their earlier decisions. In Phase 4, they were asked some general questions about how they read proofs. For example: When you read a proof is there anything different you do, say, than in reading a newspaper? How do you tell when a proof is correct or incorrect? How do you know a proof proves this theorem instead of some other theorem?

The students made judgments regarding the correctness of each student-generated "proof" four times, at the beginning and end of each of their two readings in Phases 2 and 3. At each time, there were 32 person-proof judgments. At the beginning (Time 1), these were just 46% (15 of 32) correct, but by the end (Time 4) 81% were correct. We attribute this difference to the students (at Time 4) having thought about the "proofs" several times, and perhaps, to the interviewer's no longer accepting "unsure" as a response. Most of the errors detected were of a local/detailed nature rather than a global/structural nature, with only the two students who had proved the theorem themselves observing that the converse had been proved in (d).

---

[13] The theorem was: *For any positive integer n, if n^2 is a multiple of 3, then n is a multiple of 3.*

When asked how they read proofs, the students said they attempted careful line-by-line checks to see whether each mathematical assertion followed from previous statements, checked to make sure the steps were logical, and looked to see whether any computations were left out. Several said they went through the proofs using an example. Also, for these students, a feeling of personal understanding or not—that is, of making sense or not—seemed to be an important criterion when making a judgment about correctness of a "proof." Thus, what students say about how they read proofs seems a poor indicator of whether they can actually validate proofs with reasonable reliability. While these students tended to "talk a good line," their judgments at Time 1 were no better than chance (46% correct).

On the other hand, even without explicit instruction, the reflection and reconsideration engendered by the interview process eventually yielded 81% correct judgments, suggesting that explicit instruction in validation could be effective (Selden & Selden, 2003). Indeed, several transition-to-proof textbooks include "proofs to grade,"[14] but we think it would also be helpful to have students validate actual student-generated proofs.

Knowing and Using Relevant Concepts

In addition to the ability to unpack, understand, and interpret definitions and theorems correctly, and check one's logic, one must have some relevant content knowledge. Constructing all but the most straightforward of proofs involves a good deal of persistence and problem solving to put together relevant concepts. And in order to use a concept flexibly, it is important to have a rich *concept image*, that is, a lot of examples, non-examples, facts, properties, relationships, diagrams, and visualizations, that one associates with that concept.[15]

In many upper-level mathematics courses, students are given definitions together with a few examples, after which they are expected to use these definitions reasonably flexibly. To do this, students may need to find additional examples and non-examples and to prove or disprove related conjectures, more or less without guidance. One can think of these activities as helping to build students' concept images. How does one go about building a rich concept image for a newly introduced concept? Do undergraduate students actively try to enhance their concept images, for instance, by considering examples and nonexamples?

Getting to Know and Use a New Definition

In one study conducted by Dahlberg & Housman (1997), eleven students, all of whom had successfully completed introductory real analysis, abstract algebra, linear algebra, set theory, and foundations of analysis, were presented with the following formal definition. A function is called *fine* if it has a root (zero) at each integer. They were first asked to study the definition for five to ten minutes, saying or writing as much as possible of what they were thinking, after which they were asked to generate examples and nonexamples. Subsequently, they were given functions and asked to determine whether these were fine functions and, if so, why. Next, they were asked to determine the truth of four conjectures, such as "No polynomial is a fine function."

Four basic learning strategies were used by the students on being presented with this new definition – example generation, reformulation, decomposition and synthesis, and memorization. Examples generated included the constant zero function and a sinusoidal graph with integer x-intercepts. Reformulations included $f(-1) = 0$, $f(0) = 0$, $f(1) = 0$, $f(2) = 0,\dots$, and $f(n) = 0 \ \forall n \in Z$. Decomposition and synthesis included underlining parts of the definition and asking about the meaning of "root." Two students simply read the definition – they could not provide examples without interviewer help and were the ones who most often misinterpreted the definition.

Of these four strategies, example generation, together with reflection, elicited the most powerful "learning events," that is, instances where the authors thought students made real progress in understanding the newly introduced concept. Students who initially employed example generation as their learning strategy came up with a variety of discontinuous, periodic continuous, and non-periodic continuous examples and were able to use these in their explanations. Those who employed memorization or decomposition and synthesis as their learning strategies often misinterpreted the

---

[14] Textbooks for such courses have from none to just a few to a moderate number of exercises involving critiquing "proofs." The directions vary — students may be asked to: (a) find the fallacy in a "proof;" (b) tell whether a "proof" is correct; (c) grade a "proof," A for correct, C for partially correct, or F; or (d) evaluate both a "proof" and a "counterexample." Most of these "proofs" have been carefully constructed by the textbook authors so there is just one error to detect. See, for example, Smith, Eggen, and St. Andre (1990; p. 39).

[15] The idea of concept image was introduced by Vinner and Hershkowitz (1980), elaborated by Tall and Vinner (1981), and is now a much used notion in mathematics education research.

definition, for example, interpreting the phrase "root at each integer" to mean a fine function must vanish at each integer in its domain, but that its domain need not include all integers. Students who employed reformulation as their learning strategy developed algorithms to decide whether functions they were given were fine, but had difficulty providing counterexamples to false conjectures (Dahlberg & Housman, 1997).

Thus, it seems that while students are often reluctant, or unable, to generate examples and counterexamples, doing so helps enrich their concept images immensely and enables them to judge the probable truth of conjectures.

Dealing with Various Symbolic Representations

Another aspect of understanding and using a concept is knowing which symbolic representations are likely to be appropriate in certain situations; this can be very important for success in proving. Concepts can have several (easily manipulated) symbolic representations or none at all. For example, prime numbers have no such representation; they are sometimes defined as those positive integers having exactly two factors or being divisible only by 1 and themselves. It has been argued that the lack of an (easily manipulated) symbolic representation makes understanding prime numbers especially difficult, in particular, for preservice teachers (Zazkis & Liljedahl, 2004). Similarly, irrational numbers have no such representation; thus, in proving results such as $\sqrt{2}$ is irrational or the sum of a rational and an irrational is irrational, one is led to consider proofs by contradiction — something often difficult for beginning students.

Symbolic representations can make certain features transparent and others opaque.[16] For example, if one wants to prove a multiplicative property of complex numbers, it is often better to use the representation $re^{i\theta}$, rather than $x + iy$, and if one wants to prove certain results in linear algebra, it may be better to use linear transformations, T, rather than matrices. Students often lack the experience to know when a given representation is likely to be useful.

It has been argued that moving flexibly between representations (e.g., of functions given symbolically or as a graph) is an indication of the richness of a student's understanding of a concept (Even, 1998). Also, understanding an abstract mathematical concept can be regarded as possessing "a notationally rich web of representations and applications" (Kaput, 1991, p. 61).

Bringing Appropriate Knowledge to Mind

No one questions the need for content knowledge, sometimes referred to as resources,[17] in order to solve problems and prove theorems. But students can have such resources and not be able to bring them to bear on a problem, or proof, at the right time.

Knowing, but not Using, Factual Knowledge

In two companion studies, 19 volunteer third quarter A and B calculus students, and later 28 volunteer differential equations students, took a one-hour paper-and-pencil test (without calculators) asking them to solve five moderately non-routine first calculus problems, that is, problems somewhat, but not very, different from what they had been taught. Immediately afterwards, they took a half-hour routine test, covering the resources needed to solve the non-routine problems. For example, one non-routine problem was: *Find at least one solution to the equation* $4x^3 - x^4 = 30$ *or explain why no such solution exists.* Two-thirds of the calculus students failed to solve a single problem completely and more than 40% did not make substantial progress on a single problem. Also, more than half of the differential equations students were unable to solve even one problem and more than a third made no substantial progress toward a solution. Of those non-routine problems for which the students had full factual knowledge, just 18% of the calculus students' solutions and 24% of the differential equations students' solutions were completely correct (Selden, Selden, & Mason, 1994; Selden, Selden, Hauk, & Mason, 2000).

To solve the above non-routine problem, one needs to know (1) that one might set the derivative of $4x^3 - x^4 - 30$ equal to zero to find its maximum -3 and (2) that solutions of the given equation are where this function crosses the x-axis (which it does not). Many of the students had these two resources, but apparently could not bring

---

[16] Representations can be *transparent* or *opaque* with respect to certain features. For example, representing 784 as 28^2 makes the property of being a perfect square transparent, but representing 784 as $(13 \times 60) + 4$ makes that property opaque. For more details, see Zazkis and Liljedahl (2004, pp. 165–166).

[17] Schoenfeld (1985) described good mathematical problem-solving performance in terms of resources, heuristics, control, and belief systems.

them to mind at an appropriate time. We conjectured that, in studying and doing homework, the students had mainly followed worked examples from their textbooks and had thus never needed to consider various different ways to attempt problems. Thus, they had no experience at bringing their assorted resources to mind. It seems very likely that a similar phenomenon could occur in attempting to prove theorems.

How does one think of bringing the appropriate knowledge to bear at the right time? To date, mathematics education research has had only a little to say about the difficult question of how an idea, formula, definition, or theorem comes to mind when it would be particularly helpful, and probably there are several ways. In their study of problem solving, Carlson and Bloom (2005) found that mathematicians frequently did not access the most useful information at the right time, suggesting how difficult it is to draw from even a vast reservoir of facts, concepts, and heuristics when attempting to solve a problem or to prove a theorem. Instead, the authors found that mathematicians' progress was dependent on their approach, that is, on such things as their ability to persist in making and testing various conjectures.

Our own personal experience of eventually bringing to mind resources that we had — but did not at first think of using — suggests that persistence, over a time considerably longer than that of the Carlson and Bloom interviews, can be beneficial. We conjecture that certain ideas get in the way of others, and that after a good deal of consideration, such unhelpful ideas become less prominent and no longer block more helpful ideas. This may be related to a psychological phenomenon that can take several forms; for example, in vision, if one fixates on a single spot in a picture, it will eventually disappear.

While coming to mind at the right time can be seen as an idiosyncratic, individual act, it may sometimes be related to the idea of transfer of one's knowledge. How does one come to see a new mathematical situation as similar to a previously encountered situation and bring the earlier resources to bear on the new situation?

Knowing What's Important and Useful

In addition to knowing what a proof is, being able to reason logically, unpack definitions, and apply theorems, and having a rich concept image of relevant ideas, one needs a "feel" for the content and what kinds of properties and theorems are important. Knowing what's important should go a long way towards bringing to mind appropriate resources.

Not Seeing that Geometry Theorems are Useful when Making Constructions

Seeing the relevance and usefulness of one's knowledge and bringing it to bear on a problem, or a proof, is not easy. Schoenfeld (1985, pp. 36–42) provides an example of two beginning college students who had completed a year of high school geometry and were asked to make a construction: *You are given two intersecting straight lines and a point P marked on one of them. Show how to construct, using straightedge and compass, a circle that is tangent to both lines and that has the point P as its point of tangency to one of the lines.* During a 15-minute joint attempt, they made rough sketches and conjectures, and tested their conjectures by making constructions. When asked why their constructions ought or ought not to work, they responded in terms of the mechanics of construction, but did not provide any mathematical justification. Yet the next day they were able to give the proof of two relevant geometric theorems within five minutes. Apparently, these students simply did not see the relevance of these theorems at the time.

Knowing to Use Properties, Rather than the Definitions, to Check Whether Groups are Isomorphic

In another study, four undergraduates who had completed a first abstract algebra course and four doctoral students working on algebraic topics were observed as they proved two group theory theorems and attempted to prove or disprove whether specific pairs of groups are isomorphic: $\mathbf{Z}_n$ and $\mathbf{S}_n$, $\mathbf{Q}$ and $\mathbf{Z}$, $\mathbf{Z}_p \times \mathbf{Z}_q$ and $\mathbf{Z}_{pq}$ (where p and q are coprime), $\mathbf{Z}_p \times \mathbf{Z}_q$ and $\mathbf{Z}_{pq}$ (where p and q are not coprime), $\mathbf{S}_4$ and $\mathbf{D}_{12}$. Nine times these undergraduates, who were successful in only two of twenty instances, first looked to see if the groups had the same cardinality; after which they attempted unsuccessfully to construct an isomorphism between the groups. They rarely considered properties preserved under isomorphism, despite knowing them (as ascertained by a subsequent paper-and-pencil test). For example, they all knew $\mathbf{Z}$ is cyclic, $\mathbf{Q}$ is not, and a cyclic group could not be isomorphic to a non-cyclic group, but they did not use these facts and none were able to show $\mathbf{Z}$ is not isomorphic to $\mathbf{Q}$, until afterwards. These facts did not seem to come to mind spontaneously, or in reaction to this kind of question.

In contrast, the doctoral students, who were successful in comparing all of the pairs of groups, rarely considered the definition of isomorphic groups. Instead, they examined properties preserved under isomorphism. When the groups were not isomorphic, they showed one group possessed a property that the other did not; for example, $\mathbf{Z}$ is cyclic, but $\mathbf{Q}$ is not. To prove $\mathbf{Z}_p \times \mathbf{Z}_q$ is isomorphic to $\mathbf{Z}_{pq}$, where p and q are coprime, three of them noted that the two groups have the same cardinality and showed $\mathbf{Z}_p \times \mathbf{Z}_q$ is cyclic. None tried to construct an isomorphism (Weber & Alcock, 2004).

Knowing which Theorems are Important

In comparing the proving behaviors of four undergraduates who had just completed abstract algebra and four doctoral students who were writing dissertations on algebraic topics, it was found that the doctoral students had knowledge of which theorems were important when considering homomorphisms. For example, in considering the proposition: *Let G and H be groups. G has order pq (where p and q are prime). f is a surjective homomorphism from G to H. Show that G is isomorphic to H or H is abelian,* all four doctoral students recalled the First Isomorphism Theorem within 90 seconds. In contrast, two undergraduates did not invoke the theorem, while the other two invoked its weaker form only after considerable struggle. When the doctoral students were asked why they used such sophisticated techniques, a typical response was, "Because this is such a fundamental and crucial fact that it's one of the first things you turn to" (Weber, 2001).

Another four undergraduates, who had recently completed their second course in abstract algebra, and four mathematics professors, who regularly used group-theoretic concepts in their research, were interviewed about isomorphism and proof (Weber & Alcock, 2004). They were asked for the ways they think about and represent groups, for the formal definition and intuitive descriptions of isomorphism, and about how to prove or disprove two groups are isomorphic. The algebraists thought about groups in terms of group multiplication tables and also in terms of generators and relations, as well as having representations that applied only to specific groups, such as matrix groups. Each algebraist gave two intuitive descriptions of groups being isomorphic: that they are essentially the same and that one group is simply a re-labeling of the other group. To prove or disprove two groups are isomorphic, they said they would do such things as "size up the groups" and "get a feel for the groups," but could not be more specific. In addition, they said that they would consider properties preserved by isomorphism and facts such as $\mathbf{Z}_n$ is *the* cyclic group of order n.

In contrast, none of the undergraduates could provide a single intuitive description of a group; for them, it was a structure that satisfies a list of axioms. While all four undergraduates could give the formal definition of isomorphic groups, none could provide an intuitive description. To prove or disprove that two groups were isomorphic, these undergraduates said they would first compare the order (i.e., the cardinality) of the two groups. If the groups were of the same order, they would look for bijective maps between them and check whether these maps were isomorphisms (Weber & Alcock, 2004).

It may be that undergraduates mainly study completed proofs and focus on their details, rather than noticing the importance of certain results and how they fit together. That is, they may not come to see some theorems as particularly important or useful. The mathematics education research literature contains few specific teaching suggestions on how to help students come to know which theorems are likely to be important in various situations. But, it might be helpful to discuss with them: (1) which theorems and properties you (the teacher) think are important and why, (2) your own intuitive, or informal ideas, regarding concepts, and (3) the advantages and disadvantages of various representations.

Teaching Proof and Proving

Some Suggestions Emanating from Research

One very positive finding, which was described earlier, is the remarkable sophistication of reasoning reached by some average school students who received brief interventions over a number of years (Maher & Martino, 1996a, 1996b, 1997). As described above, these students used a variety of spontaneously developed heuristics. Eventually, in order to come to agreement, these students, more or less, invented the idea of proof in a concrete case. If grade school students can be encouraged in this way, why not university students? Perhaps this could be done in part with relatively short "interventions" spread across the entire undergraduate program.

Another result is that younger students seem to prefer explanatory proofs written with a minimum of notation. This was certainly the case for U.K. Year 10 students (Healy & Hoyles, 1998). For example, instead of using mathematical induction to prove that sum of the first n integers is $n(n+1)/2$, one could use a variant of Gauss's original argument. Namely, for any n, one can write the sum in two ways as $(1 + 2 + 3 + \cdots + n)$ and as $(n + (n-1) + (n-2) + \cdots + 1)$, then add corresponding terms to obtain n identical summands equal to $n+1$, so twice the original sum equals $n(n+1)$. Hence, the original sum must equal $n(n+1)/2$ (Hanna, 1989, 1990). It seems plausible that undergraduates, and people more generally, might prefer proofs that provide insight to proofs that just establish the validity of a result.[18]

It also appears that great care should be taken to distinguish empirical reasoning from mathematical proof. Exactly how this can be done effectively is not especially clear, since merely giving high school geometry students a short instructional unit on this distinction left some of them very unclear as to the difference between empirical evidence and proof (Chazan, 1993). Perhaps secondary and university teachers need to stress this distinction often and also get students to discuss and reflect on situations where simple pattern generalization does not work.

Since current secondary teachers' conceptions of proof are somewhat limited and they sometimes accept non-proofs as proofs (Knuth, 2002a, 2002b), one way to enhance preservice secondary teachers' abilities to check the correctness of proofs might be to have them consider and discuss, in groups, a variety of student-generated "proofs," as well as having them provide feedback on each other's proofs.

In addition to explaining the difference between descriptive definitions in a dictionary and mathematical definitions, one can engage students in the defining process. For example, when using Henderson's (2001) investigational geometry text, one can begin with a definition of triangle initially useful in the Euclidean plane, on the sphere, and on the hyperbolic plane, but eventually students will notice that the usual Side-Angle-Side Theorem (SAS) is not true for all triangles on the sphere. At this point, they can be brought to see the need for, and participate in developing, a definition of "small triangle" for which SAS remains true on the sphere.

Perhaps it would also be possible to create classroom activities to improve students' ability to enhance their concept images and deal with representations flexibly. One suggestion is that upon introducing a new definition, one could ask students to generate their own examples, alternatively, to decide whether professor-provided instances are examples or non-examples, "without authoritative confirmation by an outside source" (Dahlberg & Housman, 1997, p. 298). Another possibility might be to engage students in conjecturing which kinds of symbolic representations might be useful for solving a given problem or proving a specific result. Also, one could point out that when a theorem has a negative conclusion (e.g., $\sqrt{2}$ is irrational), a proof by contradiction may be just about the only way to proceed.

For certain theorems in number theory, it has been suggested that the transition to formal proof can be aided by going through a (suitable) proof using a generic example that is neither too trivial nor too complicated (Rowland, 2002). Gauss's proof that the sum of the first n integers is $n(n+1)/2$, done for $n = 100$ is one such generic proof. Done with care, going over generic proofs interactively with students could enable them to "see" for themselves the general arguments embedded in the particular instances. If the theorem involves a property about primes, 13 and 19 are often suitable, provided the proof is constructive and that prime (e.g., 13) can be "tracked" through the stages of the argument. A generic proof, but not the standard one, can be given for Wilson's Theorem: *For all primes p, $(p-1)! \equiv p-1 \pmod p$.* That argument for $p = 13$ involves pairing each integer from 2 to 11 with its (distinct) multiplicative inverse mod 13, noting the product of each pair is congruent to $1 \pmod{13}$, and concluding that $12! \equiv 1 \times 1 \times 12 \pmod{13}$.[19] There is one caveat; there is some danger that students will not understand the generic character of the proof. In an attempt to avoid this, one can subsequently have them write out the general proof.

Some Personal Observations and Ongoing Work

We see learning to construct proofs, especially for beginning students, as composed largely of the acquisition of a complex constellation of skills, content knowledge, beliefs, and problem solving ability — much of which is best learned by doing. As a result, we think university teachers should consider including a good deal of student-student and teacher-student interaction regarding students' proof attempts, as opposed to just presenting their own or textbook's

---

[18] It has been suggested that proofs have various functions within mathematics: explanation, communication, discovery of new results, justification of a new definition, developing intuition, and providing autonomy (e.g., Hanna, 1989; de Villiers, 1990; Weber, 2002).

[19] For details of this and some other number-theoretic generic proofs, along with a description of how they were used with Cambridge University undergraduates, see Rowland (2002).

proofs. Trying to teach such a complex constellation entirely by lecture seems like trying to teach someone to tie her/ his shoelaces entirely over the telephone. It might be possible, but seems unlikely to be the most effective way.

In that connection, it might be useful, and certainly could do no harm, to discuss with students some of the difficulties mentioned above: in particular, the difference between mathematical proof and other types of arguments and the difference between mathematical definitions and everyday definitions. It might be helpful to stress that mathematicians of today see proofs as consisting of such careful deductive reasoning that, barring mistakes, the results (theorems) require no further evidence, are permanently true, and can be immediately used anywhere that the premises hold.

Another suggestion is that, when presenting proofs, one could take a top-down approach to explanation, first giving a global overview of the proof's structure to avoid the appearance of "pulling a rabbit out of a hat," followed by introducing and developing concepts as needed (Leron 1983, 1985).

Students often do not appreciate that proofs themselves can have a hierarchical structure — that there are subproofs (and subconstructions) within proofs, perhaps several levels deep. One could make students aware of this and illustrate how structure comes into thinking about how to prove a theorem. Students need to understand that proofs are not generally conceived of in the order they are written. Not realizing this may result in quite a few students not making use of the hierarchical structure of proofs in their own proving attempts and lead to some of the difficulties mentioned earlier. Students need to be encouraged to write parts of a tentative proof "out of order" (e.g., What will the last line say?), even when they sometimes resist doing so.

There seems to be quite a lot to learn about the way in which proofs are customarily written. If students were taught about this way of writing in some of their courses, they might not be so puzzled about how to begin a proof. Indeed, we take the point of view that proofs are deductive arguments *in an identifiable genre*. They differ from arguments in legal, political, and philosophical works. Within this genre, individual styles can vary, just as novels by Hemingway and Faulkner have differing styles, although their novels are easily seen as belonging to a single genre that clearly differs from newspaper articles, short stories, or poems. As part of some ongoing work, we have been collecting general features of the genre of proof. For example, definitions already stated outside of proofs tend not to be written into them. In teaching, we have found that pointing out such features, especially in the context of a student's own work, can be helpful to students.

Furthermore, we have found it useful to have students carefully examine the structure of the statement that they are trying to prove, and even to think about how a tentative proof might be structured, before launching into it. For example, consider proving the theorem (mentioned earlier): *Let f and g be functions on A. If $f \circ g$ is one-to-one, then g is one-to-one.* It would be useful for a prover to first unpack the meaning of g being one-to-one. Doing so can direct one to begin the proof by writing, "Let x and y be in the domain of g and suppose $g(x) = g(y)$." This also makes clear that the desired conclusion is "Thus $x = y$." In this way, one exposes the "real, but hidden" mathematical task, namely, to get from $g(x) = g(y)$ to $x = y$. After that, students can concentrate on how the hypothesis that $f \circ g$ is one-to-one might help.

Concluding Remarks

We have tried to provide readers with a coherent organization of some of the mathematics education research on proof and proving, but there is much more.[20] Awareness of the variety of difficulties undergraduates have with proof and proving can make one more sensitive regarding how to help them. The above pedagogical suggestions indicate some steps one might take; however, more information on "what works" is needed.

References

Carlson, M. P., & Bloom, I. (2005). The cyclic nature of problem solving: An emergent multidimensional problem solving framework. *Educational Studies in Mathematics, 58,* 45–76.

Chazan, D. (1993). High school geometry students' justification for their views of empirical evidence and mathematical proof. *Educational Studies in Mathematics, 24,* 359–387.

---

[20] Anyone wanting to delve into the considerable literature on proof and proving can go to the bibliography maintained by the *International Newsletter on the Teaching and Learning of Proof* at: www.lettredelapreuve.it/. Those with a more philosophical bent might want to consult the annotated bibliography at: fcis.oise.utoronto.ca/~ghanna/mainedu.html.

Cockcroft, W. M. (Chair). (1982). *Mathematics counts: Report of the Committee of Inquiry into the Teaching of Mathematics in Schools* (Department of Education and Science). London: HSMO.

Coe, R., & Ruthven, K. (1994). Proof practices and constructs of advanced mathematics students. *British Educational Research Journal, 20*(1), 41–53.

Dahlberg, R. P., & Housman, D. L. (1997). Facilitating learning events through example generation. *Educational Studies in Mathematics, 33*, 283–299.

de Villiers, M. (1990). The role and function of proof in mathematics. *Pythagoras*, 24, 7–24

Dubinsky, E., & Yiparaki, O. (2000). On student understanding of AE and EA quantification. In E. Dubinsky, A. H. Schoenfeld, & J. Kaput (Eds.), *Issues in mathematics education: Vol. 8. Research in collegiate mathematics education. IV* (pp. 239–289). Providence, RI: American Mathematical Society.

Edwards, B. S., & Ward, M. B. (2004). Surprises from mathematics education research: Student (mis)use of mathematical definitions. *American Mathematical Monthly, 111*, 411–424.

Even, R. (1998). Factors involved in linking representations. *Journal of Mathematical Behavior, 17*, 105–121.

Hanna, G. (1989). Proofs that prove and proofs that explain. In G. Vergnaud, J. Rogalski, & M. Artigue (Eds.), *Proceedings of the Thirteenth Conference of the International Group for the Psychology of Mathematics Education* (Vol. 2, pp. 45–51). Paris: CNRS - Paris V.

—— (1990). Some pedagogical aspects of proof. *Interchange, 21*(1), 6–13.

Harel, G., & Sowder, L. (1998). Students' proof schemes: Results from exploratory studies. In A. H. Schoenfeld, J. Kaput, & E. Dubinsky (Eds.), *Issues in mathematics education: Vol. 7. Research in collegiate mathematics education. III* (pp. 234–283). Providence, RI: American Mathematical Society.

Hazzan, O., & Leron, U. (1996). Students' use and misuse of mathematical theorems: The case of Lagrange's Theorem. *For the Learning of Mathematics, 16*(1), 23–26.

Healy, L., & Hoyles, C. (1998). *Technical report on the nationwide survey: Justifying and proving in school mathematics*. Institute of Education, University of London.

—— (2000). A study of proof conceptions in algebra. *Journal for Research in Mathematics Education, 31*, 396–428.

Henderson, D. W. (2001). *Experiencing geometry in Euclidean, spherical and hyperbolic spaces* (2nd ed.). Upper Saddle River, NJ: Prentice Hall.

Hoyles, C. (1997). The curricular shaping of students' approaches to proof. *For the Learning of Mathematics, 17*(1), 7–16.

Jones, F. B. (1977). The Moore method. *American Mathematical Monthly, 84* (4), 273–278.

Kaput, J. (1991). Notations and representations as mediators of constructive processes. In E. von Glasersfeld (Ed.), *Radical constructivism in mathematics education* (pp. 53–74). Dordrecht, The Netherlands: Kluwer Academic Publishers.

Knuth, E. J. (2002a). Teachers' conceptions of proof in the context of secondary school mathematics. *Journal of Mathematics Teacher Education, 5*(1), 61–88.

—— (2002b). Secondary school mathematics teachers' conceptions of proof. *Journal for Research in Mathematics Education, 33*, 379–405.

Leron, U. (1983). Structuring mathematical proofs. *American Mathematical Monthly, 90*, 174–184.

Leron, U. (1985). Heuristic presentations: The role of structuring. *For the Learning of Mathematics, 5*(3), 7–13.

London Mathematical Society, Institute of Mathematics and its Applications, Royal Statistical Society. (1995). *Tackling the mathematics problem*. London, U.K.: London Mathematical Society.

Mahavier, W. S. (1999). What is the Moore method? *PRIMUS, 9*, 339–354.

Maher, C. A., & Martino, A. M. (1996a). The development of the idea of proof. A five-year case study. *Journal for Research in Mathematics Education, 27*, 194–219.

—— (1996b). Young children inventing methods of proof: The gang of four. In L. Steffe, P. Nesher, P. Cobb, G. Goldin, & B. Greer (Eds.), *Theories of mathematical learning* (pp. 431–447). Hillsdale, NJ: Lawrence Erlbaum Publishers.

—— (1997). Conditions for conceptual change: From pattern recognition to theory posting. In H. Mansfield & N. H. Pateman (Eds.), *Young children and mathematics: Concepts and their representation* (pp. 58–81). Sydney, Australia: Australian Association of Mathematics Teachers.

Maher, C. A., & Speiser, R. (1997). How far can you go with a tower of blocks? In E. Pehkonen (Ed.), *Proceedings of the 21st Conference of the International Group for the Psychology of Mathematics Education* (Vol. 4, pp. 174–183). Jyväskylä, Finland: Gummerus.

Martin, G. W., & Harel, G. (1989). Proof frames of preservice elementary teachers. *Journal for Research in Mathematics Education, 29,* 41–51.

Moore, R. C. (1994). Making the transition to formal proof. *Educational Studies in Mathematics, 27,* 249–266.

National Council of Teachers of Mathematics. (2000). *Principles and standards for school mathematics.* Reston, VA: Author.

Rowland, T. (2002). Generic proofs in number theory. In S. R. Campbell & R. Zazkis (Eds.), *Learning and teaching number theory: Research in cognition and instruction* (pp. 157–183). Westport, CT: Ablex Publishing.

Schoenfeld, A. H. (1985). *Mathematical problem solving.* Orlando, FL: Academic Press.

Selden, A., & Selden, J. (2003). Validations of proofs written as texts: Can undergraduates tell whether an argument proves a theorem? *Journal for Research in Mathematics Education, 34,* 4–36.

Selden, A., Selden, J., Hauk, S., & Mason, A. (2000). Why can't calculus students access their knowledge to solve non-routine problems? In E. Dubinsky, A. H. Schoenfeld., & J. Kaput, (Eds.), *Issues in mathematics education: Vol. 8. Research in collegiate mathematics education. IV* (pp. 128–153). Providence, RI: American Mathematical Society.

Selden, J., & Selden, A. (1995). Unpacking the logic of mathematical statements. *Educational Studies in Mathematics, 29,* 123–151.

Selden, J., Selden, A., & Mason, A. (1994). Even good calculus students can't solve nonroutine problems. In J. Kaput and E. Dubinsky (Eds.), *Research issues in undergraduate mathematics learning: Preliminary analyses and results* (MAA Notes No. 33, pp. 19–26). Washington, DC: Mathematical Association of America.

Smith, D., Eggen, M., & St. Andre, R. (1990). *A transition to advanced mathematics* (3rd ed.). Pacific Grove, CA: Brooks/Cole Publishing Company.

Tall, D., & Vinner, S. (1981). Concept image and concept definition with particular reference to limits and continuity. *Educational Studies in Mathematics, 12,* 151–169.

Vinner, S., & Hershkowitz, R. (1980). Concept images and common cognitive paths in the development of some simple geometrical concepts. In *Proceedings of the Fourth International Conference for the Psychology of Mathematics Education* (pp. 177–184). Berkeley, CA.

Weber, K. (2001). Student difficulty in constructing proofs: The need for strategic knowledge. *Educational Studies in Mathematics, 48,* 101–119.

—— (2002). Beyond proving and explaining: Proofs that justify the use of definitions and axiomatic structures and proofs that illustrate technique. *For the Learning of Mathematics, 22*(3), 14–17.

Weber, K., & Alcock, L. (2004). Semantic and syntactic proof productions. *Educational Studies in Mathematics, 56,* 209–234.

Zazkis, R., & Campbell, S. (1996). Prime decomposition: Understanding uniqueness. *Journal of Mathematical Behavior, 15,* 207–218.

Zazkis, R., & Liljedahl, P. (2004) Understanding primes: The role of representation. *Journal for Research in Mathematics Education, 35,* 164–186.

9

Mathematical Induction:
Cognitive and Instructional Considerations

Guershon Harel, *University of California, San Diego*
Stacy Brown, *Pitzer College*

The principle of mathematical induction (MI) is a prominent proof technique used to justify theorems involving properties of the set of natural numbers. The principle can be stated in different, yet equivalent, versions. The following are two versions common in textbooks:[1]

Version 1: Let S be a subset of $\mathbb{N}$ (the set of natural numbers). If the following two properties hold, then $S = \mathbb{N}$.

  (i) $1 \in S$.

  (ii) $k \in S$, then $k + 1 \in S$.

Version 2: Suppose we have a sequence of mathematical statements $P(1), P(2), \ldots$ (one for each natural number). If the following two properties hold, then for every $n \in \mathbb{N}$, $P(n)$ is true.

  (i) $P(1)$ is true.

  (ii) If $P(k)$ is true, then $P(k + 1)$ is true.

Poincaré, among others, viewed the principle of MI as intrinsic to humans' intuition: "Mathematical induction … is … necessarily imposed on us, because it is … the affirmation of a property of the mind itself" (Poincaré, 1952). In fact, historical documents indicate that mathematicians employed this method of proof for at least 200 years before it was explicitly formulated by Peano. Despite its intuitive appeal and its central role within mathematics, research in mathematics education has documented that students have major difficulties understanding MI (Dubinsky, 1986; Fischbein & Engel, 1989; Movshovitz-Hadar, 1993; Reid, 1992; Robert & Schwarzenberger, 1991; Harel, 2001; Brown, 2003).

In this chapter, we explore students' difficulties with MI when taught with the standard instructional treatment and we present results from our teaching experiments, which employed alternative instructional approaches. We begin by introducing a construct central to our work, the notion of a *proof scheme*. We then describe the standard instructional treatment of MI, pointing to its possible inadequacies. Drawing from our and others' research on students' difficulties with MI when taught with the standard instructional treatment, we demonstrate how many of these difficulties are indicative of students' deficient proof schemes. Having described the standard instructional treatment and the related student difficulties, we proceed with an account of two independent, yet related, studies, Harel (2001) and Brown (2003). After which, we present a synthesis of our results in the form of a three-stage model of students' development of MI. We conclude with a summary and instructional recommendations.

---

[1] In this paper we are not concerned with *strong induction*.

The Concept of Proof Scheme

The notion of *proof scheme* was first defined in Harel and Sowder (1998), where a taxonomy of students' proof schemes was drawn from a long sequence of teaching experiments with primarily mathematics and engineering students. Later, in Harel (2007), this taxonomy was refined and expanded to reflect the historical development of proof. Within this taxonomy, proving is defined as the process employed by a person to remove or create doubts about the truth of an observation—a process that involves both ascertaining and persuading. "Ascertaining is a process an individual employs to remove her or his own doubts about the truth of an observation. ... Persuading is a process an individual employs to remove others' doubts about the truth of an observation" (Harel & Sowder, 1998, p. 241). Thus, an individual's proof scheme consists of his or her means for ascertaining and persuading. Of course, seldom do these processes occur separately: in ascertaining for oneself, one considers how to persuade others, and vice versa.

To illustrate—and hopefully help the reader appreciate the pedagogical value of— the notion of "proof scheme," consider the following story told by Hartshorne (2000, pp. 11–12) in his book, *Geometry: Euclid and Beyond*:

> I choose a point *A* on the circle, and with my compass centered at *A*, and radius *AO*, I mark off a point *B* on the circumference. Then with center *B* and radius *BO* I mark off another point *C* on the circumference. I repeat this process, always with radius equal to the radius of the original circle, to get further points *D*, *E*, and *F*. Then I draw *AB*, *BC*, *CD*, *DE*, *EF*, all of which have the same length, so that *ABCDEF* will be an equilateral hexagon inscribed in the circle.
>
>   Why does this work? How would you explain this so as to convince another person?
>
>   To get a real-life answer, I put this question to my seventeen-year-old son, then a high school senior. His first response was, "I have done it myself, so I know it works."

In this response, Hartshorne's son seems to have applied the empirical proof scheme. With this scheme one convinces oneself or attempts to persuade others about the truth of an assertion by relying on evidence from perception or examples of direct measurements of quantities, substitutions of specific numbers in algebraic expressions, etc. (Harel and Sowder, 1998). The response by Hartshorne's son seems to indicate that perception was dominant in his conviction that the above process of constructing an equilateral hexagon works.[2]

> "Yes," I said, "from a practical point of view it works. But how do you know this is an exact solution and not just a very good approximation?" After a few minutes of thought he drew the lines from *O* to *A*, *B*, *C*, *D*, *E*, *F* and then explained that *OAB* is an equilateral triangle by construction. Therefore, the angle $\angle AOB$ at the center is $60°$. The same is true for the next four triangles *BOC*, …, *EOF*. Thus we have five $60°$ angles, so the remaining angle $\angle AOF$ must also be $60°$. The triangle AOF having two sides the same and the same central angles must be the same as the triangle *AOB*, and so *FA=AB*.

The skepticism Hartshorne expressed to his son about his initial response led the son to produce a deductive proof. Usually, unfortunately, students do not respond to such skepticism in the way Hartshorne's son did. Many students do not posses a *deductive proof scheme—a scheme by which assertions are proved by the rules of deduction*—and cannot appreciate, or have difficulty answering, such skepticism; hence, they see no need to look for a deductive proof. Questions such as "How do you know this is an exact solution?" or "How do you know that a pattern derived from a finite number of cases always holds?" seem to these students contrived and artificial. We will come back to this point later. For now, let us continue with Hartshorne's story, which demonstrates the complexity of the deductive proof scheme.

> "Fine," I said, "that is very convincing, assuming that your listener knows that the angles of an equilateral triangle are $60°$, and the angle of one total revolution is $360°$. It seems your listener would have to know the theorem that the sum of the three angles of a triangle is $180°$. What if he asked you to explain why that is true?
>
>   I mentioned a proof of the sum of the angles by drawing a line parallel to one side *AB* of a triangle through the third vertex *C*. Then $\alpha = \alpha'$ because of the parallel lines, and $\beta = \beta'$ because of the parallel lines, so $\alpha + \beta + \gamma = \alpha' + \beta' + \gamma = 180$ because it is a straight angle. "But then you have to know theorems

---

[2] Of course, this is a speculation on our part since it is difficult to determine one's meaning on the basis of a single statement.

about the angles formed when a line cuts two parallel lines." There ensued a discussion about proliferation of questions, like the endless "why's of a three-year-old, and danger of getting into circular arguments."

Hartshorne continues:

> So we see that while the notion of proof as a convincing argument may work well, it depends on who your listener is, and is also subject to the danger of infinite regress if your listener is uncooperative.

The dialogue further points to the fact that the person who is presenting the proof and his interlocutor must have some common basis for determining what constitutes a convincing argument. It also points to the role of axiom systems in proving. The concept of "axiom system" has evolved in the course of the history of mathematics. What is accepted as a proof in Euclidean geometry may not be a valid proof in a geometry that is based on Hilbert's axioms.

This notion of proof scheme was the basis for the design and methodologies of the two independent, yet related, studies—one reported in Harel (2001) and the other in Brown (2003)—which we summarize below. Before describing the studies, however, it is important to first consider standard instructional treatments of MI and the difficulties that arise for students taught with such treatments.

Standard Instructional Treatments

Typically, standard undergraduate mathematics textbooks introduce the principle of MI either after a brief statement about the "need" for this method of proof or after a few examples of its application in simple problems. For example, Barnier and Feldman (1990) start the section on MI as follows:

> If consecutive odd integers, starting with 1, are added, a nice pattern emerges, namely
>
> $n=1$: $1=1$
> $n=2$: $1+3=4$
> $n=3$: $1+3+5=9$
> $n=4$: $1+3+5+7=16$
>
> where n is the number of odd numbers to be added.
>
> It appears, then, that the sum of the first n odd integers is always equal to the square of n. But how is such a statement proved? Verifying an infinite sequence of statements, statement by statement, is out of the question. Mathematical Induction is what is needed in such cases.

Though some students see that one cannot verify an infinite sequence of statements, statement by statement, many others, especially those in introductory courses, possess the empirical proof scheme, and so they generalize patterns from particular cases. For these students, the abrupt introduction of the principle of MI leaves little time for the students to see the necessity for a particular method of proof, to consider what such a method might look like, or to consider which constraints it must satisfy. Furthermore, introductions of this type do not facilitate the students' rejection of empirical approaches. As Brown (2003) noted, one consequence of this is that students in introductory proof courses often continue to use empirical approaches after having received instruction on MI and may even interpret MI as an empirical approach.

The type of MI problems and the order they are presented in standard textbooks are also of pedagogical concern. In general, induction problems can be classified into two categories, *recursion problems* and *non-recursion problems* (see Harel, 2001). "Prove that $1+3+5+\cdots+(2n-1)=n^2$ for all positive integers n" is an example of a recursion problem because, to solve it, the left-hand side of the identity must be interpreted as a recursive representation of a function. In contrast, the problem, "Prove that $n<2^n$ for all positive integers $n \geq 1$," is an example of a non-recursive problem because, to solve it, one does not need to involve a recursive representation of a function. The category of recursion problems can be further classified into two sub-categories, *explicit recursion problems* and *implicit recursion problems*, according to whether the recursive representation of a function is explicit or implicit in the problem statement. For example, in the aforementioned problem the rule, $f(n)=f(n-1)+(2n-1), f(1)=1$, is virtually explicit in the problem statement; hence, it is an explicit recursion problem. On the other hand, the Towers of Hanoi problem (see Footnote 3) is implicit, for no recursively defined function is explicitly present in the problem statement. An analysis of textbook problems shows that students' first exposure to MI is often through explicit recursion and non-recursion problems of the following kind:

(a) Identity problems (e.g., "Prove that $1+3+5+\cdots+(2n-1)=n^2$ for all positive integers n),

(b) Inequality problems (e.g., "Prove that $n<2^n$ for all positive integers $n>3$), and

(c) Divisibility problems ("Prove that $3\,|\,n^3-n$ for all positive integers n").

Implicit recursion problems, such as the Towers of Hanoi Problem, usually are not included or appear in a small number at the end of the exercises listed.

Also, textbook authors often begin with "easy" problems, such as the problem, "Prove by MI the statement, for any positive integer $n\geq 4$, $n!>2^n$." Yet many students do not view such problems as problems that require MI but rather as problems that can be solved in an "easier" way; e.g., they argue the following:

(i) If $n\geq 4$, $n!=n\times\cdots\times 4\times 3\times 2\times 1=n\times\cdots\times(2\times 2)\times 3\times 2$.

(ii) Since each of the n terms of $n\times 0°\times(2\times 2)\times 3\times 2$ is greater than or equal to 2, it follows that $n!\geq 2\times\cdots\times 2\times 2\times 2\times 2$ (n times) when $n\geq 4$.

This is not to say that problems of this type do not have a place in a chapter on MI, but to say that they are introduced to the students in a non-suitable stage in their conceptual development. More specifically, when students are asked to use MI on problems that they do not view as necessitating such a method of proof, MI is viewed by the students as a prescription to be followed; thus reinforcing the *authoritative proof scheme* (a scheme by which one determines the truth of an assertion on the basis of a blind acceptance of what her or his teacher, textbook, or classmate says) and *non-referential symbolic proof scheme* (a scheme by which one's conclusions are based on the manipulation of symbols free from meaningful referents). Thus, there is reason to consider alternatives to the standard instructional approach, for this approach is inadequate for enabling students to see the necessity of MI.

Students' Difficulties with MI

Students experience major difficulties understanding the statement of the principle of MI, when the principle is applicable, and how to apply it. To demonstrate and describe these difficulties, we draw from our own and others' research on students' mathematical behavior in dealing with problems involving MI .

Consider the following episode taken from Brown's (2003) series of studies in which students from introductory proof courses were interviewed periodically over the duration of each course. After having been taught MI in his introductory proof course, a junior mathematics major at a highly-ranked university was shown a paper folding demonstration of the theorem "The sum of the interior angles of a triangle is 180°". In this demonstration, a triangle is folded so that its three vertices fall on a point on one of its segments and its three angles form a straight angle (see the figure to the right). The student was first asked whether the demonstration was convincing and then asked how he would respond to a skeptic who still had doubts as to whether the folding approach proved the assertion. His response to the first question was "Yes, … [and] I would use this to convince my tutees." his response to the second question was "I'd tell [the skeptic] to use [mathematical] induction." When asked what he would induct on, he responded, "the number of triangles." Thus, the student did not understand that the principle of mathematical induction (MI) does not apply to uncountable sets, such as the set of all triangles in the plane, and that it is used to justify theorems involving properties of the set of natural numbers. This episode is an example of a student having difficulty understanding when the principle is applicable.

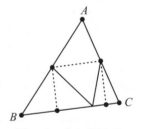

The triangle is folded along the dotted lines.

Brown (2003) also found that the students in the introductory courses often failed to apply MI to problems involving recursive relations, such as the Towers of Hanoi Problem,[3] while they frequently attempted to apply MI to

---

[3] *The Towers of Hanoi Problem*: Three pegs are stuck in a board. On one of these pegs is a pile of disks graduated in size, the smallest being on top. The object of this puzzle is to transfer the pile to one of the other two pegs by moving the disks one at a time from one peg to another in such a way that a disk is never placed on top of a smaller disk. How many moves are needed to transfer a pile of n disks?

prove assertions about the real numbers, as the following episode illustrates. Anthony was given the problem "Prove that $x + 1/x \geq 2$ for any positive real number x." Anthony began by writing:

$$S = \left\{ \forall x \left(x \in \mathbb{R}^+ \wedge x > 0 \right) x + 1/x \geq 2 \right\}$$
$$1 + 1/1 = 2 \geq 2 \Rightarrow 1 \in S$$
$$\text{Given } x + 1/x \geq 2, \text{ prove } (x+1) + 1/(x+1) \geq 2$$

Then he said:

I need to show ... (pause) ... to show that $x + 1$ is in S,

and wrote:

$$\left[x(x+1) + (x+1) + 1 \right]/(x+1) \geq 2,$$
$$x + 1/x \geq 2, \ x + 1 + 1/x \geq 2 + 1$$

Finally he asked himself:

How do I get a 1 down there (pointing to the denominator of $1/x$)?

This episode and the one before demonstrate that students have difficulties recognizing when MI is applicable. The second episode also demonstrates another difficulty, described below in more detail, where the student views MI as a mere procedure for manipulating symbols—a manifestation of the symbolic non-referential proof scheme rather than a deductive process.

Another area of difficulty for students concerns the statement of the principle of MI. Consider the following episode: "Let's prove this statement by MI," declared the instructor during a teaching experiment session of an advanced linear algebra class. As he completed the base step ($P(1)$ is true) and began the inductive step ($P(n) \Rightarrow P(n+1)$ for every positive integer n), he sensed that the class as a whole was not following him. The instructor paused, turned to the students, and asked if they remembered the principle of MI. "It is a proof with steps" was the only response to his question. "How many of you have heard the term 'Mathematical Induction' " he asked. All of the students raised their hand, confirming that they had heard the term in previous classes. Dubinsky (1986) tells of a similar experience: "If you question students—even those who have had several mathematics courses—although almost all of them will have heard of induction, not many of them will be able to say anything intelligent about what it is, much less actually use it to solve a problem" (p. 305).

Upon further analysis of this session and interviews with the students, we found that many of the students believed that the conclusion that an assertion is true for all positive integers n is based on the "fact" that it was proved for $n+1$: "We proved it for $n+1$, so we proved it for n" typified many of the students' responses in explaining the principle of MI. The students viewed the inductive step as the proof that the assertion is true for any positive integer n. Furthermore, and as a consequence, these students saw the base step to be unnecessary. For some, it was a step one performs as an initial confirmation for the truth of the assertion, i.e., one evaluates the truth of the assertion for the case $n = 1$ because it is the "simplest number." For others, it was a step needed merely to follow the rule dictated by the instructor or textbook. Thus, the central ideas behind the statement of MI eluded the students.

Many students can solve certain kinds of MI problems correctly, especially those typified by the explicit recursion problem:

Prove that $1/(1 \cdot 2) + 1/(2 \cdot 3) + \cdots + 1/(n \cdot (n+1)) = n(n+1)$ for any integer $n \geq 1$.

However, for these students, MI is merely a "proof procedure" where one takes an equation involving n and adds an expression to both sides so as to produce a similar equation with $n+1$ in place of n. Though such students may, at times, correctly apply MI, they do so without understanding what they are doing. Specifically, they do not understand the meaning of the inductive step. Namely, that, for example, one adds $1/(n+1)(n+2)$ to both sides of the above equality in order to consider the truth of $P(n+1)$ given that $P(n)$ is true.

In sum, our interpretation of students' difficulties understanding the statement of the principle of MI, when the principle is applicable, and how to apply it, is that these difficulties are, for the most part, a result of employing instructional techniques that (a) do not facilitate a need—an intellectual need—for MI and, (b) do not allow MI to arise as a proof technique.

The Studies

The notion of proof scheme was the basis for the design of the studies we report in this paper. These studies, Harel (2001) and Brown (2003), focused on students' development of MI as a deductive proof scheme. Both studies explored the conceptual changes that occurred over time as the students received one of two alternative instructional treatments.

Methodologies

Both Harel (2001) and Brown (2003) used the teaching experiment methodology. "Teaching experiment" here is as defined by Cobb and Steffe (1983): Namely, a teaching experiment consists of a series of classroom observations and individual interviews over an extended period of time (a semester, for example). Each teaching session is analyzed in terms of the classroom discourse and students' performance. The results of the analysis can, and usually do, adjust or amend the plan for subsequent lessons. Results accumulated from extensive analyses usually refine, and in some cases alter, the researchers' theoretical perspective. These analyses are then used to develop models of what students know and how their knowledge evolves.

Harel conducted a series of six teaching experiments, involving a total of 139 students. The study reported here is one of these experiments. Data were collected in the form of field notes, retrospective notes, written tests and quizzes, videotaped classroom sessions, and clinical interviews. Brown's experiments involved a year-long series of observations in introductory proof courses, including interviews with cohorts of students from each course, and three teaching experiments. Data from the observations of introductory proof courses was collected in the form of field notes, student work, and videotapes and transcripts of individual student interviews. Data for the teaching experiments was collected in the form of field notes, retrospective notes, videotapes and transcripts of each classroom session, student work from each classroom session, and videotapes and transcripts of individual interviews.

Alternative Instructional Treatments

The new instructional treatments of MI used by Harel and Brown take into consideration the deficiencies of the standard instructional treatment (as described above). They are designed (a) to help students develop the principle of MI gradually through the use of a reordered sequence of traditional MI problems and the use of some non-traditional MI problems, and (b) to facilitate the creation of a situation in which the students can both understand and appreciate the need for MI while avoiding developing a non-referential symbolic interpretation of MI (i.e., one in which MI is performed by the students as a procedure, yet the students fail to understand what they are doing).

Harel's treatment was first introduced in Harel and Sowder (1998) and a full description of the teaching experiment on MI can be found in Harel (2001). Briefly, the treatment consisted of three phases. Of particular relevance to this paper are the first two phases, before the formal statement of the principle of MI is introduced. In the first phase, students' exposure to MI is through engagement with implicit recursion problems typified by:

1. Find an upper bound to the sequence $\sqrt{2}, \sqrt{2+\sqrt{2}}, \sqrt{2+\sqrt{2+\sqrt{2}}}, \ldots$.

2. You are given 3^n coins, all identical except for one which is heavier. Using a balance, prove that you can find the heavy coin in n weighings.

The students in Harel's teaching experiments were engaged in tasks of this type for about one week, during which time they solved about twenty problems. The goal achieved in this phase was that these implicit recursion problems necessitated for the students the formation of recursive relations. For example, when solving Problem 1, a student demonstrated that the third item is less than 2 because it is the square root of a number that is smaller than 4, this number being the sum of 2 and a number that is smaller than 2. She then proceeded to argue that the same relationship exists between any two consecutive terms in the sequence. Thus, key to her argument was the recognition of a recursive relation (Harel, 2001).

In Phase 2, explicit recursion problems, such as Problems 3-5 below, are introduced.

3. Prove that for any positive integer $n \geq 1$, $\dfrac{1}{1 \cdot 2} + \dfrac{1}{2 \cdot 3} + \cdots + \dfrac{1}{n \cdot (n+1)} = \dfrac{n}{n+1}$.

4. Find a formula for the sum $a + a^2 + a^3 + \cdots + a^n$.

5. Compute the sum $1 + 3 + \cdots + 2n - 1$.

At first, the students did not appear to see any relation between this kind of problem (explicit recursion problems) and the kind of problems they worked on in Phase 1 (implicit recursion problems). They did not interpret the latter problems in terms of a sequence of propositions. But once they were explicitly asked to see if such a common structure existed, they understood the problems in these terms, i.e., they saw the similarity between the two problem sets. Accordingly, they applied the solution approaches they used to solve Phase 1 problems to Phase 2 problems. Thus, it is during Phase 2 that students realize that the same method of proof can be applied to both sets of problems. The following example (also described in Harel (2001)), illustrates how one student made this connection when solving Problem 5. The student argued:

> ...Like in Problem 1, where the relationship between two consecutive elements a_k and a_{k+1} is $a_{k+1} = \sqrt{2 + a_k}$, here the relationship is $a_{k+1} = a_k + 2k + 1$. And like in Problem 1, where we used the fact that a_1 is smaller than 2 to derive that a_2 is smaller than 2, and so on, here we use the fact that $a_1 = 1^2$ to derive that $a_2 = 2^2$, because $a_2 = a_1 + 2 \cdot 1 + 1 = 1^2 + 2 \cdot 1 + 1 = 1^2 + 2 \cdot 1 + 1^2 = (1+1)^2 = 2^2$, and in a similar way to use that $a_2 = 2^2$ to derive that $a^3 = 3^2$, and so on.

Moreover, the process of applying their method to Phase 2 problems enabled the students to recognize their solution approach as a method of proof.

Finally, in the third phase, the formal statement of the principle of MI is introduced as a refined formulation of the students' method of proof they developed and applied in the previous two phases. Upon realizing that the same method of proof—mathematical induction—could be applied to all three sets of problems, the students were assigned additional problems to help them solidify their understanding of MI.

Brown's alternative curricular treatment is based on the curricular and pedagogical approaches suggested in Harel and Sowder (1998) and Brousseau (1997). The MI problems and their sequence resemble those suggested in Harel (2001). As in Harel's alternative treatment, the introduction of the principle of MI is postponed until the students (a) have formulated a method of proof, (b) have discussed the criteria the method must satisfy, and (c) have characterized the class of problems for which the method of proof was developed. The key idea of this approach is to have the principle of MI arise as a means to solve a class of problems and as a response to a fundamental question: Let S be the set of values for which a proposition P is true. How can one show $S = \mathbb{N}$? Initially, as is the case with Harel's alternative instructional treatment, students are presented with and asked to solve implicit recursion problems and then explicit recursion problems. As the students develop solutions, specific interventions may be required to address the students' robust empirical proof schemes. These interventions are described below in detail. Having progressed through these two sets of problems, the students are then asked to describe commonalities in the tasks and their solution methods. Once the students recognize that they have considered propositions whose domain is the set of natural numbers, a whole class discussion takes place concerning the question: Let S be the set of values for which a proposition P is true. How can one show $S = \mathbb{N}$? As students characterize and refine their solution technique, they are then asked to consider how one might apply this method to other tasks, i.e., other implicit and explicit recursion and non-recursion problems. These activities allow the students to extend and refine the method of proof, ultimately preparing them for the introduction of the principle of MI.

A Three-Stage Model for the Conceptual Development of MI

With the alternative curricular approaches described above, we found that students progress through three stages when developing MI as a deductive proof scheme.

Stage 1

During Stage 1, students' responses to MI problems are mostly a manifestation of the empirical proof scheme. Students at this stage convince themselves and others of the validity of a mathematical statement by observing a pattern in the results obtained from computations—a manifestation of the *empirical proof scheme*. For example, when asked to solve the Towers of Hanoi problem (see Footnote 3), Stage 1 students may conclude that the number of moves for n disks is $2^n - 1$ on the basis of empirical observations for a few cases. For instance, a student argued that given the data in her table (see below) she knew the solution was $2^n - 1$ because $2^0 - 1 = 1$, $2^2 - 1 = 3$, $2^3 - 1 = 7$, $2^4 - 1 = 15$.

# disks	# moves
1	1
2	3
3	7
4	15

Paula's Table

Thus, in the eyes of Stage 1 students, the formula is valid because it follows a pattern of results they obtained empirically. This approach — when one generalizes from a pattern in the results — is referred to in Harel (2001) as *result pattern generalization* (RPG). Brown (2003) observed that students in Stage 1 may end up applying RPG even if initially they attend to the underlying structure of the pattern. For example, in solving the problem "Prove that $1+3+5+\cdots+(2n-1) = n^2$ for all positive integers n, some students compute as follows: $3^2 + 7 = 16$, $4^2 + 9 = 25$, $5^2 + 11 = 36$. However, rather than generalizing the pattern of this computational process, e.g., into $n^2 + (2n+1) = n^2 + 2n + 1 = (n+1)^2$), they generalize only the pattern of the computed results, focusing only on $1 = 1^2$, $1 + 3 = 4 = 2^2$, $1 + 3 + 5 = 9 = 3^2$ to conclude that $1 + 3 + 5 + \cdots + (2n-1) = n^2$.

Attending to the underlying structure of the pattern is referred to in Harel (2001) as *process pattern generalization* (PPG). While in RPG one's conviction is based on regularity in the results—obtained, for instance, by substituting numbers—in PPG one's conviction is based on regularity in the process, though it might be initiated by regularity in the results. As we will show, it is in the third stage that students shift their focus from RPG to PPG.

Stage 2

During Stage 2, students recognize the limitations of justifications based on (a) a set of particular cases, and (b) patterns derived from such cases. A student in Stage 2 will recognize, for example, the mathematical necessity of proving that the solution to the Towers of Hanoi is $2^n - 1$ after having observed that $1 = 2^1 - 1$, $3 = 2^2 - 1$, and $7 = 2^3 - 1$. In other words, Stage 2 students do not view approaches based on a result pattern generalization approach to be convincing, but rather their sense of conviction begins to rely on a pattern in the process of obtaining the results—on process pattern generalization. Consider, for example, how a student in Stage 2 solved the problem:

> Assuming no two lines are parallel and no three lines intersect at a point, how many regions are created in the plane by n lines?

The following is an outline of his solution:

(1) Let t_n be the total number of regions created by n lines.

(2) One line creates two regions, $t_1 = 2$.

(3) When a second line is added, two additional regions are created, $t_2 = 2 + 2$.

(4) When a third line is added, three additional regions are created, $t_3 = 2 + 2 + 3$.

(5) When the nth line is added to the plane, it will intersect $n-1$ lines and create n additional regions. Since we begin with one region, $t_1 = 1 + 1$. Thus, $t_n = 1 + 1 + 2 + 3 + \cdots + n$.

Thus, the reasoning exhibited by students during Stage 2 is more sophisticated than that used by Stage 1 students. Stage 2 students recognize a recurring process and generalize the pattern observed in the process—a *process pattern generalization*. Stage 1 students, on the other hand, generalize the pattern from the results they observe in particular cases—a *result pattern generalization*. Students in Stage 2 propose arguments that resemble proofs by MI. In particular, they may argue P(1), P(1) $\rightarrow$ P(2), P(2) $\rightarrow$ P(3), P(3) $\rightarrow$ P(4), so P(n) $\rightarrow$ P($n+1$), therefore for all n, P(n), or P(1), P(2), P(3), P(3) $\rightarrow$ P(4), P(4) $\rightarrow$ P(5), so P(n) $\rightarrow$ P($n+1$), therefore for all n, P(n). Students' thinking at this stage is, however, limited in two ways. First, they arrive at the implication P(n) $\rightarrow$ P($n+1$) by generalizing from a finite sequence of implications. Second, they do not recognize their use of an inductive hypothesis, the premise P(n), but rather view P(n) as a verified statement. For example, students at this stage may find their descriptions of the identified processes sufficient justification; thus avoiding mathematical formulations of these processes. As a result, the students fail to recognize their use of an inductive hypothesis. To illustrate, consider the following excerpt from a group of students as they attempt to generalize a process:

The L-Tiling Task: Let n be a positive integer. Can any $2^n \times 2^n$ grid, with one square removed, be tiled with an L-shaped tile that consists of three squares like the one on the right?

Susan: Just by visually … I mean saying that you could expand … you know like you take your eight by eight ($2^3 \times 2^3$ grid) … you use your eight … put the thing (removed square) in the corner … put the four of them together and you're always going to have that because you got two to the n (2^n), you're always, you know, *it's basically going to be like four of the one before.*
       […]

Susan: … so based on that you're always expanding in the same way and you could just do it infinitely …

Johan: Yeah

Susan: (continuing) … many times.

As illustrated by Susan's comments, these students recognized a process when constructing the $2^2 \times 2^2$ grid with four $2^1 \times 2^1$ grids, the $2^3 \times 2^3$ grid with four $2^2 \times 2^2$ grids, and the $2^4 \times 2^4$ grid with four $2^3 \times 2^3$ grids, and viewed this process as being generic, i.e., they recognized the underlying structure and viewed the examples as representative. They then argued that these constructions were sufficient evidence of a general solution. They did not see, however, the necessity of creating a representation of how one can create a tiled $2^{n+1} \times 2^{n+1}$ grid, given a tiled $2^n \times 2^n$ grid (for some n). Rather, they argued that their series of examples demonstrates that one can simply use "four of the one before." Consequently, they failed to recognize their use of an inductive hypothesis, namely, that the nth case could be tiled. In other words, the students used an inductive hypothesis without actively engaging in hypothetical thinking and, consequently, failed to fully recognize the logical structure of MI.

Recognizing and understanding the use of an inductive hypothesis is nontrivial for many students. For example, while working on Problem 5 (see the section entitled Alternative Instruction Treatments), a cohort of students began to attend to the number of tiles one would add to a square of a given dimension to create the "next" square, as demonstrated below, when developing a geometric argument.

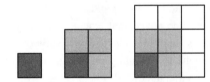

Geometric Argument for $1 + 3 + 5 + … + (2n-1) = n^2$

Once this process was generalized "take your n minus one by n minus one (square) and add two n minus one (tiles)" the students decided the statement $1 + 3 + 5 + … + (2n–1) = n^2$ was true for all n. It was at this time that the instructor pointed out to the students that they were making the assumption that the statement was true for the $(n–1)^{th}$ case. As illustrated below, the students found their use of an inductive hypothesis problematic in that they viewed the inductive hypothesis as something to be proved.

Paula: This is basically, this is the general formula we've proved (reference to "$(n–1)^2 + 2n – 1 = n^2$"). So basically, all we need to prove is n minus one squared is true for all natural numbers (reference to "$1 + 3 + 5 + … + 2(n–1) – 1 = (n–1)^2$") and that'll prove it because that's the only assumption we made.

Stage 3

During Stage 3, the student begins to produce proofs that are, in essence, proofs by MI. Their justifications are of the form: P(1) and P(n) → P(n+1) for all n, therefore P(1) → P(2), P(2) → P(3), P(3) → P(4), … , and hence, P(n) for all n. The distinction between Stage 2 and Stage 3 is this: In Stage 2 the implication P(n) → P(n+1) is generalized from a finite sequence P(1) → P(2), P(2) → P(3), P(3) → P(4) and so on—usually consisting of the first two or three implications—where students deduce P(2) from P(1), P(3) from P(2), and so on. Of particular importance is that they realize that the underlying deduction process is the same in all implication cases, and so one, in principle, proceeds in the same manner to conclude P(n) for any given n. In Stage 3, on the other hand, the implication P(n) → P(n+1) is not generalized (as in Stage 2) but proven deductively, and the sequence of cases P(1) → P(2), P(2) → P(3), P(3) →

P(4) and so on are seen as its instantiations. For example, when working on the Two-Color Problem (below), without having verified a sequence of cases, Johan argued that given an arbitrary two-colored plane, he could add a line and create a two-coloring. The implication $P(n) \rightarrow P(n+1)$ was then used to conclude P(n) for all n.

> *Two-Color Problem*: Consider any map formed by drawing n straight lines in a plane to represent boundaries. Is it possible to color the countries using two colors, if no two adjoining countries (those with a line segment as a common border) have the same color?

> Johan: Here's some number of lines, and it works and here's like a generic … a generic method of adding lines to it.
> SB: Okay.
> Johan: If this works, then whatever line we add there works also, using the same method.
> SB: Okay.
> Johan: So technically what we proved over here was … the first case was n … and adding a line was the n plus one case.

It should be pointed out that while the students in our teaching experiments developed a strong intuition about the validity of the principle of MI, our program did not proceed beyond Stage 3 where one raises the question about the status of the principle as an axiom or a theorem. To do so, we posit, one must possess the *axiomatic proof scheme*—a scheme by which one understands that, in principle, any proof must rely on a system of assertions accepted without proofs and terms without definitions.

Implications for Instruction

As defined, the notion of "proof scheme" is subjective—it can vary from person to person, and, as we have learned from history, it varied from civilization to civilization and generation to generation. It is this subjective stance that makes this notion central to the works presented in this chapter. All our studies were based on the premise that it is only when students' current knowledge is recognized that teachers can devise and implement instruction that can bring about desirable outcomes in students' learning. Despite this subjective definition of proof scheme, the goal of instruction must be unambiguous; namely, to gradually refine students' current proof schemes toward the proof scheme shared and practiced by the mathematicians of today.

For this reason, we embarked on the task of understanding current teaching treatments of MI and their potential consequences to student learning. We found that in these treatments, the three most prevalent proof schemes among students are the authoritative proof scheme, the symbolic non-quantitative proof scheme, and the result pattern generalization, which is a manifestation of the empirical proof scheme. We conjecture that these undesirable schemes are the result of employing instructional techniques that do not facilitate an intellectual need for MI and do not allow MI to arise as a proof technique from problems students understand and appreciate. Our alternative instructional treatments of MI were designed to help students develop the principle of MI gradually by facilitating the creation of situations in which the students can both understand and appreciate the need for MI.

A synthesis of our independent studies led to a three-stage model of the conceptual development of MI: In Stage 1, students prove by result pattern generalization; in Stage 2, they move to proving by process pattern generalization; and in Stage 3 their proving is consistent with the principle of MI. One critical question is how to facilitate the transition between stages. In the remainder of this chapter, we outline recommendations on how to achieve this goal.

Students often possess robust empirical proof schemes and it is this way of thinking that hinders the students' transition between stages, particularly from Stage 1 to Stage 2. The empirical proof scheme should not be underestimated, for it stems from our tendency to generalize when making sense of the world. Thus, even when one recognizes the limitations of this way of thinking within mathematics, one may continue to use it in other settings.

Our observations indicate that even in classrooms where the limitations of the empirical proof scheme have been demonstrated (e.g., one may show the students a formula that holds for numerous cases and still fails to be valid for all cases), students' use of empirical arguments persists. It is for this reason that instructors must do more than demonstrate its limitations. One approach is to have students present and then compare solutions. Facilitated by the teacher, such discussions can lead students to recognize why certain justifications, especially those of the Stage 1 kind, are insufficient. With weaker students, however, the class may be willing to accept a collection of examples as proof. In such instances, the teacher must create a situation in which the students see the necessity of rejecting arguments based on a collection of examples. For example, one may ask the students to solve the Chords of a Circle task (below):

Chords of a Circle Task: Suppose you have a circle with n points marked on the circumference. By connecting each pair of points with line segments the circle can be partitioned into a number of regions. Is there a function for calculating the number of regions?

This task satisfies two criteria: First, a pattern is easily recognized. Second, the pattern fails to hold for some n. In regard to the first criteria, when one considers the cases $n = 1, 2, 3, 4$, and 5 (see table below), one can see that the number of regions can be described with the easily recognized expression 2^{n-1}.

N	number of regions
1	1
2	2
3	4
4	8
5	16

In regard to the second criteria, the pattern fails to hold for $n > 5$. In particular, for $n = 6$, one can obtain either 30 or 31 regions depending on whether or not the chords align in such a way that three meet at a point.[4] Tasks that satisfy these criteria surprise the empirically focused student, who is often convinced of a solution's validity after testing three to four cases. For example, such students are often surprised when they do not obtain 32 regions for the case $n = 6$ and assume they have made a computational error. When repeated attempts to correct this error fail, the students are forced to both examine other possibilities and recognize that their expectation that the pattern would continue was not met.

It should be noted, however, that it is critical that the students are asked to solve this task as opposed to having it demonstrated to them, for they must actively engage in the act of assuming that a pattern exists in the results if a conflict is to arise between their results and their anticipations. As demonstrated in the excerpt below, where Jill argues with Calvin that a series of examples are insufficient, the students' experiences with the Chords of a Circle problem may instill in them a sense of skepticism about empirically derived results.

Jill: But how do you know at one point it might not … it might not happen? I understand what you're saying here, if it works for this one it's going to work for that one but it … what if at one point it doesn't? Like the circle thing?

We found that once students reject empirically-based results, they turn to the processes through which they generated their results to gain a sense of conviction, which in turn helps them transition into Stage 2. For example, when Calvin tried to convince Jill of the validity of his approach to the Towers of Hanoi Problem, he first used a series of examples. Following her remarks, he reconsidered his approach and argued:

Calvin: Suppose it (the formula $2^n - 1$) works for some stack, then we know the next stack takes (writes $(2^n - 1) + 1 + (2^n - 1) = 2^{n+1} - 1$) since we know the two k plus one formula works. So, if it works for five it works for six and we have all of these (reference to their data table) so we know it always works. We don't have to worry about the circles.

During Stage 2, the intuitive validity of a solution may hinder the students' formulation and articulation of a solution. For example, students may wish to include a series of examples demonstrating a process by which one creates successors and phrases such as "and so on …" when describing their solution. It is critical at this point, therefore, that the teacher facilitates the students' formulation of these processes through whole class discussions and activities aimed at the refinement of justifications presented by the students. One can achieve this pedagogical goal by introducing tasks that foster the students' formulation of their solution technique, or by explicitly asking students to

---

[4] The stipulation that no three chords meet at a point is often included in textbook versions of this problem. It is intentionally not included in this version of the problem so as to create a specific problem situation for the students.

describe the solution technique that was used to solve a set of problems. Once formulated, however, students may feel uncomfortable using an inductive hypothesis or they may, psychologically, view the inductive hypothesis as a verified statement. To address the issue of facilitating Stage 2 students' recognition of the need for a statement in the form of an implication, and therefore an inductive hypothesis, one can ask the students to solve tasks that satisfy the following criterion: For relatively small n, one cannot visualize or compute all of the possibilities for a particular case. The Two-Color Problem is an example of a task that satisfies this criterion (for $n = 17$, for example, P(n) concerns the set of all planes with 17-lines, arranged according to the constraints described above). In this case, unlike a mathematician, Stage 2 students often view the various possibilities as being distinct. Consequently, when the student formulates the implication P(n) $\rightarrow$ P(n+1), and assumes P(n), the premise functions as a pure hypothesis. Johan's comments (see the section entitled Stage 3) followed his realization that he could not produce a sufficiently generic sequence of the form P(1) $\rightarrow$ P(2), P(2) $\rightarrow$ P(3), P(3) $\rightarrow$ P(4), ... because each statement P(n) had numerous (in his eyes, infinitely many) possibilities. Consequently, he realized that, in order to show the statement was valid for sufficiently large n, he needed a general way to create a two-colored map with n+1 lines from any two-colored map with n lines.

Another task that facilitates Stage 2 students' recognition of the need for an inductive hypothesis is the general L-Tiling Problem (below). This task satisfies the criterion that for relatively small n, one cannot visualize or compute all of the possibilities. For $n = 5$, for example, generating a solution involves visualizing the placement of the 341 L-tiles along with the 1024 possibilities for the missing tile.[5]

(General) L-Tiling Problem: Let n be a positive integer. Can any $2^n \cdot 2^n$ grid, with one square removed, be tiled with an L-shaped tile that consists of three squares like the following?

Initially one may want to ask students to solve the modified L-Tiling Problem (below) and then, when the students are prepared to transition from Stage 2 to Stage 3, ask the students to solve the general L-Tiling Problem (above). Though the distinction between the two tasks — removing any tile as opposed to removing a corner tile — may appear trivial, it is significant to students. The modified L-Tiling Problem fosters students' use of recursive relations, which is important during Stage 1, but does not necessitate, in the eyes of the student, the use of hypothetical thinking, which is necessary to progress from Stage 2 to Stage 3. In particular, the inductive hypothesis for the general L-tiling Problem is such that one must assume the existence of a *class of objects* — the set of tiled $2^n \cdot 2^n$ grids with a tile removed. In contrast, the inductive hypothesis for the modified L-Tiling Problem simply involves assuming the existence of *an object* — a tiled $2^n \cdot 2^n$ grid with a corner tile removed. To carry out the latter, the student may simply generalize from smaller cases (e.g., the case $n = 4$), as is typical of Stage 1 students, whereas to solve the general problem the student must engage in hypothetical thinking by assuming the existence of a tiling for each member of the set of all $2^n \cdot 2^n$ grids with a tile removed. The general problem, therefore, necessitates treating the inductive hypothesis, P(n), as a hypothesis rather than as a generalization — one of the key aspects of Stage 3.

(Modified) L-Tiling Problem: Let n be a positive integer. Can any $2^n \cdot 2^n$ grid, with a corner tile removed, be tiled with an L-shaped tile that consists of three squares?

Thus, as can be seen from our recommendations, facilitating the transition from one stage to the next involves taking into consideration the student's current proof scheme and then creating a context that supports its refinement into a more desirable proof scheme. Through iterations of this process one can create a context for students to develop, understand, and appreciate the need for MI.

References

Barnier, W., & Feldman, N. (1990). *Introduction to advanced mathematics*. Englewood Cliffs, NJ: Prentice-Hall.

Brousseau, G. (1997). *Theory of didactical situations in mathematics*. Dordrecht, Netherlands: Kluwer.

Brown, S. (2003). *The evolution of students understanding of mathematical induction: A teaching experiment*. Unpublished doctoral dissertation, University of California, San Diego and San Diego State University.

Cobb, P., & Steffe, L. (1983). The constructivist researcher as teacher and model builder. *Journal for Research in Mathematics Education, 14*, 83–94.

---

[5] One does not necessarily need to visualize all of the possibilities if one recognizes that symmetry can be used to reduce the number of cases.

Dubinsky, E. (1986). Teaching mathematical induction I. *Journal of Mathematical Behavior*, 5, 305–317.

Fischbein, E., & Engel, I. (1989). Psychological difficulties in understanding the principle of mathematical induction. In G. Vergenaud, et al. (Eds.) *Proceedings of the 13th International Conference for the Psychology of Mathematics Education* (pp. 276–282). Paris, France.

Harel, G. (2001). The development of mathematical induction as a proof scheme: A model for DNR-based instruction. In S.R. Campbell and R. Zazkis (Eds.), *Learning and teaching number theory: Research in cognition and instruction* (pp. 185–212). Journal of Mathematical Behavior Monograph. Westport, CT: Ablex Publishing.

Harel, G., & Sowder, L. (1998). Students' proof schemes. In E. Dubinsky, A. Schoenfeld, and J. Kaput (Eds.), *Research on Collegiate Mathematics Education III* (pp. 234–283). Providence, RI: American Mathematical Society.

—— (2007). Toward a comprehensive perspective on proof. In F. Lester (Ed.), *Second Handbook of Research on Mathematics Teaching and Learning* (pp. 805–842). Charlotte, NC: Information Age Publishing.

Hartshorne, R. (2000). *Geometry: Euclid and beyond.* New York: Springer-Verlag.

Movshovitz-Hadar, N. (1993). The false coin problem, mathematical induction and knowledge fragility. *Journal of Mathematical Behavior*, *12*, 253–268.

Poincaré, H. (1952). *Science and Hypothesis* (pp. 1–19). New York, NY: Dover. Excerpted and Reprinted in (1983) in Benacerraf, P., & Putnam, H. (Eds.) *Philosophy of Mathematics: Selected Readings* (pp. 394–402). Cambridge, United Kingdom: Cambridge University Press.

Reid, D. (1992). *Mathematical induction: An epistemological study with consequences for teaching.* Unpublished thesis, Canada: Concordia University.

Robert, A., & Schwarzenberger, R. (1991). Research in teaching and learning mathematics at an advanced level. In D. Tall (Ed.), *Advanced mathematical thinking* (pp. 127–139). Dordrecht, Netherlands: Kluwer.

10

Proving Starting from Informal Notions of Symmetry and Transformations

Michelle Zandieh, *Arizona State University*
Sean Larsen, *Portland State University*
Denise Nunley, *Arizona State University*

In this chapter we consider the challenge of promoting students' ability to develop their own proofs of geometry theorems. We have found that students can make use of transformations and symmetries of geometric figures to gain insight into why a particular theorem is true. These insights often have the potential to form the basis for rigorous proofs. In the following classroom vignette, we see the excitement that comes from discovering an idea that seems to explain exactly why a theorem is true, followed by the realization that there is significant work to be done in order to develop a rigorous proof based on such an idea.

Classroom Vignette

Setting: A college geometry class using the Henderson (2001) text has been asked to work in groups to prove the isosceles triangle theorem (ITT). That is, given two sides of a triangle are congruent, prove that the angles opposite those sides are congruent. After about 3 minutes without much progress, the group of Alice, Emily, and Valerie burst into activity.

Alice:	The book says to use symmetries.
Emily:	Symmetries?
Valerie:	That angle equals that angle —
Alice:	Okay! Yeah! Yeah.
Valerie:	And then this angle —
Alice:	If you have, yeah! If you have, like, a bisected angle —
Emily:	You do the angle bisector —
Alice:	Yeah! And then this matches this [rotates her right hand from palm-up to palm-down across her triangle drawing] because it can lay right on top of it! [Moves her left hand to land (at word "top") on palm-up right hand.] Because then you like rotate it.
Emily:	You do a reflection over the perpendicular bisector of the angle. [Throws pencil down.]
Alice:	Yes! [Leans back in chair.] And then it proves it!
Emily:	Or, not the perpendicular, but the bisector angle.
Alice:	The, the bisector angle. But you do make it. I make it. But it is perpendicular. That's why you can do it. It's 'cause it's like this line's perpendicular here.
Emily:	It becomes perpendicular to the —

Alice: Yeah, so these are the same, and this folds right onto that.

Valerie: So, I agree with the perpendicular, but you know she's [the teacher's] going to ask, "Well, how do you know that the angle bisector is also a perpendicular bisector to the line?"

Introduction

Alice's excited declaration that, "it proves it!" indicates that she believed she had found the key to the proof. On the videotape we see Alice lean back in her chair and Emily throw down her pencil indicating their satisfaction that they had found a way to prove the theorem. Note that although we already see the students beginning to work out the details of their argument (Is the line they are interested in an angle bisector or a perpendicular bisector or both? Is the transformation a rotation, a reflection, a folding?), they are already convinced that their argument establishes the truth of the theorem. The conviction precedes the details needed to make their argument a formal proof, but it anticipates that those details can be worked out.

The discussion that follows involves students working to clarify Alice's argument for why the theorem is true and beginning to check whether the initial idea can in fact be turned into a fully justified proof. The fact that these are typical points for discussion in this class is hinted at by Valerie's phrase, "you know she's going to ask…"

However, as we will see in more detail below, Alice's confident statement that "the bisector angle … is perpendicular" was, for Alice, self-evident and did not require further explanation. This self-evidence turned out to have a coercive effect on her ability to complete the proof, despite heated discussions with Valerie and Emily who did not believe that this statement could be used in the proof without further justification.

The classroom episode beginning with our opening vignette, and described in detail below as Group 1, is illustrative of a process in which students work on developing a rigorous argument beginning with an intuitive idea. This chapter documents three examples of students engaged in the activity of proving by starting with intuitive, informal ideas and moving toward deductive proofs with claims supported by appropriate arguments. Below we will discuss several ideas from the mathematics education literature that we have found helpful for studying this type of student activity. Then we will consider in more detail three different ways in which students engaged in this activity using notions of congruence, symmetry and transformations to work toward proving the Isosceles Triangle Theorem. We will conclude with some thoughts for practice.

Explanation of Terms

Intuitions and Key Ideas

Efraim Fischbein (1982, 1987, 1999) described several kinds of intuitions, including *affirmatory* and *anticipatory* intuitions. He stated that intuitive cognitions are self-evident and coercive. With affirmatory intuitions, the self-evidence may be persistently coercive in that an individual with this intuition may not be able to consider other alternatives. Anticipatory intuitions "appear during a solving endeavor, usually, suddenly after a phase of intensive search … [and] are associated with a feeling of certitude, though the detailed justification or proof is *yet* to be found" (Fischbein, 1999, p. 34, his italics). Notice that Fischbein emphasized 'yet' because the intuitions anticipate a further refinement into the "formal, analytical, deductively justified steps of the solution" (Fischbein, 1999, p. 34).

We note that Alice's idea for the proof (that a reflection across the angle bisector will cause the triangle to land on itself) is an anticipatory intuition in that it occurs during a solving process and is associated with a feeling of certitude, not only on her part, but also Emily's, that this idea can be used to complete the proof. Alice's related idea, that the angle bisector will be the perpendicular bisector of the opposite side of the triangle, functions more as an affirmatory intuition for Alice because she finds it self-evident and, as we will see below, this has a coercive affect on her ability to complete the proof.

Earlier we stated that Alice believed she had found the key to the proof. Raman (2003, 2004) has used the phrase "key idea" in the sense of a single idea that holds the "key" to a proof. A key idea answers the question, "Why is this claim true?" in a way that connects or has the potential to connect an intuitive answer to the question with a rigorous answer to the question.

Raman's (2004) development of *key idea* was focused on the role of the key idea as a connector between what she called the public and private aspects of proof. She explained that a *procedural idea* in proof production is public in the sense that it generates a formal, deductive argument suitable for a textbook or journal article. Whereas a *heuristic idea* in proof production is private in the sense that these ideas are used "behind the scenes – for instance, as a mathematician tries to develop an intuition for why a claim is true" (p.635, Raman, 2004). In summary, Raman states:

> A key idea is a mapping between heuristic idea(s) and procedural idea(s). It links together the public and private domains, and in doing so provides a sense of understanding and conviction. The key idea is the essence of the proof, providing both a sense of why a claim is true and the basis for a formal rigorous argument. (p. 635)

Note that often in journals, mathematicians will publicly discuss the heuristic ideas that contributed to their production of a proof. Thus, private aspects of proof production may sometimes be publicly displayed (or at least described retrospectively). The real distinction between a procedural idea and a heuristic idea is that a procedural idea generates a formal deductive argument that satisfies the standards of rigor set by the mathematics community whereas heuristic ideas are more informal or intuitive ways of reasoning that need not satisfy the standards of rigor set by the mathematics community. Since Raman's (2004) notion of key idea is a bridge between heuristic ideas and (deductive) procedural ideas, the bridge may serve in both directions. Not only can one use a key idea to help move from a heuristic idea to a procedural idea, but one can also use a key idea as a way of describing heuristically, intuitively, the main ideas of a preexisting deductive proof. Fischbein (1987) described *conclusive intuitions* as summarizing "in a global, structured vision the basic ideas of the solution to a problem previously elaborated" (Fischbein, 1987, p. 64). We see key ideas as being able to serve not only as anticipatory intuitions but also, in other situations, as conclusive intuitions.

Raman's earlier papers (2003, 2004) focus on the notion of key idea in terms of already completed proofs. In more recent work, Raman and Zandieh (2007) elaborate the role of a key idea in student proving activity in terms of the powerful and problematic ways that a key idea influences students' development of a proof over the length of a class period. In this chapter we consider three proof attempts, each of which seems to start from the same affirmatory intuition or potential key idea — that one can fold or reflect an isosceles triangle over a line that bisects the angle between the two congruent sides and that doing so will cause the two angles opposite the congruent sides to land on each other, exhibiting their congruence. Alice states this more casually and kinesthetically in the opening vignette.

Alice: If you have, yeah! If you have, like, a bisected angle —
Emily: You do the angle bisector —
Alice: Yeah! And then this matches this [rotates her right hand from palm-up to palm-down across her triangle drawing] because it can lay right on top of it! [Moves her left hand to land (at word "top") on palm-up right hand.]

However, each of the three proof attempts discussed below takes a different path. In the end, Group 2 and Group 3 have partial proofs based on different key ideas even though they seem to have had similar starting points.

Transformational Reasoning

As students search for key ideas and work to relate a key idea to the arguments needed to provide a rigorous proof, they often need to develop an intuitive sense of how the system in question works. Simon (1996) explains that one can develop this sense through the use of *transformational reasoning*. For Simon, transformational reasoning is "the mental or physical enactment of an operation or set of operations on an object or set of objects that allows one to envision the transformations that these objects undergo and the set of results of these operations" (p. 201). A key to this type of reasoning is being able to deal with dynamic processes as opposed to static elements. Similarly, Harel and Sowder (1998) differentiate between arguments that call on a static visual perception to provide justification (taken as indicative of a perceptual proof scheme) and arguments based on dynamic or transformational observations (which are necessary for a transformational proof scheme). Both Simon and Harel and Sowder see transformational reasoning (observations) as goal oriented and anticipatory. However, Harel and Sowder go further when they discuss a transformational proof scheme, stating that it is a deductive argument that considers the "generality aspects of the conjecture" (p. 261).

Alice's revelation in the opening vignette is a type of transformational reasoning. Notice that Alice did not take the hint of symmetry to simply state that there was a static sense of "sameness" on either side of the figure, including the matching angles that were to be proved congruent. Instead she imagined a transformation, "Yeah! And then this matches this because it can lay right on top of it! Because you like rotate it." As she stated this, she moved her hand to indicate motion. This operation and its result provide insight into how the system works, yielding an answer to why the claim is true with some hints as to how to develop this key idea into a more rigorous proof.

Mathematizing

To further develop these intuitive beginnings into something that has more of the structure of formal mathematics, students engage in *mathematizing*. Mathematizing is a process of taking informal reasoning and making it more "mathematical" by clarifying terms or statements, converting ideas into more formal mathematical language or symbols, delineating relationships between terms or ideas and justifying statements using increasingly more formal mathematical arguments. Examples of mathematizing include defining, algorithmatizing, symbolizing, generalizing and formalizing (Rasmussen et al., 2005). Following Rasmussen et al. we believe that informal and formal are relative terms and that mathematizing is a process of moving from relatively less formal to relatively more formal reasoning.

We can see the mathematizing process beginning in the above vignette. As Alice excitedly laid out the key idea, Emily interspersed comments that sought to clarify the statement of the key idea. At the end of the vignette, Valerie reminded her group that one of their claims, "that the angle bisector is also a perpendicular bisector," was a claim that required further justification.

Mathematizing and transformational reasoning occur in many different mathematical settings. Examples from the research literature include transformational reasoning in precalculus and calculus (Carlson, Jacobs, Coe, Larsen, Hsu, 2002) and geometry and algebra (Simon, 1996); transformational observations and proof schemes in geometry, linear algebra, and number theory (Harel & Sowder, 1998); and mathematizing in geometry and differential equations (Rasmussen et al., 2005; Zandieh & Rasmussen, 2007) and elementary school arithmetic (Cobb, Gravemeijer, Yackel, McClain, & Whitenack, 1997). However, our setting was a particularly rich source of such reasoning because of four factors.

1. The content of geometry is inherently visual.
2. Our text, Henderson (2001), emphasized an approach in which students were expected to work extensively with notions of symmetry and transformation.
3. Henderson's tasks for students began very intuitively starting with asking students to look to their own personal experience for notions (including symmetries) that can be used to understand geometric constructs and make arguments about relationships in geometric settings.
4. Our pedagogical approach, following Henderson, pushed students to formalize these intuitive notions by always asking students how they knew that their intuitive insight was in fact true.

Valerie's last statement in the opening vignette, "So, I agree with the perpendicular, but you know she's going to ask ..." pointed to her awareness of the teacher's emphasis on justifying any new assertion. The teacher for this class worked to establish norms for classroom interactions that included students questioning each others' arguments and justifying all statements or assumptions that had not been previously established in class. Within this context the teacher emphasized a process of creating more formal mathematics from more intuitive notions that were personally meaningful to the students.

A Note on Symmetry

Informally, if we speak of the symmetry of a figure, we might be referring to a perception of "sameness" about parts of the figure, for example that the figure looks "the same" on either side of a dividing line. There are two different, but closely related, ways to think about mathematizing this notion of sameness. To most closely match this initial intuition one may speak of the symmetric parts of the figure as being congruent and define an appropriate sense of congruence. Depending on the definition of congruence or the method for determining congruence this would allow for a non-dynamic mathematization of sameness. Another way to consider this sameness is to define a symmetry of a figure as an isometry (a transformation that preserves distances and angle measures) that takes the figure onto itself (Henderson, 2001). This is a dynamic view of symmetry.

In asking students to prove the isosceles triangle theorem, we put them in the position of thinking about the sameness on either side of this figure. They are given that two sides are congruent (i.e., "the same") and they are asked to prove that the corresponding angles are congruent. In an exchange moments before the opening vignette, Alice, Emily and Valerie's discussion shows that they have no problem accepting that the figure has congruent angles.

Emily: Yeah, given that we have two sides congruent, we have to prove that the two angles —

Alice: OK. Prove that the two angles are congruent.

Emily: The two opposite angles —

Alice: On a plane. They just have to be.

Emily: That's just the way it is.

The students' confidence in the statement may have been due to seeing this theorem before in high school, but it was probably also due to the fact that the angles looked the same in their sketch of the figure. Later, when they read in the book the hint to use symmetries, they thought of a reflection over a dividing line. This group (Group 1) was never able to overcome the coercive affect of Alice's affirmatory intuition that the dividing line was both an angle bisector and a perpendicular bisector. As a result, we were never able to learn whether the group intended to focus on the symmetry of the figure in a static sense (by attempting to show that two parts of the figure were congruent), or in a dynamic sense (by attempting to mathematize their reasoning about a transformation). On the other hand, the proof attempts of two other groups of students (Group 2 and Group 3) may be distinguished from each other by the fact that Group 2 mathematized the sameness or symmetry by trying to work out the details of proving that the transformation makes the figure land on itself, whereas Group 3 mathematized the sameness or symmetry of the figure using congruence definitions and theorems.

In the following section we describe the proof attempts of all three groups. Each of the proof attempts starts with the notion of a reflection of the triangle across the angle bisector of the angle between the two congruent sides. However, each group works to formalize its initial intuitive idea in different ways.

Three Attempts to Prove ITT

Background

In this section, we describe three different attempts to prove ITT. These attempts took place in the context of a college geometry course. The course text (Henderson, 2001) takes the approach of asking students to look to their own personal experience for notions (including symmetry) that can be used to understand geometric constructs and make arguments about relationships in geometric settings. Our pedagogical approach, following Henderson, pushed students to formalize these intuitive notions by always asking students how they knew that their intuitive insight was in fact true. Instruction generally followed an inquiry-oriented approach, and classroom interactions fell into three main categories: whole-class discussion with the teacher in front of the class, whole class discussion with a student in front of the class, and students working in small groups.

Innovative aspects of the course included: daily use of group work for problem solving and group proof construction; student presentation of proofs to the class with subsequent questioning and critiques by other class members; student writing and rewriting of paragraph proofs closely critiqued by the teacher; and discovery activities using plastic spheres. During small group discussion the teacher usually moved from group to group. Her interactions focused both on listening and responding to questions from the students. Although "hints" were sometimes offered, the teacher tried not to "give students the answers." Much of the interaction involved listening to hear what the students were thinking. Part of this information fed into her coordination of the whole class discussion that followed.

The vignette transcribed above occurred on Day 12 (out of 28 teaching days) of the semester. Previously the students had created definitions, conjectures, counterexamples and proofs for a number of topics on both the plane and the sphere, including defining a straight line, an angle, a triangle, and exploring the side-angle-side and angle-side-angle congruence theorems. On Day 12, three of the seven groups of students were assigned to prove the isosceles triangle theorem while the other four groups were assigned to prove its converse. The groups worked on their proofs for about 30 minutes followed by whole class discussion focusing on the proof sketches written on poster paper by each group and displayed at the front of the classroom. Although only one of the three groups working on ITT was on camera during the group work, we will use data from the posters and whole class discussion to discuss the other two proofs as well.

Group 1: The Coercive Nature of Affirmatory Intuitions

From the vignette at the beginning of this chapter we saw that the group consisting of Alice, Emily and Valerie had a key idea for the proof, i.e., that folding over the angle bisector of angle B will cause angle C to land on angle A showing that the two are congruent (See Figure 1.). Note that previously in class the students had worked with several different definitions for angle congruence including that angles are congruent if they coincide exactly (excluding length of the rays) or can be made to coincide using isometries such as reflection or rotation.

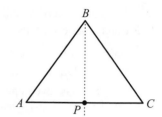

Figure 1. Sketch for ITT as labeled by Group 1.

In the next section we will see how Group 2 worked to mathematize the reflection over the angle bisector to make an argument that angles A and C (Figure 1) would coincide. Then in the discussion of Group 3 we will see how those students used the reflection as an idea that led them to a deductive proof that circumvents the details of the transformation. Group 1, however, took neither of these approaches.

Group 1 spent the entire half hour of group work circling around the issue of whether the angle bisector of angle B was the same line as the perpendicular bisector of the line segment AC. The last sentence of the opening vignette shows Valerie's concern about Alice's assumption.

Valerie: So, I agree with the perpendicular, but you know she's going to ask, "Well, how do you know that the angle bisector is also a perpendicular bisector to the line?"

Alice's immediate response was to explain how to do a ruler and compass construction to create a perpendicular to AC through B. Valerie questioned whether this line was the angle bisector of B. Alice countered by imagining reflecting over this line "since this side equals this side we can do a reflection, so that means that AB is going to lie on top of AC." (Note that although she incorrectly refers to AC, she was pointing to the correct side BC.) Valerie and Emily still did not seem convinced, so Alice introduced the idea of the midpoint of AC, labeled P, to add detail to her reflection line.

Alice's comments suggest that she had a strong affirmatory intuition that there was a line that went through P and B, was perpendicular to AC, bisects B and would reflect one side of the triangle exactly onto the other side of the triangle. She saw the line having these properties holistically without feeling the need to separate out the properties, i.e., without a need to assume one or two of the properties and then prove the other properties. When she did entertain such a separation, she stated that it must be so or that the transformation of folding showed it to be true.

Valerie: Okay so if you make it a perpendicular bisector and forget about it being an angle bisector. So if we start with a perpendicular bisector how are you going to guarantee that it even goes through B and it doesn't go—

Alice: We're saying given some point B, given vertex B where the intersection of AB and BC.

Valerie: So you're drawing a perpendicular line from B to AC. How do you know it's the midpoint?

Alice: We're going to fold it in half.

Valerie: So how do you know it's perpendicular and it's at the midpoint?

Alice: Well we're going to say that we know that P is the midpoint.

Valerie: How?

Alice: We're just going to say that.

Emily: You can't.

Alice: Well then why can't we just take it and just fold it? If we assume that AC is just a straight line, if we assume that AC is the shortest distance that connects them or even just a straight line that connects A and C together, fold that line in half then it has to be in the middle, doesn't it? If you just look only at the line to begin with just the line by itself, fold that line in half so that point A lands directly on point C that will give you the midpoint of AC which would be point P given the vertex B draw a perpendicular line that goes through B and P.

Valerie: How do you know that's possible?

Alice: Because I can do it. I can get a piece of paper and do it.

Valerie: But you can only do it because the sides and angles are congruent. You couldn't do it otherwise.

This group was never able to clarify what it is about the sides being congruent that allows this procedure (described by Alice) to work. Alice then reverted to a focus on using constructions to create the midpoint of the line. While she stepped away from the table to go get a compass, Emily made a discovery.

Emily: Oh. Oh. Oh, wait a minute. Okay. You have a triangle. You know this is equal to this. How do you draw an angle bisector? You stick them at each end and make your little hash marks and you draw a line. How do you draw the perpendicular bisector of this? Put it at each end... So you're doing both at the same time.

Emily recognized that if one constructs an angle bisector at B by using the length AB to create your hash marks, then this construction is identical to the construction for the perpendicular to AC that passes through B. Valerie agreed but was concerned about another issue.

Valerie: [Laughs] But how do you know... one thing she [the teacher] did last time though. How do we know that that really does bisect the angle? I mean, we know it does because we've done it so many times, but how do we really know that it does? That's what she's going to ask. But I totally agree with that.

Constructions had not been an intentional part of the curriculum for this class. They were only discussed briefly on one previous day of class in response to a suggestion from a student. At that point the teacher indicated that the students would need to prove that a construction did indeed construct what it was stated to construct. Although not discussed in class, it turns out that the proof that these two constructions work is usually done using the isosceles triangle theorem that these students were trying to prove.

The coercive nature of Alice's affirmatory intuitions was a source of frustration for the group. Emily and Valerie became increasingly frustrated with Alice's inability to see a need to prove that her assumptions were true.

Alice tried to explain her view again by doing the physical construction for the perpendicular bisector of AC and claiming that it will go through B, even though her construction did not. Without irony (but correctly) she blamed this on the inaccuracy of her skill with the compass.

Alice: Okay, so then the reason we know this is an angle bisector is by reflection. We're going to take AB which we know is equal, is congruent to BC, and so we're going to take AB and reflect it over the perpendicular line and it's going to lie exactly on top of BC.

Note that it is true that AB and BC can be made to lie on each other by some set of isometries because they are congruent, but it is not automatically given that this particular isometry will make them land on top of each other instead of next to each other.

Valerie remained unconvinced that Alice had proven that the constructed perpendicular bisector of AC goes through B.

Valerie: So, this is your B, A, C. So, we're finding the midpoint of that line. Let's pretend it doesn't go through B because we haven't officially proven that it does go through B. Right? Not officially.

Alice: Using the fact that you're making that, you can't just assume that that's the thing. You have to make it using these two.

Valerie: Right. But let's pretend we did. Okay?

Alice: It has to go through B.

Emily: Why?

Alice: Because your two hashes — since these are equal sides

Emily: I don't like that.

At this point Emily and Valerie buried their faces in their hands in frustration.

Alice: That's the rule! Look at it. I just did that. I mean, I just made the hashes there and there.

Valerie: Okay. Prove to me that what you did creates a perpendicular bisector.

Alice: I just did.

Emily: The only reason she said was proving it was the reflection. I don't like that. Say okay, since when we reflect it, it evens out.

Alice: That's all it is. No, it's not that I just, I didn't reflect the whole triangle. I used the given properties. I just I made the bisector here and here and actually first by doing this and doing this and then drawing the line, you actually have only made the midpoint. And that point there where it intersects *AC* is the midpoint. And since these two lines are the same it's going to go through point *B*.

Valerie: I know it does.

Emily: I understand it. I understand that. I just don't think that proves it.

Valerie: I agree.

Valerie and Emily remained unconvinced and Alice remained unable to understand their concerns. Valerie and Emily suggested that Alice write up her proof on the posterboard. Because members of the group were still writing when the class moved to whole class discussion, Group 1 did not end up posting their solution nor was it discussed with the rest of the class. For homework Emily and Valerie wrote proofs similar to that of Group 3 (see below) while Alice wrote a proof using her arguments from the small group discussion.

Group 2: Mathematizing Transformational Reasoning

In Figure 2 we reproduce the proof sketch that Group 2 wrote and posted for class discussion. This proof sketch is evidence of a process in which the students tried to directly mathematize the transformational reasoning involved in the key idea of their proof. In this case, the key idea involved a transformation in which the triangle was folded so that one side of the triangle lands on the other.

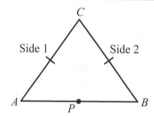

Create an angle bisector through angle *C*. If we reflect (fold) over the angle bisector then Side 1 will land on Side 2 because the angle bisector created two congruent angles. Given this, point *A* will land on point *B*. Every point on the line *AP* will land on the line *PB*. Points *A* and *B* are the same perpendicular distance from the angle bisector. Therefore the angles are congruent.

Figure 2. Group 2 ITT Proof Sketch.

The students' attempt to mathematize their key idea included explicitly describing the transformation, analyzing the results of the transformation, and providing justification for their conclusions. In the first sentence, the students described the transformation itself by specifying that they were thinking of a reflection (or fold) across an angle bisector. Then in the second sentence, the students indicated that they analyzed the results of applying this transformation and determined that "Side 1 will land on Side 2." They then further mathematized their reasoning by justifying this conclusion, citing the fact that the bisector created two congruent angles. The students then deduce from this that the points *A* and *B* land on each other. Note that the students could have supported this particular deduction more convincingly if they also cited the fact that the point *C* stayed fixed under the transformation and the fact that the two sides were the same length.

In working to further mathematize their reasoning, the students attempted to prove that one half of the base of the triangle landed on the other half. This justification attempt referred to the "perpendicular distance" between the points *A* and *B*. During the whole class discussion, one of the students in the group, Penny, explained their uncertainty with this part of the proof. In the excerpt that follows, we see that at least one other group, represented by Matt, had struggled with similar questions. Penny also indicated that she was now considering the SAS theorem as a possible way to deal with this issue.

Penny: I think with us, our main concern was once we did the angle bisector, did the angle bisector actually cut the line *AB* in half? And if you fold it over does *A* land on *B*? That's what our main concern was. And we were trying to prove that the line *PB* and *AP* are both perpendicular to the angle bisector, that was our main concern when we were doing it. We didn't even think about using the Side Angle Side. [Like Group 3 who had just presented their proof.]

Teacher: Were you able to prove that?

Penny: Um, well we just stated that we were thinking that every point on the line *AP* will—

Teacher: Where is *P*?

Penny: *P* is that little tiny blue dot.

Teacher: Oh, I see. There's *P*. I don't know if you guys saw that this is P.

Penny: So what we were saying is every point on the line *AP* will land on the line *PB* and therefore points *A*
              and *B* are the same perpendicular distance from the angle bisector.

Matt: So you guys constructed the angle bisector?

Penny: Yeah, it's like the really, really light blue one.

Matt: So that every point on *AP* will land on *PB* only if the angle bisector is perpendicular to *AB* right?

Penny& Josh: Yes.

Matt: And how do we know that it's perpendicular?

Penny: Yeah that's what we had trouble trying to—

Matt: Same here.

Note that in the discussion above, both Penny and Matt were explicitly talking about mathematizing the
transformational observation that *AP* and *PB* of the triangle landed on top of each other. In general, Group 2 examined
the details of the reflection across the angle bisector with a focus on determining whether the parts of triangle *CBP*
would in fact land on the corresponding parts of triangle *CAP* under this transformation. This can be contrasted with
the following story in which a group of students thought about a transformation, but actually ended up mathematizing
a congruence (a static view of symmetry).

Group 3: Seeing a Transformation, but Mathematizing Congruence

In their poster (reproduced in Figure 3) Group 3 referred to the same transformation of the triangle that the students in
Group 2 tried to mathematize (a fold of one half of the triangle onto the other half). However, Group 3 did not directly
mathematize this transformation or any transformational reasoning they may have used in their small group discussion
to convince themselves the theorem was true. Instead, they mathematized the symmetry of the figure as seen in the
congruence of the two half-triangles.

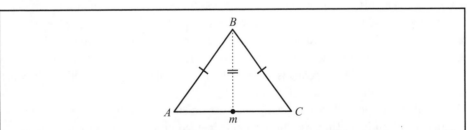

Fold the triangle in half such that the fold is through point *A* and the line *AB* lands on top of *AC*.
The fold creates two congruent angles (∠*BAM* & ∠*CAM*). Now we have two congruent triangles
because of SAS [Side-Angle-Side congruence theorem]. Therefore ∠*ACB* ≡ ∠*ABC*.

Figure 3. Poster of ITT Proof Sketch (Group 3).

In the following excerpt one of the students, Abby, described the proof and the group's thinking to the class.
Notice that their reference to a transformation in the first line was their *only* reference to a dynamic process. The rest of
their description dealt with a static figure consisting of a pair of congruent triangles that formed the larger triangle.

Abby: Okay, ours was sort of the fact that we would fold this triangle like in half sort of creating the angle
              bisector up here. And then we would know that this angle and this angle are congruent and that this line
              right here since it's the same line for both triangles that it's also like a congruent type line and by Side
              Angle Side those would be congruent too.

During the class discussion that followed, a student from another group questioned only one aspect of this proof
— the reference to the fold.

Alexis: Does that work, the first proof?

Teacher: What do you guys think? Eric, do you have a comment?

Eric: I think it works, I just wouldn't have used folded in half, I would have just said the angle bisector of angle *A*. Every other step seems to work though. Just because folding in half—

Martin: It assumes something.

Eric: It assumes that you can fold it in half. It seems like it's symmetric to me anyways.

Teacher: Well, yeah it's like saying folding "in half" sort of assumes something.

Eric: But I'm not taking anything away from them. That was a great proof.

Notice that Eric did not believe that his objection invalidated the proof and in fact he saw the reference to the transformation as being unnecessary. Furthermore, the group that presented the proof did not feel compelled to defend it against this objection. We interpret this as evidence that they did not think of their proof as mathematizing the transformational reasoning stated in the first line of the proof. Instead, the transformational reasoning seemed to play a very minor role, perhaps limited to the production of a picture that emphasized the sameness of the two sides of the figure. It was this property that the group mathematized, along with its consequences, by proving that the two "halves" of the triangle were congruent. As Eric noted, the reference to the folding could be interpreted as an assumption that the triangle could be folded in half. However, the students' proof did not depend upon this assumption but instead it consisted of a mathematization of the observation that the two sides of the figure were the same.

Discussion

In summary we can see that the idea of reflecting or folding the triangle along the angle bisector played an important, but differing role, for each of the groups. For Group 2, the notion that this transformation would cause the figure to land on itself served as the key idea for their proof. It is the key idea in the sense that it answers the question of why the theorem is true, and it is the idea that they mathematized to construct their proof.

In the case of Group 3, the folding of the triangle seemed to function as an idea that led them to see the two halves of the triangle as congruent. This latter observation then functioned as the key idea for their proof. That is, in answer to the question, "Why is ITT true?" this group could answer, "Because the angle bisector divides the original triangle into two congruent triangles." Notice that this is a static sense of sameness and can be contrasted with the dynamic transformational reasoning involved in describing a figure being folded and landing on itself.

Because we do not have video of these two groups working on their proofs, it is hard to tell to what extent the reflection functioned as an anticipatory intuition for these two groups. The retrospective view, however, does allow us to see the key idea of a completed (or nearly completed) proof. Viewing a completed proof through the lens of a key idea is brought out in Raman's (2004) work and is compatible with Fischbein's (1987) notion of a conclusive intuition as a global, holistic way to recall the essence of a proof.

On the other hand, our video of Group 1 clearly shows them starting out by using the reflection over the angle bisector as an anticipatory intuition. However, they became bogged down in trying to prove that the angle bisector and the perpendicular bisector are the same line. For this group the idea of folding the triangle led to an affirmatory intuition about the line that was extremely coercive for Alice and kept the group from moving toward a proof of ITT.

While static and dynamic views of symmetry are strongly related mathematically (the two triangle halves are congruent if and only if the one side of the triangle lands on the other after it is folded along the angle bisector) the process of mathematizing to create a more formal proof can look very different depending on which idea is used.

Implications for Practice

A primary goal of the college geometry course described in this chapter was to help students develop the ability to construct proofs. One way (but certainly not the only way) that mathematicians approach proving theorems is to first try to get a feel for why a statement is true prior to working out the details of the proof. For example, Weber and Alcock (2004) asked mathematicians how they would prove or disprove that two groups were isomorphic. The mathematicians said that they would first "size up the groups" or "see what they looked like" in order to determine whether the groups were essentially the same before working out the details of the proof. Weber and Alcock argue that it is important for students to develop the ability to use this kind of approach (in addition to learning to use more procedural approaches to proving). It is this kind of process that has been the focus of this chapter. The three proof attempts that we described in this chapter all began with students getting a feel for why the theorem was true and then trying to work out the details. These proof attempts seem to indicate that the process of developing a proof starting from an informal or intuitive idea consists of two phases:

1) *Getting a feel for why the statement is true.* This phase includes coming up with and refining an idea that seems to explain why the statement is true. When the informal idea involves symmetries and/or transformations, this may take the form of a thought experiment in which the students imagine transforming a geometric figure and drawing conclusions about the results of the transformation.

2) *Working out the details.* After the students have an idea that seems likely to work, they need to make this argument more rigorous. During this phase students may have to work to overcome the coercive nature of their intuitions so that they can dig in and question their assumptions as they fill in the details of their proof.

The examples illustrate the fact that this process can be quite complex and that supporting students as they learn to construct proofs in this way is a significant pedagogical challenge. We conclude the chapter by discussing two aspects of this challenge and sharing some thoughts about how a teacher might support students as they learn to develop proofs starting from informal or intuitive ideas.

Getting a feel for why a statement is true: Finding a potential key idea

Recall Emily and Alice's excitement when they came up with their idea for proving ITT. All teachers would like their students to experience the excitement and satisfaction that these students exhibited when they found what seemed to be the key idea for the proof. These are the kinds of moments that encourage students to keep taking on new mathematical challenges. In addition to this significant emotional benefit, this idea also represented a good foundation for developing a rigorous proof. The idea gave the students something to mathematize—a blueprint for constructing the proof if only they could work out the details. But what can teachers do to help students develop these kinds of ideas?

Simon's (1996) notion of transformational reasoning seems to describe the kind of thinking that is necessary to develop an idea like the one the students developed. Each of our three groups of students seemed to have engaged in transformational reasoning, imagining the triangle being reflected or folded across an angle bisector and then analyzing the results of this transformation. In the second example, the students' transformational reasoning provided the structure for their proof. Beginning with the image of folding the triangle in half, the students were able to work through some of the details needed to complete their proof and eventually to isolate a key remaining issue they needed to address. For the third group, the transformational reasoning seemed to play a different role, the production of a suggestive static image. While they seemed to be imagining the same fold as the other groups, what they focused on in their proof was the sameness of the two sides of the triangle — a sameness that seemed to be suggested and verified by the dynamic image of folding the triangle.

Working out the details

As Fischbein (1987) observed and our first example illustrated, intuitions can have a coercive effect. Because the approach to proving that we have been exploring in this chapter often relies on intuition, students can get stuck because they are unable or unwilling to seriously question what seems obvious to them. So while intuitions can often supply a good foundation for a proof by giving students something to mathematize, intuitions can often exert a coercive affect that may make students reluctant to dig in and work out the necessary details. How can teachers help students to overcome the coercive nature of their intuitions so that they serve as the foundation for more rigorous proofs?

The examples we have explored illustrate the potentially powerful role of discourse in overcoming the coercive nature of intuitions. Discourse played an important role in the geometry course described in this chapter. The teacher worked to create an environment where students were comfortable sharing and discussing their thinking. Students were expected to respect the thinking of their classmates, but also to ask questions and offer critiques of proposed proofs. In the three examples, we see students taking on these roles by challenging each other both in small group discussions and in whole class discussions.

In the first example we have the opportunity to watch Group 1 have numerous exchanges in which Valerie and Emily push Alice to justify her arguments more carefully. Although this was not totally successful in this case with Alice, they do force her to make further attempts to justify her claim. For Group 2 and Group 3 we do not have data from their small group work, but we see a representative from each group presenting their proof and describing issues that had not been resolved in their small group. Additionally we see other students in the class questioning claims in the proofs and commenting on the validity of the proofs. Zandieh and Rasmussen (2007) and Raman and Zandieh (2007) provide further examples of the power of classroom discourse as students move from less formal to more formal reasoning in the context of a Henderson geometry course.

Conclusion

In this chapter we have explored the process of constructing proofs starting from informal notions involving symmetry and transformations. This process is one that is often employed by mathematicians—they begin by getting a feel for whether and why a statement is true and then work out the details. We have observed that it is important for students to be able to prove theorems using this kind of approach in addition to more procedural and formal approaches.

We presented three groups of students' attempts to prove the isosceles triangle theorem. Using these examples, we elaborated some of the complexity involved in constructing proofs starting from informal or intuitive notions. In particular we observed that intuitive ideas can both aid and hinder students' efforts. Intuitive ideas can provide good starting points because they can offer insight into why the statement is true and provide an overall structure for a proof provided the students can work out the details. Intuitive ideas can hinder students because they can be coercively self evident and students may be unwilling or unable to engage in the task of examining their ideas to unearth hidden assumptions that need to be justified.

The context of transformational geometry seems to be well suited to helping students develop their ability to engage in transformational reasoning, although we point out that transformation reasoning can play an important role in other contexts as well. However, more research needs to be done to increase our understanding of how to help students learn to reason in this way. We also described the important role of classroom discourse in helping students to overcome the coercive nature of their intuitions. Challenges from other students and the teacher can bring hidden assumptions to the surface so that the process of making a proof more rigorous can continue.

References

Carlson, M., Jacobs, S., Coe, E., Larsen, S., & Hsu, E. (2002). Applying covariational reasoning while modeling dynamic events: A framework and a study. *Journal for Research in Mathematics Education, 33*(5), 352–378.

Cobb, P., Gravemeijer, K., Yackel, E., McClain, K., & Whitenack, J. (1997). Mathematizing and symbolizing: The emergence of chains of signification in one first-grade classroom. In D. Kerschner & J. A. Whitson (Eds.), *Situated cognition theory: Social, semiotic, and neurological perspectives* (pp. 151–234). Mahwah, NJ: Erlbaum.

Fischbein, E. (1982). Intuition and proof. *For the Learning of Mathematics.* 3(2). 9–24.

Fischbein, E. (1987). Intuition in science and mathematics: An educational approach. Dordecht, The Netherlands: D. Reidel.

Fischbein, E. (1999). Intuitions and schemata in mathematical reasoning. *Educational Studies in Mathematics 38*, 11–50.

Harel, G., & Sowder, L. (1998). Students' proof schemes: Results from exploratory studies. *Research in Collegiate Mathematics Education III.* A. Schoenfeld, Kaput, J., and Dubinsky, E., CBMS: 234–283.

Henderson, D. (2001) *Experiencing geometry: In euclidean, spherical, and hyperbolic space.* Upper Saddle River, New Jersey: Prentice Hall.

Raman, M. (2003). Key ideas: What are they and how can they help us understand how people view proof? *Educational Studies in Mathematics 52*, 319–325.

Raman, M. (2004). Key ideas: The link between private and public aspects of proof. In D. McDougall & J. Ross (Eds.) Proceedings of the Twenty-sixth Annual Meeting of the North American Chapter of the International Group for the Psychology of Mathematics Education, 2, 635–638.

Raman, M. & Zandieh, M. (2007). *The case of Brandon: The dual nature of key ideas in the classroom.* Manuscript submitted for preparation.

Rasmussen, C., Zandieh, M., King, K., & Teppo, A. (2005). Advancing mathematical activity: A view of advanced mathematical thinking. *Mathematical Thinking and Learning, 7,* 51–73.

Simon, M. A. (1996). Beyond inductive and deductive reasoning: The search for a sense of knowing. *Educational Studies in Mathematics, 30.* 197–210.

Treffers, A. (1987). Three dimensions. A model of goal and theory description in mathematics education: The Wiskobas project. Dordecht, The Netherlands: Kluwer.

Weber, K., & Alcock, L. (2004). Semantic and syntactic proof productions. *Educational Studies in Mathematics, 56,* 209–234.

Zandieh, M., & Rasmussen, C. (2007). *A case study of defining: Creating a new mathematical reality.* Manuscript submitted for publication.

11

Teaching and Learning Group Theory

Keith Weber, *Rutgers University*
Sean Larsen, *Portland State University*

Abstract algebra is an important course in the undergraduate mathematics curriculum. For some undergraduates, abstract algebra is the first mathematics course in which they must move beyond learning templates and procedures for solving common classes of problems (Dubinsky, Dautermann, Leron, and Zazkis, 1994). For most undergraduates, this course is one of their earliest experiences in coping with the difficult notions of mathematical abstraction and formal proof. Empirical research studies attest to students' difficulties in abstract algebra; these studies have shown that many students do not understand fundamental concepts in group theory (e.g., Leron, Hazzan, and Zazkis, 1995; Asiala, Dubinsky, Mathews, Morics, and Oktac, 1997) and have difficulty writing proofs in a group theoretic context (e.g., Selden and Selden, 1987; Selden & Selden, this volume; Hart, 1994; Weber, 2001; Harel & Brown, this volume) after completing an abstract algebra course. The purpose of this chapter is to use the research literature to illustrate some of undergraduates' difficulties in group theory, the primary content area in most abstract algebra courses, and to describe alternative forms of pedagogy that may be useful in overcoming these difficulties.

Undergraduates' Difficulties in Group Theory

In this section, we present four episodes that illustrate undergraduates' difficulties in understanding and reasoning about concepts in group theory. We will begin each section by presenting an excerpt from a clinical interview in which undergraduates were asked to describe a concept in group theory or complete a group theoretic task. We will then use these transcripts as a basis to discuss general difficulties that undergraduates have in group theory and to state some of the reasons that undergraduates have these difficulties.

Episode 1. Students' Understanding Of Cosets And Normality

Asiala et al. (1997) interviewed a group of 31 students who participated in an experimental abstract algebra course. After the course, some students met with researchers for interviews in which they discussed central group theoretic concepts. Below, one student, Carla, was asked how she would determine if K (a 4-element subgroup of S_4) was a normal subgroup of S_4.

> Carla: OK, you just have to check if when you multiplied two elements together if it is the same that you did them in the reverse order. Like if you take (1234) and compose it with (1432)[…] see if you get the same thing as when you take (1432) and compose it with (1234).
>
> I: So it's um …
>
> Carla: Commutative […]
>
> I: Now if it is not commutative.
>
> Carla: Then it is not normal.

Later in the interview, Carla realized that what she described above was a commutative subgroup. She revised her description of a normal subgroup as follows:

Carla: You take an element of S_4 and multiply it times an element in K. And if you get, if you take the element in S_4, like say the permutation (12), and multiply it times the element in K, (12)(34), if whatever you get for that is the same as when you take (12)(34) and multiply times the element in S_4.

I: So this has to be done for every element in S_4.

Carla: Right. Times every element in K.

I: And each time, element by element, they have to be the same.

Carla: Yes.

In this excerpt, Carla appears to be saying K is normal if every element of K commutes with every element of S_4. In other words, K would be a subset of the group's center. Still later in the interview, Carla was able to describe a correct method to determine if a subgroup was normal. According to Asiala et al., these interview segments (and other data) suggest that Carla has three competing notions for what it means for a subgroup H of a group G to be normal: H is commutative, every element of H commutes with every element of G, and $gH = Hg$ for every g in G. Carla's difficulties in understanding normality were not unique to her. In interviews with 20 students who recently completed a standard abstract algebra course, Asiala et al. (1997) found that most of these gave responses that revealed a poor understanding of the concept of normality and only one student was able to state and correctly interpret the definition of a normal subgroup. (See Asiala et al. (1997) for a more detailed account of their findings).

One can interpret the cause of Carla's difficulties in multiple ways. Asiala et al. propose an intriguing hypothesis. In order to understand what it means for a subgroup H to be normal, one must be able to evaluate (i.e., attach a truth value to) expressions of the form $aH=Ha$ and then must imagine checking the truth values for all a in the group G. However, in order to determine whether $aH=Ha$ is true, one must think of the cosets aH and Ha as mathematical objects that can be compared with one another (Dubinsky et al., 1997). If one can only see the symbol aH as a prompt to compute a coset, but not as the result of applying this process, one will have difficulty understanding what it means to evaluate $aH = Ha$. A later interview segment suggests that Carla does not yet understand cosets as objects:

I: When you take an element of S_4 and multiply it by everything in K, you said you get a set [...] Does that set have a name?

Carla: Probably. Well it should be the same as K.

I: Well try it... Take your (12) and multiply it by everything in K.

Carla: OK (pause) Guess you don't (pause) You get that, unless I did something wrong.

I: It's different from K.

Carla: Right.

I: Does it have a name?

Carla: I don't know. I don't think so.

I: Could you write a notation for using that single element from S_4 and K?

Carla: Well it'd be like (12)K.

I: OK, and does (12)K, does that have a name?

Carla: It just tells me that you take (12) times every element in K. And you get that subset.

I: What is a coset?

Carla: I don't know. That's what it is, huh? (They laugh) It's when you take an element and you multiply it times every element in the subgroup. That would be a coset.

From this interview segment, it is clear that Carla is still developing her understanding of cosets. In particular, she did not initially see the symbol (12)K as representing the object that is the result of coset formation. Asiala et al. (1997) hypothesize that this weak understanding of cosets may have contributed to her unstable understanding of normality. Later in this chapter, we describe instruction designed to help students construct an understanding of cosets and normal subgroups.

Episode 2. Undergraduates' proving processes

Weber (2001) asked undergraduates who had just completed an abstract algebra course to think aloud as they proved non-trivial group theoretic propositions. One of these propositions was:

Let G be a group of order pq where p and q are prime. Let f be a surjective homomorphism from G to H. Prove that either H is abelian or isomorphic to G.

In some respects, this is not a difficult theorem to prove. Its proof is short and is a relatively direct consequence of applying a weak form of the first isomorphism theorem (i.e., $|G|/|\ker f| = |f(G)|$). An excerpt of one student's response is given below.

"I'm not quite sure what to do here. Well, injective, if G and H have the same cardinality, then we are done. Because f is injective and f is surjective. G is isomorphic to H ... OK, so let's suppose their cardinalities are not equal. So we suppose f is not injective. Show H is abelian. OK f is not injective so we can find a distinct x so that $f(x)$ is not equal to $f(y)$"

The student then made a series of valid inferences, concluding with $xy^{-1} \in \ker f$, before giving up on the problem. Earlier observations of this student indicated that he possessed an accurate conception of proof and he was able to prove more basic statements. Further, a subsequent paper-and-pencil test revealed that this student could state and apply the facts that were required to prove the proposition. When the interviewer told the student specifically which facts to use, he was able to construct a proof. Weber concluded that this student's initial inability to prove the proposition was not due to a lack of factual or procedural knowledge, but his inability to apply the knowledge that he had in a productive manner.

This student's performance was typical of the undergraduates whom Weber observed. In two studies, Weber (2001, 2002) asked eight undergraduates to think aloud while proving five statements. There were 25 instances in which the undergraduates were aware of the facts necessary to prove a particular statement. In only eight of these instances could the undergraduates construct a proof without prompting from the interviewer. Like the student in the excerpt above, most of the participants in Weber's study took a seemingly reasonable approach to the problem and drew valid inferences, but these inferences were often not useful for proving the proposition that the students were attempting to prove.

Weber's explanation for why the undergraduates could not prove these statements was that they lacked effective proof-writing strategies and heuristics. Based on his observations of doctoral students proving the same statements, Weber concluded that these doctoral students were aware of what theorems in group theory were important and when they were likely to be useful. To illustrate, when proving the proposition above, all four doctoral students immediately used the first isomorphism theorem. When the interviewer later asked them why they had done so, participants responded that this theorem is often useful when one is reasoning about homomorphisms, especially when the homomorphism in question is surjective. The undergraduates in Weber's study did not have this strategic knowledge. Weber suggests that as a result, the undergraduates' choice of which facts to use was more random and they were less likely to make use of important theorems. The undergraduates' proof attempts largely consisted of unpacking definitions and pushing symbols. In a subsequent teaching experiment, Weber demonstrated that strategic knowledge can be taught to undergraduates, and that such instruction improved their ability to construct proofs substantially. Details of this study can be found in Weber (2006), but for the sake of brevity, this teaching experiment will not be reported in this chapter.

Episode 3. Undergraduates' methods for reducing abstraction

Hazzan (1999) interviewed students from a standard lecture course in abstract algebra. In the following excerpt, Tamar is explaining why Z_3 is not a group.

Tamar: [Z_3] is not a group. Again I will not have the inverse. I will not have one half. I mean, I over...I mean, if I define this [Z_3] with multiplication, I will not have the inverse for each element. [...] It will not be in the set. [...] I'm trying to follow the definition. [...] What I mean is that I know that I have the identity, 1. What I have to check is if I have the inverse of each...I mean, I have to see whether I have the inverse of 2 and I know that the inverse of 2 is ½. [...] Now, one half, and on top of that do mod 3[...], then it is not included. I mean, I don't have the inverse of... My inverse is not included in the set. Then it's not a group. I do not have the inverse for each element."

Tamar appears to be interpreting the binary operation to be multiplication, rather than the operation that is traditionally associated with the group Z_3 (addition modulo 3). Hazzan suggests that Tamar made this interpretation in order to *reduce abstraction*. In group theoretic contexts, the notation used for the binary operation of the group is usually the same as multiplication for standard number systems. Tamar attempted to make the question that she was asked more concrete by retreating to a familiar context (in this case, numbers and multiplication). Selden and Selden (1987) also found that students tend to reason about abstract algebraic systems as if they were working with familiar number systems.

Hazzan describes other techniques by which students reduce abstraction. For instance, she found that when students were presented with tasks in abstract algebra, they tended to avoid using their conceptual knowledge of the relevant mathematical objects by relying upon well-known (canonical) procedures. An example of this is provided in Hazzan (2001). In this article, she illustrates how when students were asked to produce an operation table for a group of order 4, some students relied on a procedure that (some of them admitted) they did not really understand:

Step 1:   Choose an identity element and fill in the first row and column accordingly.

Step 2:   Choose a result for one of the remaining boxes without violating the law that each element must appear exactly once in each row and column.

Step 3:   Fill in the remaining boxes without violating the law that each element must appear exactly once in each row and column.

Note that this procedure does not work in general (it is based on the cancellation law which is not equivalent to the definition of a group). From Hazzan's perspective, it is significant that the students relied on this canonical procedure rather than their conceptual knowledge. In particular, students *did not* use the fact that the desired group must either be isomorphic to Z4 or the Klein-4 group.

Similarly, Findell (2002) found that abstract algebra students use operation tables to "mediate" abstraction. He describes a student who recognized isomorphisms between groups of order four but relied on a procedure involving renaming elements and re-ordering operation tables. Findell described this student's methods as "largely external, in the sense that it was based *in* the table and in procedures that required the operation table be present rather than in reflection on the binary operation." (p. 241, emphasis is Findell's).

Episode 4. Undergraduates' understanding and reasoning about group isomorphisms

Weber and Alcock (2004) presented four undergraduates who had recently completed an abstract algebra course with five pairs of groups and asked them to think aloud while attempting to prove or disprove that the pairs of groups were isomorphic. One such question asked participants to prove or disprove if (**Q**,+) was isomorphic to (**Z**,+), which seems to be a relatively straightforward proof once one recognizes that the second group is cyclic but the first group is not. One undergraduate's response is given below:

"I think **Q** and **Z** have different cardinalities so, no wait, **R** has a different cardinality, **Q** doesn't. Well, I guess we'll just use that as a proof. Yeah so I remember seeing this on the board […] There's something about being able to form a homomorphism by just counting diagonally [The student proceeds to create a complicated bijection between **Z** and **Q** by using a diagonalization argument—similar to the one used by Cantor to prove that **Z** and **Q** shared the same cardinality] Yeah I don't think we're on the right track here. What you are describing here is a bijection, but not a homomorphism".

This undergraduate's behavior was typical of the participants in Weber and Alcock's study. Collectively, the undergraduates were only able to prove two of the twenty propositions that they were given. This outcome was despite the fact that, for the majority of these propositions, the undergraduates possessed the factual knowledge required to write these proofs. In nearly all cases where the participants made progress on proving or disproving that two groups were isomorphic, they first determined if the groups in question were equinumerous. If they were, the participants then attempted to form a bijection between the two groups. In most cases, this approach was unsuccessful, leading the participants to abandon their proof attempt as they did not know how to proceed.

Weber and Alcock hypothesized that the undergraduates used this strategy because they did not intuitively understand isomorphic groups as being algebraically the same, but understood isomorphic groups at a purely formal level (i.e., groups G and H are isomorphic if there exists a mapping from G to H that is both bijective and a homomorphism). Interviews with (other) undergraduates corroborated this explanation; these undergraduates indicated that they had no informal ways of thinking about groups other than by reciting the group axioms. Further, they could offer no intuitive description of what it meant for two groups to be isomorphic. One undergraduate remarked, "my intuition and formal understanding of isomorphic groups are the same."

Weber and Alcock also interviewed doctoral students and mathematicians and found a sharp contrast between their responses and those of undergraduates. Mathematicians indicated that they thought of isomorphic groups as being "essentially the same" or that one group was a "re-labeling of the other group." When asked how they would prove or disprove that two groups were isomorphic, the mathematicians indicated that they would "size up the groups" and "see what they looked like." When doctoral students were observed doing these tasks, they would uniformly begin by determining which properties both groups shared and which properties they did not. If one group had a property that the other did not share, the doctoral students would immediately deduce that the groups were not isomorphic. If they did share important characteristic properties (e.g., if the groups were equinumerous and cyclic), they would use this as a basis for proving the groups were isomorphic.

Innovative Approaches To Teaching Abstract Algebra

In the last section, we highlighted some of the difficulties that undergraduates have in reasoning about group theoretic concepts. In response to these findings, a growing number of researchers have designed and evaluated innovative pedagogy to help overcome these difficulties (Leron and Dubinsky, 1995; Hannah, 2000; Larsen, 2004; Weber, 2006). In this section, we will discuss two of these teaching approaches. Due to space limitations, we will not be able to give a comprehensive description of each teaching method. Instead, we will attempt to provide the reader with the theoretical ideas behind the teaching approaches, a general description of how each approach is implemented, and some evidence suggesting that these approaches can be effective. Throughout our descriptions, we will provide references where the coverage of these teaching approaches is more thorough.

Learning abstract algebra by using ISETL to program group theoretic concepts

Leron and Dubinsky (1995) contend that students' difficulties in learning abstract algebra are largely due to their inability to understand fundamental processes and objects. They further argue that the lecture method may be insufficient to overcome these difficulties, because for most students, simply "telling students about mathematical processes, objects, and relations is not sufficient to induce meaningful mathematical learning." Leron and Dubinsky propose an alternative instructional approach that has been developed through two cycles of experimentation, analysis, and modification. The first cycle was described and reported in Dubinsky et al. (1994) and the second in Dubinsky (1997).

This instructional approach is based on APOS theory, a general theory of mathematical learning. (The acronym APOS stands for Action-Process-Object-Schema). Put simply, Leron and Dubinsky believe that group theory (and most of mathematics) can be understood in terms of actions and processes that can be applied to mathematical objects. A concept can be understood in terms of an *action*, or a set of explicit mechanical steps that can be applied to mathematical objects in response to an external cue. By reflecting on the application of an action, the student may conceive of the actions as a *process*, or a mathematical transformation that links particular inputs to particular outputs. That process can itself become a subject of mathematical investigation or can be the input of other mathematical processes. When students can conceive of the process in this way, they are said to have constructed an *object* conception. A *schema* is a coherent collection of processes, objects, and other schemas that is invoked to deal with a mathematical situation.

In group theory, the concept of coset can involve an algorithm (*action*) that a student can apply when he or she is given an element of a group and a (finite) subgroup of that same group. A coset can be thought of more abstractly as a general *process* that maps a group element and a subgroup to a subset of the group (a process conception of coset). Cosets can also be thought of as *objects* that can be compared with one another (e.g., they do have the same cardinality and left cosets may or may not be equal to corresponding right cosets) or may be the inputs to subsequent mathematical actions (such as applying binary operations to cosets of normal subgroups, a critical idea that must be grasped to understand quotient groups). A more complete discussion of these ideas is given in Dubinsky and McDonald (2001).

The instructional approach that was described by Leron and Dubinsky (1995) relies on computer programming activities to promote students' conceptual development. To encourage the development of process conceptions, students are asked to program a computer to perform that process. To encourage the development of object conception of this process, students are then asked to have this process serve as an input for other processes. All programming activities are done collaboratively with other students. In a typical computer activity, a new concept is explained to students (in non-technical language) and students are asked to write a computer program that can check whether a given input is a member of that concept. For instance, early in their course, students are asked to write a program that could determine whether a binary operation on a set of elements satisfied the closure and identity properties. An example of student-generated code is provided below:

```
is_closed := func(G,o);
    return
      forall a,b in G| a .o b in G;
  end;

has_identity := func(G,o);                $G has left identity
    return
      exists e in G|(forall a in G|e .o a = a);
  end;
```

(*Note:* Here G denotes a set of elements and o denotes a binary operation.)

The computer language used by students, ISETL (Interactive Set Language), has a syntax similar to the syntax of formal mathematics (see Leron and Dubinksy, 1995 for more examples of ISETL code). Hence, in writing these programs, students are to some extent engaged in the process of defining the concept. Leron and Dubinsky concede that students find some of the programming tasks to be quite difficult. However, they argue that the difficulties that students experience are usually due to the complexity of the underlying mathematical ideas, not the syntax of ISETL. So when students struggle to write the programs, they are actually grappling with the difficult mathematical ideas of group theory.

Other activities ask students to explore particular group theoretic situations by making conjectures, using the computer to test these conjectures, and forming explanations for why some of these conjectures are true. For instance, after students in Leron and Dubinsky's (1995) class wrote programs that could determine if a set and a binary operation on a set satisfied each of the group axioms, they were asked to "Explore the modular systems Z_n, both with and without zero, relative to addition and multiplication mod n. Formulate some conjectures, test them, and try to come up with some explanations" (p. 230). These activities provide students with the opportunity to assess and refine their understanding.

Between these activities, the class may meet as a whole to discuss important or troublesome ideas or the teacher may provide a (short) lecture in which formal definitions are given, formal proofs are provided, or key ideas from the previous activity are summarized. While these discussions and lectures are important, it is primarily the computer activities themselves that are expected to lead to students' learning.

Leron and Dubinsky list four benefits of their instructional approach:

- First, by programming a computer to do a process, the students are necessarily describing and likely discussing and reflecting upon the process that they are programming. This provides a powerful opportunity for the students to construct a *process* understanding of the concept they are studying. "If the students are asked to construct the group concept on the computer by programming it, there is a good chance that a parallel construction will occur in their mind" (Leron and Dubinsky, 1995, p. 230).

- When students are asked to anticipate the output of a computer command, this provides them with the opportunity to evaluate and, if necessary, refine their understanding of group theoretic ideas and discuss this unanticipated result with their classmates. This does not help students' progress through the Action-Process-Object-Schema learning trajectory per se, but allows them to address misconceptions that they may have developed.

- In struggling to define group theoretic concepts by programming them on a computer, students will develop an experiential basis to understand the more abstract and formal treatment of these concepts that they will encounter later in the course. Students often find the formalism of abstract algebra courses to be strange and intimidating. With an experiential basis, the formalism can be viewed as a formalization of previous experience.

- ISETL allows functions, processes, and sub-routines to be inputs into (and outputs of) other processes. By seeing how a computer program acts upon processes, students may come to construct these processes as mathematical *objects*. Understanding concepts such as group and coset as mathematical objects is essential for understanding fundamental concepts in group theory, but such an understanding proves very difficult for most students to achieve (Dubinsky et al., 1994). Programming in ISETL may facilitate this transition.

To teach the specific concept of normality, students are first asked to construct a program called PR(a,b), where each of the input variables *a* and *b* can either be a group element or a subgroup of the group. This program can be used to calculate specific cosets. Asiala et al. (1997) report that students have serious difficulty in writing this program. However, they also argue that in their struggles to produce the program, they construct an understanding of what it means to be a coset, and hence are less likely to develop the misunderstanding that Carla exhibited in Episode 1.

Students are then given five relations and are asked to determine whether given groups and subgroups satisfy these relations. These five relations are all equivalent definitions of normality, but students are not initially aware of this. Writing a program that will perform these checks requires students to enter expressions such as: For all *a* in G| PR(a, H)=PR(H, a).

In Dubinsky et al.'s (1994) framework, producing such code should theoretically help solidify students' understanding of cosets as objects (since they are using the coset expressions PR(a,H) and PR(H,a) as an input into the process of comparing cosets) and should require students to realize the iterative role of the variable *a* in coset construction. After checking which groups and subgroups satisfied the relations (different but equivalent normality conditions), the students were asked to state which relations were equivalent and to justify their hypothesis. The equivalence of some definitions was explicitly proved by the instructor, but most equivalences were proven by the students. Asiala et al. (1997) found that the majority of students who engaged in these activities were able to accurately explain what it meant for a subgroup to be normal and did not experience the difficulties that Carla experienced in Episode 1.

Learning group theory by re-inventing concepts

Introduction. Larsen (2002, 2004) reports an ongoing effort to develop an approach to teaching abstract algebra in which the formal mathematical concepts are developed beginning with students' informal knowledge and strategies. This approach is being developed through a series of developmental research projects (Gravemeijer, 1998). Developmental research projects cycle through two related phases, called the developmental phase and the research phase. During the developmental phase, activities and instruction are developed based on an evolving instructional theory. The research phase involves the analysis of classroom activity as the evolving instructional approach is implemented. This analysis then informs the next development phase by informing both the development of the instructional theory and the design of specific instructional materials.

While developmental research results in the production of an instructional sequence, the goal is to produce something more generalizable than a specific sequence of instructional activities—a local instructional theory. (Note that the word local refers to the fact that the instructional theory deals with a specific mathematical topic.) The purpose of the local instructional theory is to provide a rationale for the instructional activities. Drawing on the local instructional theory, a teacher can adapt the instructional approach to his or her specific situation. Local instructional theories feature three key ingredients. The following description of these ingredients is adapted from Gravemeijer (1998).

- Students' informal knowledge and strategies on which the instruction can be built.
- Strategies for evoking these kinds of informal knowledge and strategies.
- Strategies for fostering reflective processes that support the development of more formal mathematics based on these kinds of informal knowledge and strategies.

These components of a local instructional theory will provide the framework for our discussion of Larsen's instructional approach. For the interested reader, we note that this description of a local instructional theory draws on the theoretical perspective of *realistic mathematics education* (See Gravemeijer, 1998).

The instructional approach is still under development as Larsen continues to cycle through phases of development and research. The ongoing work is concerned with (among other things) the challenges of adapting the instructional approach to a traditional classroom setting. Here we will focus on Larsen's first teaching experiment (Cobb, 2000) as it was particularly important to the development of the local instructional theory. This is because one of the participating students produced an unexpected solution strategy that turned out to be particularly productive in terms of supporting the development of the group concept. Analysis of this first teaching experiment guided the development of activities designed to evoke this particular strategy and activities designed to promote reflection on this strategy to support the development of the formal group concept.

Description of the Teaching Experiment. The two students who participated in the teaching experiment were Jessica and Sandra. Neither student had received prior instruction in abstract algebra, but both had recently completed a transition-to-proof course in which they learned to construct basic set-theory proofs. The students met with Larsen for seven sessions, each of which lasted between ninety minutes and two hours. Each of these sessions was videotaped and all of the students' work was collected.

The first task given to the students was to describe and symbolize the symmetries of an equilateral triangle. Jessica and Sandra developed the following table (Figure 1) to illustrate the six symmetries of an equilateral triangle.

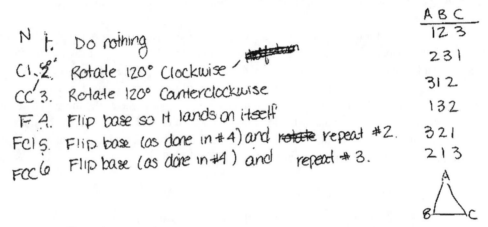

Figure 1. The students' table illustrating the symmetries of an equilateral triangle.

Calculating combinations of symmetries using rules: A spontaneous solution strategy. After the students identified the six symmetries of an equilateral triangle, they were asked to compute each composition of two symmetries. Larsen expected that the students would eventually develop an operation table to record their results (determined by manipulating a cardboard triangle that was provided). The students' table in Figure 1 makes this approach quite feasible since they could start with their triangle in the standard position (denoted by 123), manipulate their triangle, and then look up the result. However, although the students did develop an operation table to record their results, their method for determining these results was unexpected. Jessica immediately started making observations about the relationships between various symmetries and then used these observations to calculate the combinations.

Identity property of N:

Jessica: So if we do "do nothing" and one of these other ones, it's gonna be the same thing.

Inverse relationship between CL and CC:

Jessica: If you go *CL CC* that gives the same…
Sandra: Right because you're just doing the opposite.

Jessica then began using her observations to calculate more complex combinations and, after hearing Jessica's explanation, Sandra is able to perform a similar calculation herself.

Calculation of the combination F FCL

Jessica: If you combine *F* and *FCL* you're just going to get clockwise. Cause you're going to flip and then you're going to flip again and then so those cancel each other out so you're going to have these left over.
Sandra: You're going to flip…
Jessica: You're going to flip and then you're going to flip it again so both of those flips cancel each other out so this is all your going to have left is this move, clockwise.

Sandra: Do we want to assume the opposite for the other one?

Jessica: Yeah.

Sandra: So if we do flip as our first term and flip counterclockwise for our second term then you get counterclockwise.

This calculation makes implicit use of the associative law, uses the identity property, and takes advantage of the fact that F is its own inverse. The students proceeded to fill out their table, relying primarily on calculations of this type. Although Larsen did not anticipate that the students would use calculations to determine their results, after Jessica started doing it, Sandra seemed to think that this was what they were supposed to do.

Jessica: I'm just noticing these rules.

Sandra: That's great. That's what we're supposed to be doing. This [filling in the table] is just the monkey work.

This strategy that Jessica spontaneously brought to the task was very exciting from an instructional design perspective. It clearly represents a good foundation for the development of the group concept as it involves the use of group axioms—the identity property of N and associativity are used fairly explicitly along with particular inverses. As it turned out, it was fortunate that Jessica participated in this teaching experiment because it does not appear that her approach to the task is typical. The participants in the second and third teaching experiments of Larsen's dissertation study did not spontaneously use calculations to compute combinations of symmetries. Additionally, in subsequent uses of the activity in three group theory classes (approximately 70 students), no student has spontaneously used this approach to the extent that Jessica did.

Evoking the strategy of using rule-based calculations to compute symmetries. One of the important goals of the retrospective analysis of the teaching experiment data was to understand what conditions promoted Jessica's use of rule-based calculations. The analysis revealed that a number of factors seemed to support this.

The students' use of the compound symbols *FCL* and *FCC*. Jessica and Sandra used the term "compound moves" to refer to *FCL* and *FCC* since each was expressed in terms of two simpler movements. Note that it is not necessary to express these two symmetries as compound moves—they can be expressed as simple reflections (generally they are expressed this way in abstract algebra texts). It is important that these compound moves were included because otherwise all of the compositions of two symmetries would have been expressed as strings of only two symbols, leaving no room for intermediate steps between expressing the combination to be determined and expressing the result. It is the strings of three and four symbols that create the opportunity to perform calculations by grouping elements that cancel (e.g., *F FCL*). Subsequent work with the equilateral triangle activity suggests that students who have familiarity with symmetry tend to not use compound symbols. In Jessica's case, she discovered these symmetries first as compound movements and in fact had difficulty seeing them as simple movements.

Based on these observations, newer versions of the equilateral triangle sequence include a task in which students are asked to express all of the symmetries using combinations of flips across the vertical axis (denoted by F) and 120 degree clockwise rotations (denoted by R). The result of this is that the students convert their six symbols to a new set of symbols that includes what Jessica and Sandra called compound moves. Since in a typical classroom, different groups of students are likely to generate different symbols, this task also sets the stage for the adoption of a common set of symbols for the entire class to use.

While a set of symbols that contains compound symmetries may make it more likely that students will begin to perform rule-based calculations, it may also be necessary to employ other pedagogical strategies to promote the development of this approach. For example, the teacher could identify students who do use rule-based calculations spontaneously and then ask those students to share the strategy with the class. Another approach would be to present the students with a combination that lends itself to easy calculation. The first combination that Jessica calculated, F *FCL*, is an example of such a combination. The students could be asked whether they actually need to manipulate the triangle find the result of this combination. Either strategy should set the stage for asking the students to create a set of rules that can be used to calculate all 36 combinations. Jessica and Sandra compiled such a set of rules. An early version of their list is shown in Figure 2. (Note that the students had refined their notation to express *CL* as R and *CC* as R^{-1}.)

Developing the formal group concept

The list of rules in Figure 2 does not include all of the group axioms and also includes more rules than are necessary to completely determine the group of symmetries of an equilateral triangle. Here we will discuss some strategies for developing the formal group concept based on this starting point.

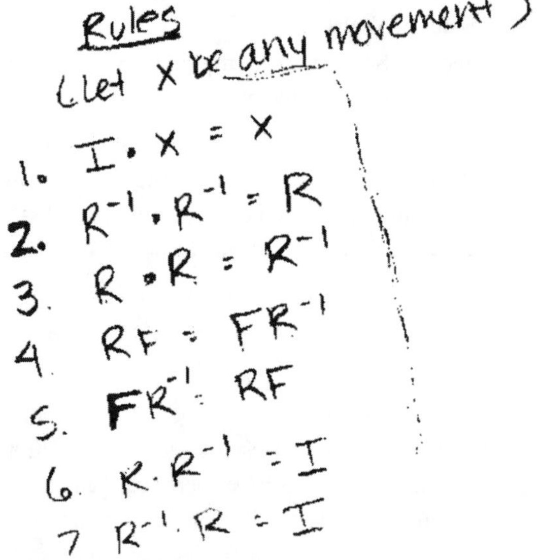

Paring down the rules to a minimum set. Since many of the calculations needed to compute the 36 combinations of two symmetries require the use of grouping (associativity), it is not difficult to raise this property to a more explicit status (a discussion can be started by asking students to explain and justify their use of grouping). With the addition of the associative law, the list of rules in Figure 2 is sufficient to calculate the 36 compositions. In order to highlight the key structural features of this group of symmetries, it is helpful to pare this list down. The students can be asked whether any of the rules are unnecessary in the sense that they can be obtained using other rules. This task gives students a chance to begin using the group axioms in proofs as they reduce their list of rules. When Jessica and Sandra had eliminated all of their unnecessary rules, they were left with the rules in Figure 3.

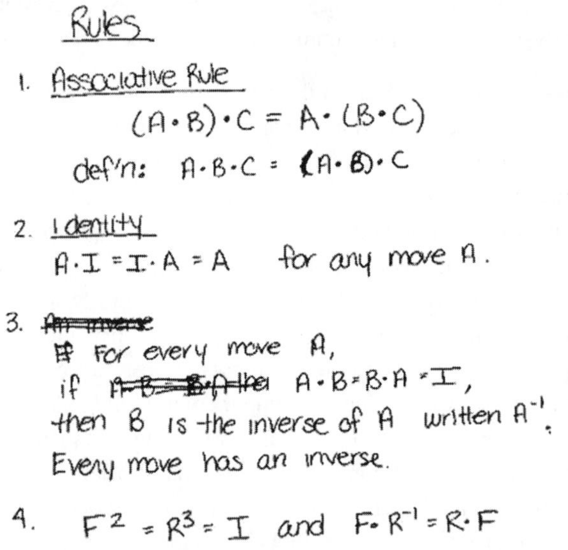

Rules
(Let X be any movement)

1. $I \cdot X = X$
2. $R^{-1} \cdot R^{-1} = R$
3. $R \cdot R = R^{-1}$
4. $RF = FR^{-1}$
5. $FR^{-1} = RF$
6. $R \cdot R^{-1} = I$
7. $R^{-1} \cdot R = I$

Figure 2. The students' list of rules for calculating combinations of symmetries.

Rules

1. Associative Rule
$$(A \cdot B) \cdot C = A \cdot (B \cdot C)$$
def'n: $A \cdot B \cdot C = (A \cdot B) \cdot C$

2. Identity
$$A \cdot I = I \cdot A = A \quad \text{for any move } A.$$

3. ~~An inverse~~
∄ For every move A,
if ~~A·B = B·A then~~ $A \cdot B = B \cdot A = I$,
then B is the inverse of A written A^{-1}.
Every move has an inverse.

4. $F^2 = R^3 = I$ and $F \cdot R^{-1} = R \cdot F$

Figure 3. Reduced list of rules for the symmetries of an equilateral triangle.

Establishing closure and the existence of inverses as rules. Jessica and Sandra paid special attention to inverses in their work. They figured out what the inverse of each element was and listed these inverses. And they included the existence of inverses on their list of rules. These particular students were unusual in this regard. It is not necessary to observe that each element has an inverse in order to calculate the 36 combinations. So, attention to the existence of inverses does not naturally arise from the activity of calculating combinations. However, while completing their operation tables, students often notice that each symmetry occurs exactly once in each operation table. Students can be asked whether they can easily prove this property using their set of rules, or whether they should add additional rules. To make their proofs work, students will need to assume closure and the existence of inverses.

Formulating a definition of group. Once the list of rules has been refined to include all of the group axioms and only a minimal set of relations for the group of symmetries of an equilateral triangle, it is useful to have students explore other systems including other groups of symmetries and arithmetic groups. The group axioms then come to the fore as the set of invariants across these various systems, setting the stage for the formulation of a definition of group. Figure 4 shows some of Jessica and Sandra's work as they formulated their definition.

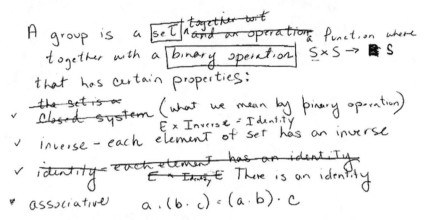

Figure 4. Working to formulate a definition of group.

Learning group theory by reinventing concepts: A discussion

The instructional approach described above is one of reinvention. First the students are engaged in activities designed to evoke powerful informal understandings or strategies. Then the students are engaged in activities designed to support reflection on these informal notions in order to promote the development of the formal concepts. In the case of the group concept, the key informal strategies are the development of an operation table to record results and the use of rule-based calculations of symmetry compositions. These two strategies were the basis for the development of the formal group concept and eventually the isomorphism concept (see Larsen 2002, 2004).

Certainly it can be challenging to teach abstract algebra via reinvention. It clearly takes longer to have students reinvent the group concept in the manner described above than it does to write the definition on the board and give a few examples. We would not recommend that every concept in abstract algebra be learned through reinvention. However, it may be worthwhile to teach some fundamental concepts in this way. Here we will discuss some of the possible benefits.

Freudenthal (1991) argues that, "knowledge and ability, when acquired by one's own activity, stick better and are more readily available than when imposed by others" (p. 47). In the instructional approach described here, the students actively participate in developing the symbols and notation systems that they use. Thus it is expected that these symbols and notation systems are meaningful and useful to the students. Additionally, since the formal notions are developed from the students' informal ideas, it seems likely there will be a stronger connection between the students' informal understandings and the formal concepts. Thus it seems likely that the students will be able to rely on their informal understandings to help them develop formal arguments in much the same way as the mathematicians described by Weber and Alcock (2004).

We see some evidence of these phenomena in Jessica's proof of her conjecture that the identity element of a group is unique (see Figure 5). When asked to prove this conjecture, she argued that it was a consequence of each element appearing exactly once in each row and column of the operation table. She was able to use this idea and a generalized operation table to develop her proof.

Jessica's approach can be contrasted with phenomena illustrated in Episode 3 and Episode 4. First, Jessica was able to use her intuitive understanding of groups to develop a formal proof. She thought about why the identity had to be unique, and realized that if it were not unique then an element would have to appear twice in a row of the operation table. She was able to express this situation using equations, and then work out the details of the proof. Second, Jessica's use of the operation table contrasts sharply with that of the student described by Findell (2002) whose reasoning was largely external and required the presence of the entire table. Jessica modifies the table for her purposes—using arbitrary elements and only including aspects that she needs to support her reasoning. This suggests that she was using the table as a tool to support her reasoning and not merely as a crutch for recalling the steps of a

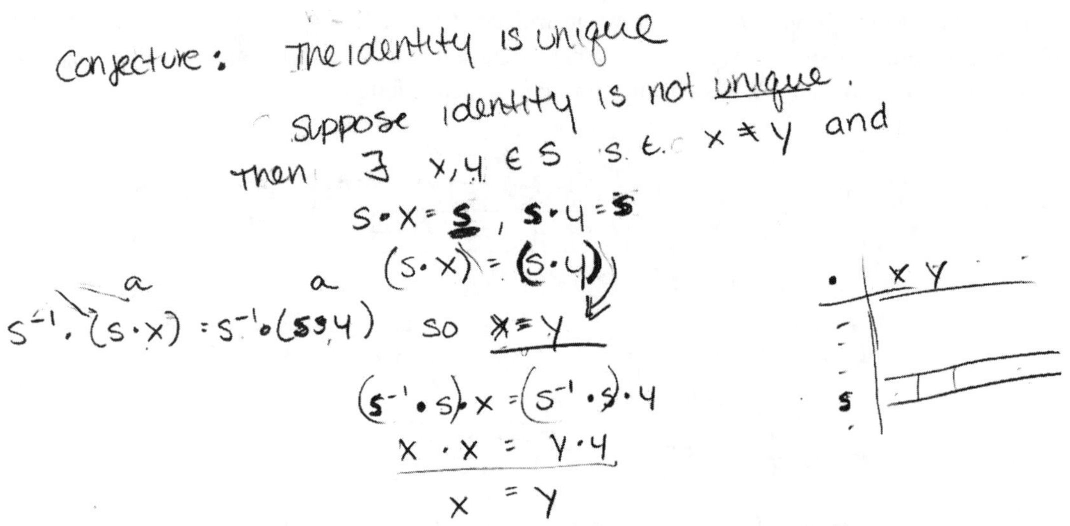

Conjecture: The identity is unique

Suppose identity is not unique.

Then $\exists\ x, y \in S$ s.t. $x \neq y$ and

$$s \cdot x = s,\ s \cdot y = s$$

$$s^{-1} \cdot (s \cdot x) = s^{-1} \cdot (s \cdot y)$$

$$a \qquad a\quad (s \cdot x) = (s \cdot y)$$

so $x = y$

$$(s^{-1} \cdot s) \cdot x = (s^{-1} \cdot s) \cdot y$$

$$x \cdot x = y \cdot y$$

$$x = y$$

·	x	y
s		

Figure 5. Jessica's proof that the identity element is unique.

procedure.

Conclusions

In the first section of this paper, we described research findings that highlight some of the difficulties that students have with understanding and reasoning about group theoretic concepts. Several researchers have argued that highly formal lecture-based abstract algebra courses may contribute to these difficulties, since they deny students the opportunity to engage in activities that might be used in developing conceptual understanding and effective reasoning processes (e.g., Leron & Dubinsky, 1995). Such activities include attempting to define central group theoretic concepts, reflecting on the processes used in group theoretic calculations, forming and evaluating conjectures, and participating in meaningful, informal mathematical activities that students can later use to understand the more formal treatment of abstract group theoretic concepts.

In the second section, we described two different, innovative instructional approaches that have been used to engage students in these activities and to have students develop a stronger, more meaningful understanding of group theory. While these instructional approaches are different in important ways, they do share some fundamental properties. Both approaches involve actively engaging students with the important concepts of group theory. Furthermore, both approaches devote considerable time to developing an informal or experiential basis for understanding these ideas. Special emphasis is given to connecting the formal treatment of these ideas to this experiential basis. In the instructional approach developed by Dubinsky and his colleagues, this is done by asking students to implement the informal ideas in the ISETL environment - using a computer programming language that is much like the language of formal mathematics. In the case of the instructional approach developed by Larsen, this is done by having students develop important ideas informally in the context of geometric symmetry and then formalize their informal ideas and methods to reinvent the formal theory.

References

Asiala, M., Dubinsky, E., Mathews, D.W., Morics, S., and Oktac, A. (1997). Development of students' understanding of cosets, normality, and quotient groups. *Journal of Mathematical Behavior*, 16, 241–309.

Dubinsky, E. (Ed.). (1997). An investigation of students' understanding of abstract algebra (binary operation, groups and subgroups) and The use of abstract structures to build other structures (through cosets, normality and quotient groups) [Special Issue]. *The Journal of Mathematical Behavior*, 16(3).

Dubinsky, E., Dautermann, J., Leron, U., and Zazkis, R. (1994). On learning the fundamental concepts of group theory, *Educational Studies in Mathematics*, 27 (3), 267–305.

Dubinsky, E. and McDonald, M. (2001). APOS: a constructivist theory of learning in undergraduate mathematics education. In D. Holton (ed.) *The teaching and learning of mathematics at the undergraduate level*. Dordrecht: Kluwer.

Cobb, P. (2000). Conducting teaching experiments in collaboration with teachers. In A. Kelly & R. Lesh (Eds.), *Handbook of Research Design in Mathematics and Science Education* (pp. 307–334). Mahwah, NJ: Erlbaum.

Findell, B. (2002). The operation table as metaphor in learning abstract algebra. In D. Mewborn (ed.) *Proceedings of the 24th Conference for the North American chapter of the Psychology of Mathematics Education*. 233–245. Athens, GA.

Freudenthal, H. (1991). Revisiting mathematics education: The China lectures. Dordrecht, The Netherlands: Kluwer.

Gravemeijer, K. (1998). Developmental research as a research method. In A. Sierpinska & J. Kilpatrick (Eds.), *Mathematics Education as a Research Domain: A Search for Identity* (pp. 277–296). Dordrecht, The Netherlands: Kluwer.

Hannah, J. (2000). Visual confusion in permutation representations. *Research in Collegiate Mathematics Education* IV, 8, 188–209.

Hart, E.W. (1994). A conceptual analysis of the proof-writing performance of expert and novice students in elementary group theory, In J. J. Kaput and E. Dubinsky (Eds.) *Research Issues in Undergraduate Mathematics Learning: Preliminary Analysis and Results*, Washington, D.C.: MAA.

Hazzan, O. (1999). Reducing abstraction levels when learning abstract algebra concepts. *Educational Studies in Mathematics*, 40, 71–90.

Hazzan, O. (2001). Reducing abstraction: The case of constructing an operation table for a group. Journal of Mathematical Behavior, 20, 163–172.

Larsen, S. (2004). Supporting the guided reinvention of the concepts of group and isomorphism: A developmental research project (Doctoral dissertation, Arizona State University, 2004) Dissertation Abstracts International, B 65/02, 781.

Larsen, S. (2002). Progressive mathematization in elementary group theory: Students develop formal notions of group isomorphism. Proceedings of the Twenty-Fourth Annual Meeting of the North American Chapter of the International Group for the Psychology of Mathematics Education. Athens, Georgia. 307–316.

Leron, U. and Dubinsky, E. (1995). An abstract algebra story. *American Mathematical Monthly*, 102, 227–242.

Leron, U., Hazzan, O., and Zazkis, R. (1995). Learning group isomorphisms: A crossroad of many concepts. *Educational Studies in Mathematics*, 29, 153–174.

Selden, A. and Selden, J. (1987). Errors and misconceptions in college level theorem proving. In *Proceedings of the 2nd International Seminar on Misconceptions and Educational Strategies in Science and Mathematics*. Cornell University, NY.

Weber, K. (2001). Student difficulty in constructing proofs: The need for strategic knowledge. *Educational Studies in Mathematics*, 48, 101–119.

Weber, K. (2002). The role of instrumental and relational understanding in proofs about group isomorphisms. In *Proceedings from the 2nd International Conference for the Teaching of Mathematics*. Hersonisoss, Greece.

Weber, K. (2006). Investigating and teaching the thought processes used to construct proofs. *Research in Collegiate Mathematics Education*, 6, 197–232.

Weber, K. and Alcock, L. (2004). Semantic and syntactic proof productions. *Educational Studies in Mathematics*, 56, 209–234.

12

Teaching for Understanding: A Case of Students Learning to Use the Uniqueness Theorem as a Tool in Differential Equations

Chris Rasmussen, *San Diego State University*
Wei Ruan, *Purdue University Calumet*

Students in many undergraduate mathematics courses tend not to readily and appropriately use theorems as tools for making arguments and solving problems (Schoenfeld, 1989; Hazzan & Leron, 1996). Students' reluctance to use theorems as tools is a problem that is not only cognitive in nature (that is, the difficulty is in how students conceptualize particular mathematical ideas), but also social in nature (that is, the nature of class discussion, the interpretation of tasks and ideas, etc.). In this chapter we highlight results from a classroom-based research program in differential equations that has resulted in some positive progress on the problem of students' reluctance to use theorems as tools for reasoning and solving problems.

The main result of the analysis of student learning and use of the uniqueness theorem for first order differential equations is the delineation of four interrelated cognitive and social factors that help account for why students actually made progress in using the uniqueness theorem as a tool for making arguments and solving problems (Rasmussen, 2004). The intention is that readers might, after understanding the details of this particular case, find the four factors useful more generally as an orienting framework for thinking about ways in which they can promote their students' use of theorems as tools for reasoning in other content areas. Thus, even if one does not regularly teach differential equations, this chapter intends to offer useful information for those who want their students to develop and use formal mathematics with understanding.

We begin by describing the results of one of the individual student problem solving interviews conducted as part of the study. The interview occurred just prior to students' first day of instruction in differential equations, and hence responses to the tasks posed in the interview are not a result of any instruction in differential equations. The research purpose for conducting the interviews was to glean insights into what informal or intuitive thinking students have about the issue of uniqueness before receiving any formal instruction. The task shown in Figure 1 was posed to Joe, who turned out to be one of the most mathematically able students in the class. All student names in this chapter are pseudonyms.

Joe rewrote the rate of change equation as $dL/dt = -0.1L + 7$ and determined that dL/dt is -3 by inserting the value of 100 for L. He then represented this work graphically and orally addressed the question posed.

A scientist developed the rate of change equation $dL/dt = -0.1(L - 70)$ in order to make predictions about the future temperature for a particular hot liquid (where L is the temperature of the liquid). Suppose one vat of this hot liquid has initial temperature of 100°F and a different vat of this hot liquid has initial temperature of 120°F. According to the rate of change equation, will there be a time when the two vats have the exact same temperature?

Figure 1. Interview task

Joe: Let me put it like this. This would be temperature versus time. [Sketches a graph similar to that shown in Figure 2]

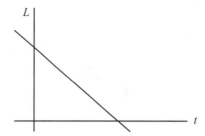

Figure 2. Joe's sketch of temperature versus time.

Joe: Given what I know about physics, I would say yes because at some time they both would have to reach room temperature. And they both should stop cooling at that point.

Interviewer: That connects with our everyday experience, right?

Joe: That connects with our everyday experience. But when I look at this equation [circles the rate of change equation $dL/dt = -0.1L + 7$] mathematically speaking, I see this [points to the rate of change equation] as telling me that it's [gestures with his hand in a way that re-creates his graph of temperature versus time] just going to get colder and colder and colder, infinitely colder. So I would say, no, there is no way for the second one to catch up with the first because at no point is this going to stop decreasing, it's just going keep getting colder and colder and colder.

Interviewer: So could you draw me a graph of the temperature versus time for both vats?

Joe: I would expect it to be a parallel line. That it's constantly lagging by a constant amount of whatever 20 degrees difference ends up causing. That it will always be that far behind.

There are two things to notice in Joe's response. First is the fact that Joe makes two conclusions, one based on his expectations of the real world situation and one based on his interpretation of the mathematical equation. The former conclusion is an empirically based argument, and the latter is a mathematically based argument. Second is the way in which he thinks about rate of change. In particular, Joe appears to think about rate of change as a constant, non-varying quantity. He does not immediately view dL/dt as a quantity that changes as the temperature of the vat, L, changes. Effective reasoning in differential equations requires one to coordinate changes in quantities such as dL/dt and L, and represents a sophisticated form of covariational reasoning (see Oehrtman, Carlson, and Thompson, this volume).

As will be detailed in the main section of this chapter, shifts in the source of one's arguments (empirical versus mathematical) and shifts in how one thinks about a central idea, such as rate of change, are two significant factors that contributed to students' progress on using a theorem (in this case, the uniqueness theorem) as a reasoning tool.

In the sections that follow we first provide some background on the research approach and the mathematical goals of the project. We then describe and illustrate the social and cognitive factors that help account for students' progress in using the uniqueness theorem as a reasoning tool. We conclude the chapter with some reflections on instructional design and how it relates to the social and cognitive factors.

Background

Advances in technology and an increased interest in dynamical systems are prompting new directions in many first courses in differential equations. These new directions include qualitative, geometric, and numerical analyses as complements to the traditional emphasis on analytic techniques. The MAA Notes #50, *Revolutions in Differential Equations*, edited by Kallaher (1999) and the special issue of the *College Mathematics Journal* edited by West (1994) offer excellent examples of these new directions. At the same time, research is beginning to illuminate students' thinking about central ideas and methods of analysis associated with these new directions. A recent review of the literature by Rasmussen & Whitehead (2003) highlights the primary findings to date, including a delineation of students' strategies, understandings, and difficulties with (a) coordinating algebraic, graphical, and numerical representations, (b) creating and interpreting various representations including phase portraits and bifurcation diagrams, and (c) making warranted predictions about the long-term behavior of solution functions. Noticeably missing are research reports on students' use of and thinking about the uniqueness theorem. In general, mathematics education research has concentrated on proof construction (see Selden & Selden, this volume) rather than on theorem use. The chapter reported here makes a contribution to this important, yet under-researched area of mathematical reasoning.

Originating in the seventeenth century as a technique for solving geometrical and mechanical problems, the study of differential equations initially centered on attempts to find analytic solution techniques. As scientists moved toward increasingly analytically intractable differential equations arising in physical and graphical situations, they became motivated to ask the questions of first existence and then uniqueness of solutions. The first to draw attention to the unspoken assumption that there exists a solution to a given differential equation was Cauchy, who in the 1820s gave a rigorous proof for the existence of a solution. Several decades later, the Lipschitz condition provided the first guarantee of unique solutions to first order ordinary differential equations (Boyer & Merzbach, 1989; Kline, 1972).

The following is a formal statement of the uniqueness theorem in terms of the Lipshitz condition for the differential equation of the form $dy/dt = f(t, y)$:

> If $f(t, y)$ is continuous on $a < t < b$, $c < y < d$, and there exists a constant L such that $|f(t, y) - f(t, z)| \leq L|y - z|$ for all $t \in (a, b)$, and all $y, z \in (c, d)$, then the initial-value problem $dy/dt = f(t, y)$ with $y(t_0) = y_0$ where $t_0 \in (a, b)$ and $y_0 \in (c, d)$ has at most one solution for all $t \in (a, b)$ such that $y(t) \in (c, d)$.

Often the uniqueness theorem (together with the existence theorem) is stated for students in terms of continuity of f and $\frac{\partial f}{\partial y}$ in the hope that they would use it as a tool for making arguments and solving problems. The extent to which this happens, however, is typically not encouraging (Raychaudhuri, 2007). Students usually do not understand the theorem and hence do not use it as a tool for making arguments or solving problems. We conjectured that one reason for this discouraging result is that the more typical requirement of continuity (rather than Lipschitz) does not, from students' perspective, express mathematical relationships that are relevant to them. Although students are familiar with continuity, why continuity of f and $\frac{\partial f}{\partial y}$ are needed appears to be a mystery. We have found, however, that the Lipschitz condition is much more likely to connect to students' thinking about relevant mathematical relationships when given problems that raise their intellectual curiosity about the issue of uniqueness.

In the research reported here we set the following two instructional goals for the uniqueness theorem: First, students would come to view the issue of uniqueness as personally relevant (historically this took a very long time). Second, the Lipschitz condition for uniqueness of solutions would be a relevant, formal description of students' manipulation of vectors and their corresponding explanations for why graphs of solutions do or do not intersect. Students in our study essentially framed their observations and analysis in terms of the condition $|\frac{\partial}{\partial y} f(t, y)| \leq L$ rather than in terms of $|f(t,y) - f(t,z)| \leq L|y - z|$. The teacher of course must play an active and direct role in connecting these two statements for students.

Two forms of evidence indicate that we made reasonably good progress in achieving these two goals. First, a review of student work after instruction on the uniqueness theorem revealed several instances of students spontaneously invoking the uniqueness theorem to make arguments, even though the task did not specifically request such arguments. Second, a quantitative analysis of students' understanding in more traditional classes versus project classes found that students who received instruction similar to that described here used the uniqueness theorem as a tool for solving problems significantly more so than students who received more conventional instruction (Rasmussen, Kwon, Allen, Marrongelle, & Burtch, 2005).

Research Approach

This chapter is based on analysis of data collected during a 15-week introductory course in differential equations conducted at a mid-sized public university. The data collected included classroom video recordings of each class session from two cameras, copies of students' written class work, copies of exams and homework, and video recordings of individual student problem solving interviews. The project team consisted of the teacher, who was an experienced research mathematician with expertise in partial differential equations (and the second author of this chapter), and two mathematics education researchers who attended each class session, taking field notes and listening carefully to student thinking. The research method used is one referred to in mathematics and science education as *design research* (Cobb, 2000; Kelly, 2003). Central characteristics of design research include (1) the study of learning and teaching while simultaneously designing teaching tasks and teaching interactions; (2) the use of continuous cycles of design, testing, analysis, and redesign; and (3) the creation of knowledge for practitioners and other researchers in mathematics education.

Typical class sessions consisted of cycles of small group work on challenging problems and whole class discussion of students' work on these problems, however tentative. Sharing, presenting, and discussing student work was typical in this class, and the teacher regularly sought to support and encourage students in their efforts to explain and justify their thinking. The course materials, largely developed in previous design research projects, were modified as needed as the semester progressed. An essential characteristic of the course was the creation of a learning environment in which students could reinvent important mathematical ideas and methods as they solved problems and explained their reasoning (Freudenthal, 1991; Gravemeijer, 1999; Rasmussen, Yackel, & King, 2003).

At the completion of the semester, the research team analyzed the data through cycles of reviewing videorecordings, creating transcripts, and writing interpretive notes. This iterative process resulted in the identification of four overarching cognitive and social factors that help explain why students were successful in using the uniqueness theorem as a tool for reasoning and solving problems. Sample episodes from this iterative process are used in this chapter for illustrative purposes. To sharpen the discussion even further, we focus primarily on the reasoning of three students, Bill, Adam, and Joe.

In general, the analysis draws on theories of learning in which students' mathematical reasoning is viewed as constrained and enabled by both their current understandings and the nature of the learning environment (Cobb & Yackel, 1996). As emphasized by Piaget (1970), learning is a process involving a constant interaction between the learner and her environment. This involves the integration of things to be known with existing ways of thinking and the reorganization of these ways of thinking as students participate in and form the patterns of argumentation that become routine in their mathematics classrooms. Given this theoretical orientation, we were interested in changes in student thinking and shifts in what counts as an acceptable justification. In keeping with an emphasis on framing the evolution of thinking in the context of classroom learning, this chapter describes important cognitive and social factors that facilitated students' progress in developing and ultimately using the uniqueness theorem as a tool for reasoning and solving problems.

Using Theorems as Tools

The framework for capturing progress in students' use of the uniqueness theorem as a tool is traced in terms of four interrelated social and cognitive factors. We characterize a factor as social when we want to emphasize the process of interaction between students, the teacher, and the mathematics. The two social factors we identified are the negotiation of acceptable justifications and the familiarization with mathematical terminology and meaning. On the other hand, we characterize factors as cognitive when we want to emphasize particular conceptions that students engage. The cognitive factors we identified are the engagement of an intuitive theory and a conceptual reorganization about rate of change.

In this chapter we illustrate and clarify the social factor pertaining to justification and both cognitive factors, while only briefly discussing the social factor of familiarization with mathematical terminology and meaning. Whenever possible, we point to connections between these factors and the role of the teacher in their evolution. The concluding section relates the evolution of these factors to the overall instructional design.

Negotiation of Acceptable Justifications — A Social Factor

Analysis of the classroom videotape data points to an important interplay between empirically based justifications and justifications based on mathematical relationships. Justifications were deemed empirical if they were (1) based on observed or imagined graphs or (2) based on an imagined, real-world phenomenon. By design, instructional sequences drew heavily on geometric approaches and the framing of problem situations in terms of real world phenomena. Thus, it is perhaps not surprising that students' justifications were, at least initially, grounded in observed or imagined graphs or imagined real world events. What is significant is that there was a shift in the nature of students' justifications over the course of the semester — from those with an empirical basis to those with a basis in mathematical relationships.

For example, prior to work on the sequence of tasks dealing explicitly with the uniqueness theorem, the class discussed whether or not an imagined solution graph, one that was initially increasing, would ever actually reach or touch the equilibrium value of 12.5. Under scrutiny was the differential equation $\frac{dP}{dt} = 0.3P\left(1 - \frac{P}{12.5}\right)$, which was intended to model population growth. Tangent vector fields generated by a computer program (without curve sketching capabilities) and analysis of the differential equation itself were the primary mathematical ideas and tools available to students. The following whole class excerpt succinctly captures the interplay, and tension, between empirically based and mathematically based justifications.

Bill:	What Jeff and I were thinking was that eventually, ideally, this would seek an equilibrium, fluctuating up and down around 12.5. But obviously, you can't have half a deer running around, so, you know, it's gonna at some point go above 12.5, then it goes in negative, in the uh slope, so it'll drop below 12.5. Then, then you're back positive, and it'll, so it'll be rising and falling up and down around, around 12.5.
Joe:	How do you rise up and down when you have a zero tangent?
Adam:	Maybe theoretically, but that's not what our equation's saying. Our equation's saying that uh 12.5 is gonna be the limit. It's gonna go up, it's gonna, it's gonna be what's it called, asymptotic to 12.5 or, I think that's —
Joe:	Well, it's not actually [inaudible] It's when P to the 12.5 is one.
Adam:	Yeah. Yeah.
Teacher:	Okay. What's so important for this 12.5? It seems that some of you think it's positive if P is less than 12.5?
Stds:	Yes.
Joe:	It's positive if it's less than 12 and a half. It's negative if it's greater than 12 and a half, and it's zero at 12 and a half. And so I have a problem with it being able to fluctuate around 12.5 because if you have a zero. If you had zero change,
Bill:	Okay, well, my thinking was —
Joe:	I mean, it doesn't change over time no matter [inaudible].
Bill:	Well, my thing was that you talk about a population, you're talking about a population, you have to have whole numbers.

Bill explained that his group's initial idea was that the graph would oscillate around 12.5 with decreasing amplitude. No justification for why such asymptotic behavior might be the case was offered. Bill said that he and Jeff then rejected this conclusion that the population would settle down to 12.5 because "you can't have half a deer." Joe and Adam immediately rejected any kind of oscillation based on the mathematical relationship between the slope of a graph as dictated by the differential equation (in particular there should be a tangent with zero slope at 12.5) and the shape of the graph. This clarification was, in part, solicited from the teacher when he asked, "What's so important about 12.5?" In response to Adam and Joe's point, Bill then clearly stated that his reasoning was based on the need to have "whole numbers" due to the population setting. Bill's justification falls within the realm of empirically based justifications, while Joe and Adam's justification falls with the realm of justifications based on mathematical relationships.

As the discussion continued, Joe, Adam, and a third student, Jake, argued further against Bill's conclusion.

Joe:	What you think a population would be doesn't mean that that's what that equation is going to do.
Adam:	We're talking about a model here.
Jake:	Yeah, that's just a representation, I mean like that's like a thousand times 12.5, or three thousand times 12.5.
Teacher:	So it's like a very large number and uh,
Adam:	So the fluctuation wouldn't really, you wouldn't see it.
Joe:	Yeah.
Adam:	It's just a model!
Jake:	It would be at equilibrium at 12.5. It would level off as it approaches 12.5, the rate of change.

Teacher: Yes, certainly we cannot have half fish, or half deer. But you're saying that if we have a huge number for population then, then this, although it's not really a smooth curve, but for the model, we have a, we have a smooth curve. Is that what you're saying?

Jake: Yes.

Teacher: Yes.

In addition to Adam's argument that the differential equation is something other than an exact fit to the population setting ("it's just a model"), Jake argued that 12.5 could very well be 12,500, for example. The teacher clarified for the class that in terms of actual population values, which would be discrete, the solution graphs of interest to the class are continuous. The previous excerpts point to how norms for argumentation are co-constructed between students and the teacher. They are not rules set out in advance by the teacher. We also point out that, although the teacher's voice is less prominent in these excerpts than the students', he played an essential and proactive role in shaping the classroom discussion and norms for justification. Among other functions, he was the one who selected and made possible the conversation about whether or not a solution graph would touch 12.5, he was the one who worked to set up a classroom environment in which students felt safe to voice their ideas, even if they turn out to be rejected, and he was the one who at once honored Bill's conclusion ("Yes, certainly we cannot have half a fish, or half a deer") while implicitly reinforcing the need for conclusions to be based on mathematical relationships.

As the semester progressed, justifications based on mathematical relationships and concepts such as rate of change became more and more routine, even though the problems posed to students were often framed in terms of imagined real world settings in which prediction of future quantities was important. The significance of this social factor in the evolution of using the uniqueness theorem as a reasoning tool in relation to the cognitive factors is that it (a) creates opportunities for the teacher to become aware of students' intuitive theories, and (b) makes explicit discussion about central mathematical ideas (in this case rate of change) more viable, which in turn opens spaces for students to refine and reorganize their conceptions of these ideas.

Engaging Intuitive or Informal Theories — A Cognitive Factor

Students exhibited an intuitive theory that non-equilibrium solution functions will approach equilibrium solution functions asymptotically (Rasmussen, 2001). Students' intuitive theory about asymptotic behavior in this classroom took on one of two forms, either oscillations with decreasing amplitude toward a fixed value or strictly increasing/ decreasing behavior toward a fixed value. Both of these intuitive theories were evident in the excerpts provided in the previous section. Recall the following statement made by Bill: "What Jeff and I were thinking was that eventually, ideally, this would seek an equilibrium, fluctuating up and down around 12.5." Adam, on the other hand, assumed that the graph of the solution would approach 12.5 asymptotically in a strictly increasing manner. Recall Adam's statement that the graph is "gonna go up, it's gonna, it's gonna be what's it called, asymptotic to 12.5." Later in this same excerpt Jake also stated that the graph "would level off as it approaches 12.5."

Which of the two forms of asymptotic intuition was engaged appeared to depend on the imagined real-world setting. For settings in which oscillation of quantities was reasonable, we saw both types of asymptotic intuition. In other settings, like the one discussed in the next section, only strictly decreasing asymptotic intuition was engaged.

Although asymptotic behavior is the outcome in many cases, it is not always the case (e.g., consider solutions to $y' = -y^{1/2}$). From a student's perspective, such intuitive theories make sense because they originate in extensive mathematical experiences. In his seminal work on intuition, Fischbein (1987) characterized intuitions as self-evident statements that exceed the observable facts. Being self-evident, justifications often do not accompany statements that engage intuitive theories, as was the case in the previous excerpts. When pushed for justification, students tended to provide circular arguments. For example, as the conversation about solutions to $\frac{dP}{dt} = 0.3P\left(1 - \frac{P}{12.5}\right)$ continued, the teacher asked students what should be the graph of the solution if "we base it just on the differential equation model?" To which Bill responded,

Bill: Then I agree that P approaches 12.5, and as it gets closer and closer to 12.5 the rate of change will get smaller and smaller and yeah, I don't think it would ever reach 12 and a half. I would just keep getting closer and closer, but never quite make it.

Teacher: Why do you think that?

Bill: Because, because, the closer it gets, the rate of change keeps decreasing. You know, never going to zero. But it keeps, it keeps holding it back. The rate of change does not let it get to 12 and a half.

From an instructional design and teaching perspective, awareness of students' intuitive theories is important because it informs subsequent work with students in efforts to promote cognitive refinements and reorganizations (see next section). Students' intuitive theories, although often not generalizable to all situations, can serve an important function in the learning process. In particular, when students encounter instances that conflict with their intuitive theories, that is, when they encounter disequilibrium, they are more likely to become explicitly aware of—and search for—refined and reorganized conceptions. The point is not to replace the intuitive theory, but rather to use students' intuitive theories as opportunities for refining and reorganizing their thinking about a central mathematical idea, such as rate of change. Consistent with Brousseau's (1997) theory of cognitive obstacles, engaging intuitive or informal theories points to how such theories can function not only as a constraint, but also as a resource.

Reorganizing a Central Mathematical Idea — A Cognitive Factor

Students' thinking about rate of change as expressed in a differential equation was initially either a "descriptor" (that is, an adjective) that characterizes the slope or steepness of an observed or imagined solution curve, or a "controller" of the direction a solution graph should take. Bill's previous excerpt speaks to both of these ways of thinking about rate. Specifically, Bill had a solution graph in mind (one that is increasing toward 12.5) and he used rate of change as a descriptor of the graph. "… as it [the graph] gets closer and closer to 12.5, the rate of change will get smaller and smaller …". Here rate of change is an adjective for an already imagined graph. Rate of change is used to describe qualities of the graph. Bill also used rate of change as a mechanism or controller for how the graph should proceed. For example, in this same excerpt, Bill stated, "But it [the rate of change] keeps, it keeps holding it [the graph] back. The rate of change does not let it [the graph] get to 12 and a half." In this last quote, rate of change acts as control mechanism for how the graph will unfold, rather than as a descriptor of an already unfolded graph. The descriptor and controller metaphors for rate offer specificity on how students coordinate a quantity and its rate of change.

In subsequent problems students compared graphs of solutions to two different differential equations and reasoned about the rate of change of the rate of change to account for why one set of solutions touched an equilibrium solution and the other set did not. That is, the issue of uniqueness was a concern for students, even though the theorem had yet to be introduced. In accomplishing this comparison goal, rate became an object with its own properties that needed to be described.

To clarify, consider the analogy of a red ball. Initially the ball is the primary object of interest. Red is an adjective or a descriptor that characterizes the ball. This is analogous to an imagined solution function and its slope or rate. Now imagine there is another ball. The new one is a deep, intense red and the old is a pale, pinkish red. Through comparison we develop a need to describe redness itself. So we might say, "That is a deep red" or an intense red. Now deep is a descriptor (i.e., adjective) modifying red. The focus of our attention has shifted to red. The ball is in our subsidiary awareness while our focal awareness is on the nature of redness. This is analogous to shifting one's attention from an imagined graph where rate or slope is a descriptor of that graph or controller of a graph to analysis of rate of change with its own properties that need to be described. For example, as one student commented, "The rate of change of the rate of change increases as y approaches zero." In this quote, rate is not a property of a solution graph, but rather an object analyzed for its own properties.

The differential equations under scrutiny were $dh/dt = -h$ and $dh/dt = -h^{1/3}$, both of which were offered as models for the height of a descending airplane and both of which have constant solutions $h(t) = 0$. Students used software that provided a tangent vector field in which the user could activate and drag a vector that continuously changes direction per the differential equation. We found it important for the software not to actually sketch solution curves because this would take away from a focus on changes to the tangent vector. A snapshot of the click and drag tangent vector field, with activated vector shown near the point (.08, .27) is given in Figure 3.

Students used the software to make conjectures as to if one, both, or neither differential equation predicts a landing for the plane. The issue of uniqueness therefore became relevant to students since landing would mean that two solution graphs touch (in particular there would be two different solutions that meet at the solution $h(t) = 0$). As illustrated in the following quote, Adam used rate as descriptor and rate as controller to argue that graphs for $dh/dt = -h$ would not touch zero.

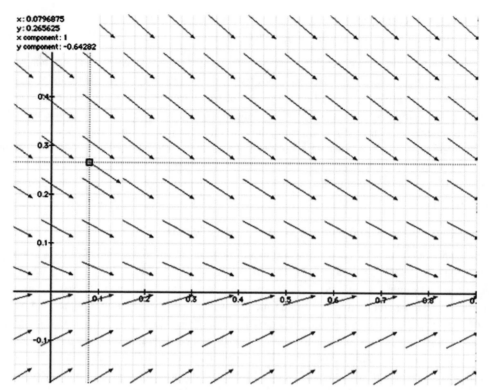

Figure 3. Tangent vector field for $dh/dt = -h^{1/3}$.

Adam: If you plot your vector field or whatever, your slope is gonna gradually taper off. As your number gets smaller and smaller, your slope's gonna get smaller and smaller [rate as descriptor]. So there's no way you're ever gonna be able to get to zero on your height because your slope is gonna slow it down [rate as controller].

Adam also stated that the other differential equation is "kind of the same as the first one." It is likely that his conclusions for both differential equations engaged asymptotic intuition. Other students claimed that $dh/dt = -h^{1/3}$ predicts that the plane would touch ground. The justification for this claim was the observed vector field, in which students experienced and observed vectors change very quickly from a nonzero slope to a zero slope. To clarify, imagine dragging the activated vector shown in Figure 3 down to the h-axis, moving parallel to the dh/dt axis. What students observed is that, as they moved the tail of the vector onto the h-axis, the vector makes a sudden and abrupt change to the horizontal position. They referred to this sudden change in slope as "snapping to zero." This sudden change in slope does not occur for vectors associated with $dh/dt = -h$.

Relying on observed vectors snapping to zero is a form of empirical justification. Accounting for the snap then became a topic of conversation and further analysis. Part of this analysis involved finding analytic solutions, which yielded conclusive evidence that solutions to $dh/dt = -h$ do not touch zero while solutions to $dh/dt = -h^{1/3}$ do touch zero. The analytic solutions did not, however, offer students insight into why solutions were or were not unique. This insight was gleaned by reexamining the "snapping" of tangent vectors near zero (or not snapping) in light of how the rate of change changes. That is, framing the snapping in terms of the rate of change of the rate of change.

Adam: Okay, um on our slope field it looks like the rate of change of our, the differential equation, is going down kind of slow for the $-h^{1/3}$. Then after it passes, what is it, one, it snaps. It starts snapping to zero [Lynn: Why?] Why? When you re-write the derivative of your differential equation. So it's negative one third, uh, 'h'. Well, next to the three, pull out your, yeah, there you go. Yeah, when you look at that [$d(-h^{1/3})/dh$] of we can see that, um, as h gets smaller, it just blows up. But that means that the, when you're at zero it's undefined. But as you go to zero it's getting bigger and bigger and bigger. As you get close to zero, you get a really big number. So as you go to zero it's not defined.

Teacher: What's the 'it' that is undefined?

Joe: The rate of change of the rate of change becomes infinitely large. As you approach zero the rate of change of the rate of change goes to infinity.

This cognitive reorganization in rate (from rate as descriptor or controller to rate as object with its own properties that need to be described) created an opportunity for the teacher to then relate student thinking to the formal, symbolic expression $\left|\frac{\partial}{\partial y} f(t, y)\right| \leq L$, a short step away from conditions for uniqueness. From the students' perspective, this "formalization" was a recognizable recasting of their analyses of the rate of change of the rate of change. Put another way, the observation that vectors for one of the differential equations snapped to zero functioned as a key idea (Raman, 2003) that provided a way to connect observations, predictions, and intuitions to formal, conventional mathematical expressions. In general, awareness of how key ideas function is important in mathematical reasoning, from using theorems as tools, to creating and using definitions, to proving. We hope that this chapter offers readers some insights into the processes that enable movement between students' ways of thinking and more conventional or formal expressions of that thinking.

The Familiarization with Mathematical Terminology and Meaning — A Social Factor

Expressing students' thinking about the rate of change of the rate of change in terms of the criteria $\left|\frac{\partial}{\partial y} f(t, y)\right| \leq L$ was critical to formally stating the uniqueness theorem. The process of becoming increasingly more familiar with the theorem, however, continued even after the introduction of the formal statement by the teacher. This process involved application of the theorem in other settings, a revisiting of how to interpret ideas such as boundedness and partial derivatives, and what one can logically infer (or not) when the conditions of the theorem are not met. For the most part, the teacher gave clear and explicit instruction on these issues. Because the issues were genuine concerns of students, the insertion of new, conceptual information from the teacher appeared to be effective. We characterized these conversations between the teacher and students as a social factor because it is the teacher who takes the lead in familiarizing students with conventional terminology and the intended mathematical meaning. We think that it is important for the teacher to find those times when the telling of new information (see Marrongelle and Rasmussen, this volume) can help students further organize and sharpen their understandings. In addition to reinventing mathematical ideas with the guidance and support of the teacher, students are also expected to adopt conventional terms and meanings and thus part of a teacher's job is to familiarize students with the language of mathematics.

To recap, the social and cognitive factors that help account for the process by which students came to use the uniqueness theorem as a tool are: (1) The negotiation of what counts as an acceptable explanation, (2) The engagement of an intuitive theory, (3) A cognitive reorganization about a central idea, and (4) The familiarization with mathematical terminology and meaning. These four factors offer a way to conceptualize the process of developing formal mathematics and use of formal mathematics, such as theorems, in a way that values both cognitive and social realities of classroom life. As such, we hope this particular example inspires readers interested in engaging their students in formal mathematics in other content areas.

Reflections on Instructional Design

We describe the sequence of tasks by which students came to use the uniqueness theorem as a tool for reasoning and solving problems in terms of four types of tasks: Prediction, Exploration, Mathematization, and Generalization. Prediction tasks required graphical predictions for the behavior of solution functions. We saw examples of this with the differential equations $dP/dt = 0.3P(1 - P/12.5)$, $dh/dt = -h$, and $dh/dt = -h^{1/3}$. Exploration tasks used technology so that students could graphically explore tangent vector fields and make inferences about solution function behavior. Whether or not graphs of solutions with positive initial conditions touched the zero solution was, for many students, an open question. Exploration type tasks tended to lend themselves to varied and useful conjectures by students.

Mathematization tasks required students to pursue analytic work to solve the differential equations, and to make comparisons between the analytic solutions and their graphical predictions. Because of the analytic solutions, students knew that solutions to $dh/dt = -h$ with positive initial conditions never touched the constant solution $h(t) = 0$, while just the opposite was the case for solutions to $dh/dt = -h^{1/3}$. The analytic solutions revealed the difference between solutions to the two differential equations. It did not, however, offer students insight into *why* there was this difference.

In need of further insight, students returned to exploration tasks in which the teacher directed their observations to how a vector changes as they dragged it toward $y = 0$. It is at this point that students noticed that the vector "snaps" onto the horizontal line $y = 0$. Some students made the connection that this snap indicates that the solution curve makes a kind of "jump" to get to the constant solution curve. Further mathematization of this connection resulted in symbolically expressing the snap in terms of the partial derivative of the rate of change function $f(t, y)$. Finally, generalization tasks, which came after this re-exploration and further mathematization, led to a generalized statement of the uniqueness theorem. Students further applied and used the uniqueness theorem to reason about solutions to a variety of different differential equations.

The boundaries between prediction, exploration, mathematization, and generalization tasks are of course not strict. Student work on a particular task, for example, might involve prediction as well as exploration. We also find it useful to plan for cycles of exploration and mathematization tasks because useful conclusions can often then be drawn, resulting in more robust generalizations.

To conclude the chapter we offer Figure 4 as a way to relate the four types of instructional tasks to the cognitive and social factors that contributed to students' use of the uniqueness theorem as a tool for reasoning and solving problems. The placement of the rectangles representing the cognitive and social factors under the various types of tasks indicates that we saw evidence of these factors when the class was engaged in the corresponding tasks. For example, there were instances of negotiating what counts as an acceptable mathematical argument at various times throughout the prediction, exploration, and part way into the mathematization tasks. Reorganization of student thinking about the central idea of rate of change, on the other hand, occurred primarily during the exploration and mathematization cycles. The familiarization with conventional terminology occurred near the end of the prediction-exploration-mathematization-generalization sequence.

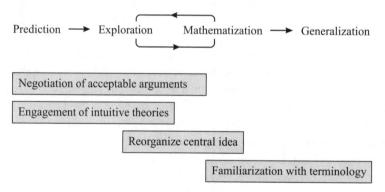

Figure 4. Relating cognitive and social factors to instructional design.

The overlap in the rectangles that represent the social and cognitive factors and the cyclical nature of the exploration-mathematization tasks point to the complexity of the learning environment. Our intention is that Figure 4 puts some structure or organization to this complexity. Teaching for understanding is indeed a complex endeavor, especially when teaching and instructional design explicitly attend to (1) social interactions in which meanings are established, (2) emerging and negotiated norms for convincing arguments, (3) student conceptions that can be built on and extended, and (4) processes in which students can connect more formal mathematical developments to their personal experience.

Finally, helping students become users of theorems as tools is of course made easier by using well thought out and research-based instructional tasks. As reported elsewhere (Rasmussen & Keynes, 2003) a similar sequence of prediction, exploration, mathematization, and generalization tasks proved useful for student reinvention of analytic solutions to systems of linear differential equations. It is important to keep in mind, however, that the tasks themselves do not guarantee that students will actually be engaged in predicting, exploring, mathematizing, and generalizing in ways that lead them to ownership of formal mathematics. The teacher has primary responsibility for enlisting the social and cognitive factors in ways that make this possible.

References

Boyer, C. & Merzbach, U. (1989). *A history of mathematics*. New York: Wiley.

Brousseau, G. (1997). *Theory of didactical situations in mathematics*. Dordrecht, The Netherlands: Kluwer Academic Publishers.

Cobb, P. (2000). Conducting classroom teaching experiments in collaboration with teachers. In R. Lesh & E. Kelly (Eds.), *New methodologies in mathematics and science education* (pp. 307–334). Mahwah, NJ: Erlbaum.

Cobb, P. & Yackel, E. (1996). Constructivist, emergent, and sociocultural perspectives in the context of developmental research. *Educational Psychologist, 31*, 175–190.

Fischbein, E. (1987). *Intuition in science and mathematics.* Dordrecht, The Netherlands: Reidel.

Freudenthal, H. (1991). *Revisiting mathematics education.* Dordrecht, The Netherlands: Kluwer Academic Publishers.

Gravemeijer, K. (1999). How emergent models may foster the constitution of formal mathematics. *Mathematical Thinking and Learning, 2*, 155–177.

Hazzan, O. & Leron, U. (1996). Students' use and misuse of mathematical theorems: The case of Lagrange's theorem. *For the Learning of Mathematics, 16*, 23–26.

Kallaher, M. J. (Ed.) (1999). *Revolutions in differential equations: Exploring ODEs with modern technology.* Washington, DC: The Mathematical Association of America.

Kelly, A. E. (2003). Theme issue: The role of design in educational research. *Educational Researcher, 32*, 3–4.

Kline, M. (1972). *Mathematical thought from ancient to modern times.* New York, NY: Oxford University Press.

Piaget. J. (1970). *Psychology and epistemology.* New York: Viking Press.

Raman, M. (2003). Key ideas: what are they and how can they help us understand how people view proof? *Educational Studies in Mathematics, 52*, 319–325.

Rasmussen, C. (2001). New directions in differential equations: A framework for interpreting students' understandings and difficulties. *Journal of Mathematical Behavior, 20*, 55–87.

Rasmussen, C. (2004). The evolution of formal mathematical reasoning: a case study of the uniqueness theorem in differential equations. In McDougal, D. (Ed.), *Proceedings of the Twenty-Fourth Annual Meeting of the North American Chapter of the International Group for the Psychology of Mathematics Education.* (Vol. 2, pp. 959–968). Toronto, Canada.

Rasmussen, C. & Keynes, M. (2003). Lines of eigenvectors and solutions to systems of linear differential equations. *PRIMUS, Volume XIII*(4), 308–320.

Rasmussen, C., Kwon, O., Allen, K., Marrongelle, K. & Burtch, M. (2006). Capitalizing on advances in mathematics and K–12 mathematics education in undergraduate mathematics: An inquiry-oriented approach to differential equations. *Asia Pacific Education Review, 7*, 85–93.

Rasmussen, C. & Whitehead, K. (2003). Learning and teaching ordinary differential equations. In A. Selden & J. Selden (Eds.), *MAA Online Research Sampler.* (http://www.maa.org/t_and_l/sampler/rs_7.html)

Rasmussen, C., Yackel, E., & King, K. (2003). Social and sociomathematical norms in the mathematics classroom. In H. Schoen & R. Charles (Eds.), *Teaching mathematics through problem solving: Grades 6–12* (pp. 143–154). Reston, VA: National Council of Teachers of Mathematics.

Raychaudhuri, D. (2007). A layer framework to investigate student understanding and application of the existence and uniqueness theorems of differential equations. *International Journal of Mathematical Education in Science and Technology, 38*, 367–381.

Schoenfeld, A. (1989). Explorations of students' mathematical beliefs and behavior. *Journal for Research in Mathematics Education, 20*, 338–355.

West, B. (Ed.). (1994). Special issue on differential equations [Special issue]. *The College Mathematics Journal, 25*(5).

Acknowledgements. Support for this paper was funded in part by the National Science Foundation under grant No. REC-9875388. The opinions expressed do not necessarily reflect the views of the foundation. The authors thank Karen Allen, Kevin Dost, and Jennifer Olszewski for their assistance in the data collection and analysis.

Part II
Cross-Cutting Themes
a. Interacting with Students

13

Meeting New Teaching Challenges: Teaching Strategies that Mediate between All Lecture and All Student Discovery

Karen Marrongelle, *Portland State University*
Chris Rasmussen, *San Diego State University*

A growing number of postsecondary mathematics educators are exploring teaching strategies other than lecture (Holton, 2001). The motivations for such change include personal dissatisfaction with student learning, students' poor retention of knowledge, student dissatisfaction with their undergraduate experiences in science, mathematics, and engineering (National Science Foundation, 1996; Seymour & Hewitt, 1997), as well as efforts to rethink core courses such as calculus, linear algebra, and differential equations. As postsecondary educators make changes to their practice they often struggle with many of the same issues that K–12 mathematics teachers encounter as they attempt to change their practice. In this chapter we address one of these issues, namely the role of teacher lecture (or telling) and strategies that teachers can use to balance student discovery and teacher telling.

Navigating a new terrain of teaching practice is particularly tricky for any teacher, elementary or university, who may never have experienced as a learner an approach to teaching other than lecture and demonstration. For example, some teachers believe that changes in practice must be dramatic and involve a total abandonment of lecture (where the teacher has all the responsibility for developing the mathematics) to a form of practice that leaves students to discover all ideas and techniques for solving problems. These are two ends of a continuum from all student discovery to all teacher telling. How, why, and when a teacher positions him or herself along this continuum is a source of tension for teachers. This was the case for Heaton (2000), an accomplished fourth-grade elementary school teacher, university teacher educator, and educational researcher. In the quote below, Heaton reveals her own doubts about her new approach to teaching mathematics.

> I had stopped all telling and eliminated any type of evaluation of students' answers. I tried to do nothing but ask questions and remain neutral…. I accepted all individual answers but was at a loss for how to move forward. I was beginning to feel that math needed to be more than just a time to share ideas…I had begun to feel as if I ought to be doing something more with responses. I was a *teacher*. I was supposed to *teach*. (p. 61, italics in original)

Heaton's account underscores two important points for teachers interested in creating classrooms in which students are more actively involved in the building and creating of ideas and methods for solving problems. First, changing teaching practice does not mean wholesale abandonment of past practice. This means that teachers must go beyond simply replacing old strategies with new ones; rather teachers must determine how to integrate new strategies into their existing repertoire. Second, the role of the teacher needs to include more than bringing tasks to the classroom and standing back as students solve problems.

But how might a teacher continue to develop mathematical ideas with students when students' meaningful constructions appear to be inadequate for the bigger picture the teacher envisions? Reflecting on her experiences with third graders, Ball (1993) posed the question in the following way:

> How do I create experiences for my students that connect with what they now know and care about but that also transcend their present? How do I value their interests and also connect them to ideas and traditions growing out of centuries of mathematical exploration and invention? (p. 375)

In this chapter we address this challenge by first highlighting two studies from the literature on the role of telling in teaching. Then we describe our research in which we developed two teaching strategies for navigating the all discovery — all telling continuum. Finally, we present two different examples of these teaching strategies and highlight how they can be used to engage students' current thinking while moving forward the mathematical agenda (Rasmussen & Marrongelle, 2006). For each example we offer commentary on the specific teacher actions that we believe help teachers move the class forward and navigate the all discovery — all telling continuum.

Navigating Along the All Discovery–All Telling Continuum: Lessons from Research

The previous quotations provide evidence that teachers struggle with the action of telling. We have noticed in our own teaching and in our work with other university mathematics faculty a tendency to make radical changes in practice, such as avoiding telling students information and abandoning many past practices. Effective and lasting shifts in teaching practice, however, rarely involve such drastic changes. Certainly the teacher action of telling has been downplayed as transmission models of teaching give way to models influenced by contemporary views of learning (Lobato, Clarke, & Ellis, 2005; National Council of Teachers of Mathematics, 2000). Research into mathematics teachers' struggles, on the other hand, provide insight into the more nuanced changes that teachers make as they move along a continuum from all student discovery to all teacher telling (Chazan & Ball, 1999; Fennema & Nelson, 1997; Heaton, 2000; Herbst, 2003; Lampert, 2001; Leinhardt & Steele, 2005; Lobato et al., 2005; Nelson, 2001).

In order to promote purposeful movement along the all discovery—all telling continuum, we explicitly name two teaching strategies: transformational record and generative alternative. By naming these teaching strategies, and thus making the work of teaching more explicit, teachers can begin to identify and plan to use such strategies in their instruction.

We begin by examining the study conducted by Chazan and Ball (1999) that examines a teacher's struggles with how to intervene when a class discussion begins to move in the direction of a debate void of reflection on important mathematical ideas. We juxtapose this research with a review of Lobato, Clarke, and Ellis's (2005) recasting of the telling action of teaching while keeping student thinking at the forefront of teachers' decision making.

Chazan and Ball (1999) analyze a case study of a high school algebra class during a lesson in which students debated how to average pay bonuses across 10 people when one person received a bonus of $0. The authors use this case study to illustrate problems with characterizations of teaching that emphasize what teachers ought *not* to do, in particular that teachers ought not to tell. In the case under study, students engaged in a debate about whether or not to include the $0 entry in the calculation of the average. One student suggested calculating the average by adding up the values and dividing by 9 (ignoring the person who received the $0 bonus). Another student argued – by analogy to computing grade averages – that the $0 must count in the computation of the average because to do otherwise would ignore a person. Other students chimed into the discussion agreeing either that you included or didn't include the $0 in the calculation of the average. The teacher characterized his feelings about this discussion as follows:

> I was enjoying the discussion and appreciating students' engagement, when I began to grow uneasy. I wondered about where the class would go with the disagreement over the zero. Now that the views had been presented, would students be willing to reflect on their own views and change them or would each argue relentlessly for his or her own view? Would they be able to come to some way to decide whether these averages were correct? (p. 3)

In other words, the teacher was worried that the discussion was degenerating into simple position taking, rather than leading to a resolution for why a particular approach was mathematically appropriate. Thus the teacher was faced with a decision about what to do next in the class — allow the discussion to continue when it was unlikely that a mathematically grounded solution would result (one extreme of the all discovery — all telling continuum) or tell the students how to compute the average (the other extreme of the all discovery — all telling continuum). Rather than position himself on one or the other extreme, the teacher positioned himself in the middle of this continuum and redirected the discussion to the interpretation or meaning of $500 or $555.56 (the results obtained by calculating the average using 10 and nine people, respectively). A *redirect* is an example of a teaching strategy that takes charge of the mathematical agenda by moving students away from a calculation to the fundamental mathematical issue under discussion.

The teacher's dilemma of how much to tell, what to tell, and when to tell is salient in the previous example. Other researchers have taken up the particular strategy of telling in light of contemporary theories of learning and the emphasis placed on non-telling strategies by reform-based policy documents. Typical conceptions of telling involve teacher telling as a one-directional action: the teacher tells and the student listens and tries to make sense of the information. Such typical ideas about telling tend to ignore the role of the student in the learning process. In particular, conceptualizing telling as simply a one-directional action fails to take into account the manner in which students interpret teacher utterances. Many teachers have experienced instances where they have told students some bit of information (be it directions about a class activity, a fact or definition, or an explanation of a process to solve a problem) and students didn't "get" it. One way to interpret students not "getting" what a teacher told them is that we cannot assume students understand what teachers tell them in the way that teachers might anticipate.

Lobato, Clarke, and Ellis (2005) offer an alternative way to think about telling in which the teacher follows up her/his utterances with other actions designed to make explicit or further students' thinking. These researchers start from the premise that teachers have experience teaching students by telling or lecturing — that is they already know *how* to tell — but struggle with *when* to tell, *what* to tell, and *why* to tell (p. 109). By focusing on the role or function of telling, rather than telling as a form of instruction itself, the researchers describe a 'reformulation' of telling that is fundamentally concerned with student interpretation or understanding of the ideas being told. This is similar to Heaton's (2000) revelation that telling has a place in new forms of teaching.

Perhaps showing or telling did have a place in this teaching. I could decide when and why to do it. The telling I did…was a move in response to a child's understanding. This is quite different from the kind of telling I did in my past practice, independent of students' understanding (p. 64).

Lobato, Clarke, and Ellis describe a sequence of teaching actions in which teacher utterances or telling of information are followed up with (and often preceded by) instances of the teacher soliciting student thinking. The authors detail the following teacher actions, all of which are strategies that teachers can use to navigate along the all discovery — all telling continuum:

- describing a new concept,
- summarizing students' ideas in order to provide a new idea,
- providing information for students to test their ideas,
- presenting a counterexample,
- presenting work from another (possibly hypothetical) student,
- engaging in Socratic questioning, and
- presenting a new representation.

For example, the authors present an episode in which the teacher describes a new concept in order to support a student's learning of division as partitioning. In this particular episode, a student was working on a problem in which she was asked to find the rate at which a faucet was leaking given that 16 oz. of water collected in 24 minutes. The student attempted to answer this question by dividing 24 by 16, resulting in an answer of 1.5, which she could not appropriately interpret. Based on prior experiences with this student, the teacher attempted to assess and move forward this student's thinking about the connection between division and partitioning by posing a simpler problem (How far does a duck travel in 1 second given that the duck travels 7cm in 3 seconds?). The student correctly computed the distance traveled in 1 second as 7/3cm, dividing both 7cm and 3 seconds by 3. The teacher then asked the student to partition a line segment representation of the event to represent her calculation (the student had previously used line segment representations in her work). The student proceeded to partition the line segment into seven equal parts instead of three equal parts and located her 'answer' of 7/3 by estimating on the portioned line segment. In response to this student's action, the teacher showed the student her own representation of the connection between division by

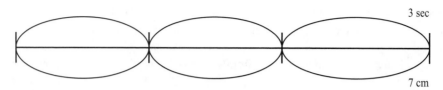

Figure 1. Teacher's representation of the connection between division by 3 and portioning into three equal pieces.

three and partitioning the segment (see Figure 1). In her explanation of her diagram, the teacher said, "One thing that dividing by 3 does is split this number line or drawing into three equal parts" (p. 119). The teacher then asked the student some questions to connect her previously calculated answer of 7/3 to the teacher's representation.

The authors point out that the teacher did not "provide [the student] with a step-by-step method for determining which number to divide" (p. 119), rather, she described an idea: the connection between division and partitioning. Additionally, rather than assume that the student constructed the same meaning of the representation as the teacher, she posed a follow-up task to ascertain the student's thinking. As it turned out, the student was not able to successfully complete the task, indicating that she did not construct an understanding as hoped for by the teacher. We stress that the follow-up task probing the student's interpretation of the teacher's utterances was equally as important as the teacher's utterances themselves.

Rethinking the role of telling in teachers' repertoires of teaching actions is an important line of inquiry in mathematics education research. Too often, reform-based teaching recommendations are interpreted as implying that acceptable teacher actions are only those toward the all student discovery end of the continuum. We do not concur with such interpretations. Teachers naturally struggle with questions of telling, both in terms of breaking away from traditional conceptions of telling and resolving new ideas of teaching in terms of new conceptions about telling.

Rethinking the role of teaching builds on contemporary views of learning, which emphasize that learning is a process involving constant interaction between the learner and her environment (Blumer, 1969; von Glaserfeld, 1995). Such perspectives on learning recognize that learning entails a process of becoming a member of a mathematical community — that is, developing ways of communicating, reasoning, and providing arguments to defend ideas — and a process of active individual participation (Cobb & Bauersfeld, 1995; Cobb & Yackel, 1996). In particular, we take symbolizing and participating in argumentation as learning and as such developed teaching strategies to promote symbolizing and argumentation.

Developing Teaching Strategies for Navigating the Continuum

As teachers of undergraduate mathematics, we are keenly aware of the tension between, on the one hand, fostering a classroom environment in which students' ideas are valued and supported, and on the other hand, moving forward students' thinking. Our work as teachers of undergraduate mathematics and researchers of undergraduate mathematics learning and teaching provides us with opportunities to carefully investigate the types of teacher interventions that build on students' ideas and further students' mathematical reasoning. In this section we describe a research project in which we identified two strategies that teachers use to connect to student thinking while moving forward their mathematical agenda; in other words, strategies for navigating the all discovery–all telling continuum.

The basis for the teaching strategies we describe below is rooted in Freudenthal's (1991) instructional design theory of Realistic Mathematics Education (RME). Part of the intent of RME is to offer heuristics to guide the creation of activities in which students develop important mathematical ideas and methods by solving a series of interesting problems. This intention is captured in Freudenthal's (1991) adage that, first and foremost, mathematics is a human activity. Freudenthal's adage has mostly been used to refer to how students should experience mathematics learning. We think that this adage applies equally well to the role of teachers. The point we want to bring out is that an important part of mathematics teaching is responding to student activity, listening to student activity, notating student activity, learning from student activity, etc. In this sense, mathematics teaching is a human activity about human (i.e., student) activity (Rasmussen & Marrongelle, 2006).

The research we report here is part of a larger design research project in differential equations. Design research consists of a cyclical process of designing instructional activities and researching the learning and teaching that takes place in conjunction with classroom implementation of the activities. An important feature of this cyclical process is the capacity to generate, test, and refine hypotheses about learning and teaching. In our design research project, we worked with a variety of undergraduate differential equations teachers who were implementing course materials inspired by the instructional design theory of RME. Instruction in these classes generally followed an inquiry approach in which important mathematical ideas and methods emerged from students' problem solving activities and discussions about their mathematical thinking. As such, these classes provided an appropriate setting to examine strategies teachers used as they navigated the all discovery — all telling continuum. The data we collected included: video-recordings of each class session, copies of students' written work, video-recorded interviews with individual students, and audio-recorded weekly project meetings that included the classroom teacher and at least one other researcher who attended each class session.

During our review of classroom videorecordings and audiorecordings of project meetings, we developed a hypothesis about two types of strategies the differential equations teachers used to navigate the all discovery — all telling continuum. One strategy we identified was based on our observation that teachers used notations, diagrams, or other graphical representations initially to record student thinking and then later students used the teacher's record to solve new problems. We call this strategy a *transformational record* (Rasmussen & Marrongelle, 2006), as the teacher's record of students' thinking is transformed by students as a means for further reasoning.

A second strategy we identified was based on our observation that teachers introduced alternate symbolic expressions or graphical representations for the purpose of promoting student explanation and justifications for the validity of these alternatives. We call this strategy a *generative alternative* (Rasmussen & Marrongelle, 2006), as the alternative representation offered by the teacher generates students' explanations and justifications and allows for progress to be made through logical reasoning.

We identified occurrences of the transformational record strategy when (1) some form of notation (typically informal or unconventional notation) was either used by a student in whole class discussion or introduced by the teacher to record or notate student reasoning and (2) this notational record was then used by students in achieving subsequent mathematical goals. We also developed two criteria for identifying episodes of generative alternatives. Students in the differential equations classes were often asked to make graphical or symbolic predictions, and on numerous occasions the teacher invited students to consider alternatives, either student-generated or invented by the teacher, to one or more of their predictions. We coded such episodes as examples of generative alternatives when the alternatives functioned to (1) contribute to the classroom expectation that students provide explanations for their responses and (2) elicit or generate justifications for why students believed particular graphs or symbolic expressions to be mathematically correct or incorrect. That is, these alternatives generated explanations and justifications[1].

In the remainder of this chapter, we illustrate the transformational record strategy with an example from one of our differential equations project classes and we illustrate the strategy of generative alternative from a senior capstone course for prospective secondary mathematics teachers taught by the second author. We encourage readers to imagine situations in which they could use either type of strategy in other content areas.

Transformational Record: Developing and Using Slope Fields

As the person who knows the discipline, a teacher has the obligation of enculturating students into the language and conventional representations of the broader community while at the same time honoring and building on student contributions. In other words, a teacher frequently needs to navigate the all discovery — all telling continuum in her work for the purpose of ensuring that students are fluent in mathematics vocabulary and conventions. Similar to Lobato et al.'s (2005) claim that teacher telling needs to be followed up with tasks to ascertain students' thinking, we demonstrate how the results of teacher telling (in this case the introduction of a mathematical convention) can be used by students to further their thinking about mathematical ideas. However, instead of introducing the mathematical idea and then posing a task to ascertain students' thinking, the teacher in this example initiates a class discussion around one student's idea and then, when he feels the class has reached consensus, introduces a conventional mathematical notation to fit the class's idea. Students demonstrate that the mathematical convention fits with their idea when they use the notation to further their thinking.

---

[1] The distinction between explanation and justification can be subtle. Explanation includes describing how one solved a problem, describing one's thinking, or clarifying another person's solution. Justification involves providing reasons why a solution is correct or incorrect or why an argument is valid or invalid.

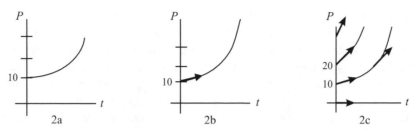

Figure 2. Population versus time graphs and records of student reasoning about rate.

The example we discuss occurred on the second day of a differential equations class in which students were asked to make predictions about the shape of a population versus time graph for a single species of fish that reproduces continuously and has unlimited resources. No differential equation was provided. Students worked on the task in small groups, after which the instructor led a whole-class discussion of students' responses to the task. The typical response to this task was an exponential or quadratic-like shaped graph (see Figure 2a) positioned above the t-axis.

The first topic of conversation initiated by the teacher about students' graphical predictions was whether or not the initial slope at $P = 10$ and $t = 0$ should be zero or have a positive value. The first student to speak up on this issue argued that the slope should be positive. The teacher received a "correct" response and could have moved on with the lesson, exercising a move situated more toward the all telling end of the continuum. However, he led a whole class discussion about the initial slope issue that lasted over eight minutes representing a slide along the all discovery — all telling continuum.

The fact that the teacher led a discussion of a student's initial response is significant for two reasons. First, it allowed for students to express alternative viewpoints and in the process important mathematical issues and interpretation were discussed, such as exponential growth, the existence of the fish pond prior to time $t = 0$, and the meaning of continuous reproduction. Second, it enabled other students in the class to take ownership of the positive slope idea. This is important because the research literature suggests that when students take ownership of mathematical ideas, they are more likely to develop deep understandings of them rather than think of them as memorized facts (c.f., NCTM, 2000).

Thus, at the end of the eight minute discussion, the teacher drew a tangent vector with positive slope at the point where $P = 10$ and time $t = 0$ (see Figure 2b), and his tangent vector served the function of recording the reasoning that emerged from the ideas put forth by students during class discussion. The teacher's move along the continuum toward the all discovery end, evidenced by his choice to give students the opportunity to discuss their ideas rather than immediately draw a tangent vector on the chalkboard, led to an idea that was more representative of the whole class's thinking, rather than the thinking of one student. In this case, the teacher's proactive role in the notating process was far more complex than simply providing a record that fit with one student's thinking for the purpose of introducing a particular conventional mathematical representation.

Subsequent topics of conversation about students' graphical predictions for population over time included how the rate of change at this initial point compared to the rate of change at a later time, how (and why) the rate of change would compare to the other rates if the initial population was greater than what was originally sketched (e.g., if at $t = 0$, $P = 20$), and what the rate of change would be if the population at time $t = 0$ was $P = 0$. As was the case with the teacher's initial tangent vector record of students' reasoning, these additional conversations provided further opportunities for the teacher to continue to record students' reasoning with additional tangent vectors, such as that shown in Figure 2c. We refer to Figure 2c as an *emerging tangent vector field* because it is a consequence of classroom discourse and it is beginning to resemble what an expert in differential equations would recognize as a slope field (or tangent vector field).

The class discussions surrounding students' graphical prediction for the population versus time scenario and the teacher's resulting records of this reasoning resulted in important consequences. One important consequence is that they provided an opportunity for the teacher to record students' reasoning in a way in which the conventional inscription of a slope field began to take form. Thus, the teacher was able to connect conventional mathematical ideas with student thinking. The point we want to make next is that the slope field, which was initially the teacher's record of students' reasoning, subsequently became a means for students to reason about the symbolic form of the rate of change equation. Thus, not only did students demonstrate their understanding of the mathematical ideas behind the teacher's record, but a transformation in the teacher's record took place as well.

A critical idea that facilitated this transformation emerged out of students' arguments for why, if the initial population is 20 at time $t = 0$, the initial slope would be the same as the slope directly to the right on the 10-curve (the solution curve starting at $t = 0$, $P = 10$; see Figure 2c). Students' reasons for this horizontal invariance in slope relied on their imagery of the scenario. The basic argument put forth by the class was that it didn't matter whether you called time zero Wednesday, Friday, or Labor Day, the population of fish would be increasing at the same rate for a given initial population. What mattered was the number of fish; what you decided to call time zero is arbitrary. As shown in Figure 2c, this line of reasoning was recorded by the teacher with two different tangent vectors that had the same slope at $P = 20$ but at two different t-values.

All the mathematical work up to this point occurred without the symbolic expression for the differential equation. Developing the symbolic expression was the next task in the sequence of instructional activities. The task began with the teacher inviting students to consider whether the rate of change, $\frac{dP}{dt}$, should depend explicitly on just the population P, just time t, or on both P and t. That is, if $\frac{dP}{dt} =$ 'something', what should the "something" consist of? Should it contain just P, just t, both P and t? Conceptually, this tends to be a challenging task for students for two reasons. First, students need to explicitly distinguish between the rate of change of a quantity and the quantity itself. Second, P stands for both an unknown function and a variable in the rate of change equation. Reasoning about what the explicit variables are in a rate of change equation involves conceptualizing rate as a function, which is cognitively more complex than conceptualizing rate of change as the slope of the tangent line at a point (Rasmussen, 2001).

During the whole class discussion, one student, Bill, pointed to the emerging tangent vector field (which is still on the chalkboard) to support his argument. In other words, the teacher's previous record of student reasoning shifted function and served as a means for students to reason about why the rate of change equation should depend explicitly on just P.

Bill: Ok. We're trying to find what the rate of change is. This differential should tell me the rate of change. That's the question. The something that is the right side of this, uh, the graph, or the right side of the [rate of change] equation. When we looked at our [P vs. t] graphs, we all agreed that, when the population reaches a certain size, all the rates of change are going to be the same. Doesn't matter what time they reach that, that change.

Notice that Bill's argument relies on the previous conclusion that "we all agreed that, when the population reaches a certain size, all the rates of change are going to be the same." This statement is significant because horizontal invariance of slopes now becomes the basis from which Bill argues that the rate of change equation should only depend on P, supporting our claim that the teacher's initial record is transformed into a means for reasoning about a different mathematical idea. As Bill continues to discuss his reasoning, the instructor, without comment, adds additional tangent vectors to the graph on the board, as illustrated in Figure 3. These additional markings serve to record Bill's thinking and to further develop the emerging tangent vector field into a slope field.

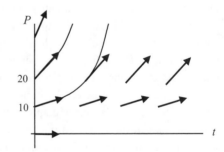

Figure 3. Emerging tangent vector field

As Bill articulates in his argument, the idea and record that for a set population value the rate of change is invariant across time, is transformed into a means for reasoning that the differential equation should explicitly depend only on P and not on time t.

The teacher's proactive role in initiating and refining the tangent vector field record of student reasoning also included making important decisions about when to withhold making a record of student reasoning. As illustrated in this example, the teacher initially chose not to make a tangent vector record of the reasoning expressed by the first student. This created an occasion for the classroom community to develop a line of reasoning for which the emerging tangent vector field was a fit. This emerging tangent vector field then served as a means for students to reason about what the differential equation should explicitly depend on. We propose that the students likely would not be able to reason in the ways that they did when they were attempting to develop the symbolic form of the differential equation if the teacher did not allow adequate discussion time before drawing the tangent vector on the chalkboard. His decision to allow for adequate discussion time stemmed from his concern that more than just one or two students think deeply about the mathematical ideas and from his desire to foster a classroom environment in which students routinely explained and justified their thinking.

Generative Alternative: Opportunities for Justification that Clarify Underlying Concepts

Sometimes the tension that occurs as a teacher navigates the all discovery — all telling continuum arises from obvious classroom situations. For example, the teacher in the study by Chazan and Ball (1999) faced a palpable dilemma in his classroom as he deliberated on how to ensure that students were discussing mathematically significant ideas while at the same time supporting students' engagement in the class discussion. Other times, the teacher must leverage her judgment and experience when making instructional decisions with less than obvious data. For instance, in the previous example, a student responded to the differential equations teacher's first question with a correct response. The teacher's strategic decision to hold back from drawing a tangent vector resulted in more students taking ownership of the concepts underlying the creation of a slope field. Consequently, students were positioned to provide sound arguments for what the symbolic form of the differential equation would look like. In this next example, we further highlight the proactive role of the teacher in connecting student thinking to the intended mathematical ideas.

This next example, which illustrates the strategy of generative alternative, occurred during a unit on function in a course for prospective high school mathematics teachers. Both teaching strategies of transformational record and generative alternative are grounded in the research literature. Both strategies were derived from work in differential equations (Rasmussen & Marrongelle, 2006), but the power of such research is that these strategies can be used by the reader in his or her own classroom in content areas other than differential equations. To illustrate this point, we use an example from a class other than differential equations in which the teacher was consciously aware of the generative alternative strategy and used the strategy intentionally to create a situation which pressed students for justification. Thus, although the example is not from the published research literature, it is grounded in the ideas from the literature.

Students were given the Bottle Problem (Oehrtman, Carlson, & Thompson, this volume), as shown in Figure 4. The teacher used the problem to highlight the covariational nature of functions and to emphasize that one does not necessarily have to have an algebraic rule to discuss a functional relationship.

Students worked on the problem in class and through cycles of small group work and whole-class discussions. During these sessions, the students gave sound reasons explaining why the portion of the graph (up to the beginning of the neck) would look similar to that shown in Figure 5. Most students also reasoned appropriately that the graph corresponding to the neck of the bottle would be linear. Based on prior research with this problem (Carlson, Jacobs, Coe, Larsen, & Hsu, 2002), the teacher anticipated that students would find it a challenge to determine the slope of the graph that depicts the transition from the round portion of the bottle to the neck of the bottle. Indeed, many students had not carefully considered the transition point at the beginning of the neck of the bottle. The teacher initiated three options: graph B (linear graph with same slope at the transition point), graph C (linear graph that is steeper than the slope at the transition point), and graph A (linear graph with slope less than that at the transition point) as shown in Figure 5. Students' assignment for the next class was to write up a convincing argument for or against each alternative.

By providing different graphs to consider, the teacher offered an occasion for students to provide explanations and justification for why they favored one option over another. They had to listen to their classmates' justifications and give reasons why they agreed or disagreed. The invitation for the students to consider different graphs opened up

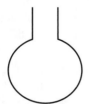

Imagine this bottle filling with water. Sketch a graph of the height of water as a function of the amount of water that is in the bottle.

Figure 4. The bottle problem.

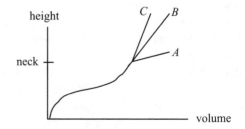

Figure 5. Alternative graphs to consider.

an opportunity for them to further explore relationships between the diameter of the neck of the bottle and the rate of change of the height of the water with respect to volume, as the bottle fills up.

We point out that the teacher assumed more responsibility for the direction of the mathematical content when he directed the class to consider the three possible graphs. The teacher's actions here are significant because he did not have to wait for students to offer diverse ideas; rather he was able to draw upon his content expertise to further question students' thinking. This is consistent with the notion that teachers' mathematics expertise is an essential aspect of making pedagogical decisions (Ball, Lubienski, & Mewborn, 2001; Fennema & Franke, 1992).

In response to this task, one student made connections to the formula for the volume of a cylinder in his response. In the quote below, the student indicated that v_2 represents the volume of water exactly at the neck of the bottle transition point and v_f stands for the final volume of water when the bottle is completely filled. Similarly, h_2 and h_f stand for the respective height of water at the respective locations.

> Now as the volume increases at a constant rate from v_2 to v_f, the height will grow at a constant rate as well. More specifically, $V = \pi r^2 h \Rightarrow h = \frac{1}{\pi r^2} \cdot V$. This begins at height h_2 [the neck of the bottle], so we can see from the function $h(v)$ above that h will grow at a constant rate of $\frac{1}{\pi r^2} \cdot V$. Since πr^2 stays constant from h_2 to h_f, height CANNOT accelerate or decelerate its growth rate, and therefore from h_2 on, the slope is constant.

This student reasoned within the context of the bottle problem, using his knowledge of the volume formula for a cylinder. We argue that because he had to consider multiple graphs, this student broadened his reasoning to include connections between graphical and symbolic representations of functions.

This task also provided an opportunity for the teacher to gain insight into student thinking that he may not have been privy to otherwise. Consider the following response to the Bottle Problem task from another student who argues for Choice C:

> I would pick (C) to be the extension of the graph because once the water is being poured at a constant rate at that point the volume is going to increase since that portion of the bottle is the thinnest, therefore, the water will overflow at a faster rate.

We see from this student's response that he is correct in reasoning that the volume is going to increase and that the height of the water in a thin bottle will increase at a faster rate than a stouter bottle ("the water will overflow at a faster rate"). However, this student did not seem to make a connection between the diameter of the bottle at the "top" of the spherical portion and the diameter of the cylindrical portion.

The teacher's proactive role in providing three graph options for the Bottle Problem provided opportunities for students to further their reasoning about rate of change and make connections between graphical and symbolic function representations. For some students, the opportunity to provide arguments for or against the graphical choices gave the teacher insight into their thinking that otherwise might not have been exposed.

Conclusion

We illustrated two strategic decisions by teachers in the previous examples. The differential equations teacher led a whole class discussion about slope before he recorded the students' thinking. His decision to lead a whole-class discussion about slope enabled his vector notation to emerge as a record of students' collective thinking. We refer to his strategic move as a *transformational record* (Rasmussen & Marrongelle, 2006). Transformational records can support a teacher's proactive role in furthering students' mathematical reasoning in ways that are increasingly compatible with the reasoning and symbolizing of the broader mathematical community.

The teacher of the capstone class introduced the class to three possible graphs of the height of the water versus the volume for the bottle's neck. We refer to his strategic move as a *generative alternative* (Rasmussen & Marrongelle, 2006). The generative alternative construct serves the dual function of furthering students' mathematical reasoning and contributing to the ongoing constitution of the norms for explaining and justifying one's thinking, listening to and attempting to make sense of others' thinking, and responding to challenges and questions. In this second example, the generative alternative offered an occasion for students to provide explanations and justification for why they favored one option over another and how they made sense of the connections between rate of change and the diameter of the bottle.

Some readers may have recognized features of their own teaching in the examples presented. We do not claim that the *ideas* of recording students' ideas with informal or formal notation and presenting students with alternative examples are novel; master teachers have been practicing these strategies for years. Rather, the innovation in transformational record and generative alternative is that by giving these strategies names, we are making the work of teaching more explicit. Thus, a teacher can begin to name the strategies in her/his repertoire and thoughtfully and consciously plan for the use of such strategies. This has significant impact for the work of teaching, impact that goes beyond what transpires in the classroom. For example, as a teacher plans her lesson, she might devise alternatives to have on hand in the case where students do not bring out a particular mathematical idea. Alternatively, as a teacher hypothesizes about potential student misconceptions that might surface during a lesson, he can leverage the tool of generative alternative to surface or address misconceptions.

The strategies of transformational record and generative alternative offer teachers a way to think about how she or he can slide along the all discovery—all telling continuum. In the two examples provided, we see the teacher act proactively to provide information, promote student reasoning, and offer occasions for students to explain and justify. Through such actions, teachers are able to engage students' current thinking while moving forward the mathematical agenda. We invite readers to reflect on how they could use these two teaching strategies in an intentional way that allows them to express their expertise, but in a way that mediates between all lecture (or teacher telling) and all student discovery.

References

Ball, D. L. (1993). With an eye on the mathematical horizon: Dilemmas of teaching elementary school mathematics. *Elementary School Journal, 93*(4), 373–397.

Ball, D. L., Lubienski, S. T., & Mewborn, D. (2001). Research on teaching mathematics: The unsolved problem of teachers' mathematical knowledge. In V. Richardson (Ed.), *Handbook of research on teaching* (4th ed., pp. 433–456). New York: Macmillan.

Blumer, H. (1969). *Symbolic interactionism.* Englewood Cliffs, NJ: Prentice-Hall.

Carlson, M. P., Jacobs, S., Coe, E., Larsen, S., & Hsu, E. (2002). Applying covariational reasoning while modeling dynamic events: A framework and a study. *Journal for Research in Mathematics Education, 33*(5), 352–378.

Chazan, D., & Ball, D. (1999). Beyond being told not to tell. *For the Learning of Mathematics, 19*(2), 2–10.

Cobb, P., & Bauersfeld, H. (1995). *The emergence of mathematical meaning.* Hillsdale, NJ: Erlbaum.

Cobb, P., & Yackel, E. (1996). Constructivist, emergent, and sociocultural perspectives in the context of developmental research. *Educational Psychologist, 31*, 175–190.

Fennema, E., & Franke, M. L. (1992). Teachers' knowledge and its impact. In D. A. Grouws (Ed.), *Handbook of research on mathematics teaching and learning* (pp. 147–164). New York: Macmillan.

Fennema, E., & Nelson, B. S. (1997). *Mathematics teachers in transition.* Hillsdale, NJ: Lawrence Erlbaum.

Freudenthal, H. (1991). *Revisiting mathematics education, China lectures.* Dordrecht, The Netherlands: Kluwer Academic Publishers.

Heaton, R. (2000). *Teaching mathematics to the new standards: Relearning the dance.* New York: Teachers College Press.

Herbst, P. G. (2003). Using novel tasks in teaching mathematics: Three tensions affecting the work of the teacher. *American Educational Research Journal, 40*(1), 197–238.

Holton, D. (Ed.). (2001). *The teaching and learning of mathematics at the university level* (Vol. 7). Dordrecht, The Netherlands: Kluwer Academic Publishers.

Lampert, M. (2001). *Teaching problems and the problems of teaching.* New Haven, CT: Yale University Press.

Leinhardt, G., & Steele, M. D. (2005). Seeing the complexity of standing to the side: Instructional dialogues. *Cognition and Instruction, 23*(1), 87–163.

Lobato, J., Clarke, D., & Ellis, A. (2005). Initiating and eliciting in teaching: A reformulation of telling. *Journal for Research in Mathematics Education, 36*(2), 101–136.

National Council of Teachers of Mathematics. (2000). *Principles and standards for school mathematics*. Reston, VA: The Council.

National Science Foundation. (1996). *Shaping the future: New expectations for undergraduate education in science, mathematics, engineering, and technology*. Arlington, VA: National Science Foundation.

Nelson, B. S. (2001). Constructing facilitative teaching. In J. Warfield (Ed.), *Beyond classical pedagogy* (pp. 251–273). Mahwah, NJ: Lawrence Erlbaum.

Rasmussen, C. (2001). New directions in differential equations: A framework for interpreting students' understandings and difficulties. *Journal of Mathematical Behavior, 20*, 55–87.

Rasmussen, C. L., & Marrongelle, K. A. (2006). Pedagogical content tools: Integrating student reasoning and mathematics in instruction. *Journal for Research in Mathematics Education 37*(5), 388–420.

Seymour, E., & Hewitt, N. (1997). *Talking about leaving: Why undergraduates leave the sciences*. Boulder, CO: Westview Press.

von Glasersfeld, E. (1995). *Radical constructivism: A way of knowing and learning*. Bristol, PA: The Falmer Press.

Acknowledgement. Support for this paper was funded in part by the National Science Foundation under grant No. REC-9875388. The opinions expressed do not necessarily reflect the views of the foundation.

14

Examining Interaction Patterns in College-Level Mathematics Classes: A Case Study

Susan Nickerson and Janet Bowers
San Diego State University

While discussing the pedagogical challenges of teaching an undergraduate discrete math course, one of our colleagues recently lamented that

> Students are ill-prepared for this course…but this ill-preparation is a curious issue. I think it has more to do with the way they learned mathematics than with the content of the previous courses.

In this chapter, we propose a response to his comment. In particular, the goal of our discussion is to illustrate that the ways in which teachers and students interact can profoundly affect the attitudes students form *as well as* the content they learn.

Why Study Interaction Patterns?

This view of the importance of interaction styles is consistent with a conclusion reached by Stigler and Hiebert (2004) regarding their recent international study of teaching patterns across the world:

> A focus on teaching must avoid the temptation to consider only the superficial aspects of teaching: the organization, tools, curriculum, content, and textbooks. The cultural activity of teaching — the ways in which the teacher and students interact about the subject — can be more powerful than the curriculum materials that teachers use. … We must find a way to change not just individual teachers, but the culture of teaching itself. (p. 16)

In short, not just *what* we teach, but *how* we teach and communicate with students — what we call interaction patterns — appears to have great influence on student learning. In this chapter, we describe several interaction patterns that educational researchers have identified. These constructs can be seen as analogous to the measures of central tendency that statisticians use to describe the general shape of a data distribution in that the constructs don't completely describe the classroom culture but still provide information about the 'shape.' We then illustrate how two novel interaction patterns that emerged in one college mathematics classroom supported not only students' strong conceptual development, but their development of a revised view of the nature of mathematics itself.

Past Studies of Classroom Interaction Patterns

Many prominent theories in educational research are based on the view that learning is an inherently social process (cf., Cobb, 1996; Lave, 1997; Vygotsky, 1978; Au, 1993; Arcavi, Kessel, Meira, & Smith, 1998; Boaler, 2000; Cazden,

1988; Puro & Bloome, 1987; Stephan & Rasmussen, 2002; Wood & Turner-Vorbeck, 2001). Therefore, examining the ways in which various educative discourses affect student learning is critically important. One of the pioneers in the study of classroom dynamics was Mehan who analyzed the patterns used during classroom interaction from the point of view of the function that it played during a lesson (Mehan, 1979). His research, which has since been confirmed by many other studies (c.f., Dillon, 1990; Medina, 2001), revealed that the predominant pattern found in most classrooms involves three distinct moves: First, a teacher *initiates* a question, next, a student *responds*, and finally there is an *evaluative* interaction. As Medina (2001) describes, Mehan went on to pose four distinct types of questions:

1) Choices, which are those that dictate the student agree or disagree with a statement provided by the teacher;

2) Products, which require students to provide factual responses;

3) Processes, which call for students' opinions or interpretations; and

4) Metaprocesses, which are elicitation questions that ask students to reflect upon the process of making connections between a question and a response to formulate the grounds of their reasoning.

These four sub-categories enable researchers to further distinguish the nature of discussions. For example, researchers have documented that in the majority of mathematics classrooms observed, teachers incorporated either choice or product elicitations. The distinction between these first two elicitations and the second two (process elicitations and metaprocess elicitations), can be seen as parallel to the distinction between calculationally-oriented teachers and conceptually-oriented teachers developed by Thompson, Philipp, Thompson, and Boyd (1994). These researchers claimed that there was a direct correlation between different teachers' views of mathematics and the types of interaction patterns that emerge in their classrooms. They distinguished between calculationally-oriented teachers, who expect their students to offer explanations in the form of procedural descriptions, and conceptually-oriented teachers, who tend to encourage rich mathematical discourse. Applying Mehan's four sub-categories to these two distinctions, it follows that calculationally-oriented teachers often follow the Initiate, Respond, and Evaluate (IRE) pattern and utilize either choice or product elicitations. In contrast, conceptually-oriented teachers often use process or metaprocess elicitations. For example, Thompson et al. report that during rich exchanges in which real mathematical proofs and refutations are offered (c.f., Lampert, 1990; Lakatos, 1976), students offer explanations that go beyond procedural descriptions of steps taken to compute an answer. In other words, whereas the patterns of interaction in the classroom of a calculationally-oriented teacher's classroom tend to focus on the means of computing an answer, the interaction patterns that emerge in conceptually-oriented teachers' classrooms begin with answers, but also include justifications for the answer or explanations of the process couched in contextually-relevant ways. The value of such interaction patterns is that they encourage students' serious justification of their thinking and hence can support students' efforts to develop increasingly sophisticated ways of reasoning.

Other researchers have documented other types of interaction patterns that rely on recall rather than student-initiated reasoning. For example, Wood (1994) described the *funneling* and *focusing* patterns, both of which aim to increase student participation in the development of an answer from mere recall and evaluation (as in the IRE patterns) to more detailed explanations of how an answer could or should be computed. The funneling pattern involves having the teacher ask a series of directed questions designed to narrow the students' responses until the correct one emerges. The focusing pattern is similar in that the teacher again uses a series of narrowing questions, but they are designed to scaffold the student's effort to arrive at a final answer. One similarity between these two patterns and the IRE pattern is that the students' role is generally one of giving single-word answers or following a fill-in-the-blank script that the teacher has laid out for formulating a justification. However, unlike the IRE patterns, teachers' goals when using the funneling and focusing patterns are to use a series of probing questions to scaffold students' explanations.

In summary, the conclusions reached by Wood and her colleagues as well as many other discourse pattern research groups (c.f., Steinbring, 1989; Wood, 1999; Stigler & Hiebert, 2004) support Mehan's general assertion that the preponderance of discourse in most mathematics classrooms can be described as teacher-generated questions with a paucity of student-initiated comments. Thus, it seems logical to explore the degree to which this pattern (or any of its derivatives) is effective, or if there are other more effective patterns that can promote and support students' deep conceptual understandings.

The Relation between Interaction Patterns and Student Learning

Although Mehan's (1979) goal for analysis was to document discourse form and function, he did not attempt to make any type of evaluative judgment on the degree to which any particular type of teaching led to more productive learning. However, other researchers have suggested that the IRE patterns (especially those involving the first two types of elicitations) can actually be deleterious to student learning. Their arguments are based on the view that if students in class are not required to think on their own (beyond recitation of known information) or engage in serious justifications of their own thinking, they do not learn the art of argumentation itself, which is a cornerstone of "doing mathematics" (Cobb & Bowers, 1999; Lampert, 1990; Voigt, 1995). A second argument against the IRE pattern itself is that a teacher may interpret various students' participation in these ritual discourse patterns as evidence that they are making sense of the mathematics when, in reality, they are only learning to play academic charades.

A third theoretical argument regarding IRE patterns (at least choice and product elicitations) is that the overall importance of personalizing new knowledge may appear to be devalued. In other words, some contemporary learning theories maintain that learning is both a cognitive and social process. On the one hand, it is a cognitive process by which students "actively construct their ways of knowing as they strive to be effective by restoring coherence to the worlds of their personal experience" (Cobb, 1996, p. 34). Thus, from a cognitive perspective, learning and understanding are *constructed*. If students believe that they are expected to merely recall memorized facts, then they are acknowledging that they have not truly constructed understandings. On the other hand, learning is also viewed as a social process wherein "an individual's mathematical activity, for example, is profoundly influenced by his or her participation in encompassing cultural practices such as completing worksheets in school, shopping in a supermarket, selling candy on the street, and packing crates in a dairy" (Cobb, 1996, p. 34). Thus, from a social perspective, if learning is seen as the activity of filling in worksheets as opposed to personally constructing arguments, then the value of mathematical argumentation – and the personalization of knowledge — is implicitly devalued.

These theoretical arguments suggest that empirical evidence is critically needed to explore the relation between interaction patterns and student learning. In their recent work, Wood and her colleagues (Wood & McNeal, 2003; Wood & Turner-Vorbeck, 2001) have developed a two-dimensional framework that considers (1) students' responsibility for thinking and (2) students' responsibility for participation. Within this framework, the authors distinguish between Conventional Classroom Cultures and Reform Classroom Cultures. In Conventional Classroom Cultures, students are responsible for recalling answers and prescribed procedures. This is indicated by teacher prompts such as "What is the answer? Two plus three is _____" (Wood & McNeal, 2003).

In contrast, Wood and McNeal note that in Reform Class Cultures, students are responsible for recognizing, building, and constructing arguments. Teacher prompts for mathematical thinking include phrases such as

- "Any comments on the answer or method?"
- "Is there a different way that you could do this?"
- "How do you know that? Why do you think that?"

These empirical results illuminating the distinction between conventional and reform-based classrooms enabled the authors to draw some conclusions on the relation between the various cultures and students' mathematical thinking. Their analysis revealed that

> [t]he frequency and complexity of teacher prompts for mathematical thinking progressively increased across the types of class cultures with the conventional environment, both textbook and problem solving, being predominantly situations of prompts for recall for children… [In contrast] In inquiry/argument classes, teacher prompts for mathematical thinking were most frequent questions focused on synthesis building. (p. 439)

The work of Wood and her colleagues provides empirical evidence to support the claim that some interaction patterns *are* positively correlated with the emergence of deeper understandings and students' efforts to offer more conceptually-oriented thinking. In particular, while the interaction patterns found most predominantly in conventional classrooms are marked by questions requiring only fact recall (hence resemble IRE patterns), the patterns that are found in reform-based cultures (such as Strategy Reporting and Inquiry/Argument) are more highly correlated with sophisticated thinking as evidenced by the proofs and justifications students give when explaining answers and making conjectures.

Examining the Teacher's Role in Establishing Productive Interaction Patterns

Given the empirically-based correlation between interaction patterns and student learning, the next logical question to discuss is the role that the teacher can play in establishing "productive" patterns. Our view is that although the role of a teacher is certainly critical, the process by which a classroom community develops specific modes for inquiry is *not* one-way. In fact, we maintain that interaction patterns must be *mutually negotiated* through successive iterations of give-and-take between the teacher and students. [1] This mutual negotiation process is of particular interest as a subject to be included in this volume because, if college-level teachers can become aware of the implicit messages that students may be forming during classroom discussions, they might develop a better understanding of the views of mathematics their students are forming and, ultimately, address the situation our colleague was describing.

In the remainder of this chapter, we describe excerpts from a case study that illustrate the teacher's role in establishing productive interaction patterns. In particular, we describe how a conceptually-oriented teacher guided his class in the process of developing two patterns of interaction that represent distinct departures from the IRE and funneling and focusing modes.

Case Study

The motivation for this case study was to study the expertise of one mathematics teacher who was internationally known for both his mathematical theories and his pedagogical skills. At the university in which he worked, he was extremely well respected by students and colleagues alike. Hence, our goal for studying his practice was to answer a simply-stated research question: "How does he succeed in helping so many people understand and appreciate advanced mathematics in such a deep way?" We refined our questions more specifically to focus on two areas: (1) How does he use technology to enhance his students' imagery, and (2) How does he encourage students to form strong mathematical arguments? Thus, unlike large, quantitative studies that involve pre-determined hypotheses and many classrooms in order to minimize any one teacher's contributions, the goal of our case study was to examine this one particular teacher's expertise (see Bowers & Nickerson, 2001, for a full description of this study).

Case Study Methodology

In order to address our basic question of "How does he do it?" we chose to conduct a naturalistic inquiry (c.f., Moschkovich & Brenner, 2000) in which we, as researchers, were nonparticipating observers (and videographers) during each class session. Although we were clearly visible in the classroom, we did not participate in any discussions or interact with the students during class. After class, the teacher debriefed us in order to clarify our working hypotheses regarding the goals of his instruction. But, as the teacher has verified, we did not affect his instructional trajectory for ensuing classes in any significant way. To ensure truth value for the analysis, we observed all the classes during the semester and triangulated our hypotheses with each other and with the teacher. This is critically important as it was only through these triangulation discussions that our efforts to paint a picture of what we saw through empirical observation could be verified. During our analysis phase, we compared sets of field notes and videotapes and conferred with some of the participants via email or face-to-face discussion to ensure further validity (cf. Cobb & Whitenack, 1996).

Setting

The class that served as a setting for the study was a mathematics course for upper-division mathematics majors and graduate students that the teacher himself had designed and molded over the past nine years. The course was specifically designed to help prospective (and practicing) secondary mathematics teachers think about mathematics in a deeper way. In our discussions with the teacher both before and during the course, we learned the teacher had two goals. His mathematical goal was to enhance the students' images of the concept of functions. Because the students were future (and practicing) teachers, he also had a pedagogical goal of challenging the students' beliefs about effective teaching.

---

[1] We use the word "negotiation" here to denote a social process by which the students and teacher talk and react to each other's actions and words, as opposed to the connotation of two opposing parties bargaining or bartering for disparate agendas.

Phases of Analysis

In the first phase of our analysis, we collected three types of data: classroom observations and artifacts, student questionnaire data, and follow-up student interview questions. The classroom data included two sets of field notes from each of the 29 class sessions, videorecordings of 17 class sessions, and photocopies of students' work (including written reports, weekly homework assignments and reflection pieces, and written examinations). Although it may appear redundant, we chose to each take separate field notes of the whole-class interaction patterns for several reasons: First, the mathematics was sufficiently complex as to warrant close attention, and second, the discussion was often fast-paced and overlapping—thus, having two people record the action was often helpful in later interpretation phases. Approximately halfway through the semester, we asked participants to complete an anonymous written survey. We completed the data collection by surveying teachers who taught the students during the following semester. All of these types of data from varied perspectives served as a form of triangulation.

After running through the notes for a first interpretive pass, we engaged in a second phase of analysis wherein we decided that the construct of interaction patterns would enable us to best address our second question regarding how the teacher was able to encourage his students to form strong mathematical arguments. To this end, we developed a methodological framework to help us identify various patterns with consistency and reliability.

1. A pattern must be **inclusive** enough to characterize how all participants engaged in the discourse.
2. A pattern must be **descriptive** enough to be distinguished from other patterns previously defined.
3. A pattern must be **repeated** several times so that its robustness can be verified.

We then each set off with our own notes and copies of the videotapes to identify various patterns. Once these were initially selected, we met with each other and with the teacher to refine our constructs and triangulate our conclusions.

Findings

After conferring with the teacher a number of different times, we developed a strong sense of his mathematical goal, which was based on Thompson's (1994) contention that students' difficulties with the Fundamental Theorem of Calculus can be traced to their impoverished images of rate. More specifically, the teacher emphasized that students construct an understanding of a function as a relationship between covarying magnitudes, and algebraic expressions as descriptions of that relationship. He also emphasized that students think of graphs as sets of points (as opposed to lines or "wires" or moveable objects), so that the coordinates of each point are viewed as a record of the magnitude's value at any given moment during their simultaneous variation.

The teacher's pedagogical goal was to challenge his students' beliefs about effective teaching as being a practice of providing clear examples and rules to follow (cf., Ambrose, 2004; Ball, 1990; Richardson, 1996). His approach for reaching this goal was to model mathematical justifications that were conceptual rather than calculational in nature, and encourage his students to describe justifications of their thinking that went far beyond simply answering a question. To this end, he developed and implemented a carefully crafted series of instructional activities that were devised to both support the conversations the teacher hoped to have with students and to initiate questions that would demand further investigation.

Identification of ERE and PD patterns

Based on the research literature mentioned above, we believed that we would find one interaction pattern that would characterize the ways in which this conceptually-oriented teacher and his class interacted. To our surprise, we identified two recurring interaction patterns, one that occurred when the teacher initiated a new idea or concept, and one that emerged as the students became more familiar with the idea.

The ERE Pattern. The pattern that was evident each time the teacher introduced a new activity. In these cases, the teacher *Elicited* observations, students *Responded,* and the teacher *Elaborat*ed on their comments. We have called this trend an ERE pattern to acknowledge its relation to Mehan's (1979) IRE (Initiate, Respond, Evaluate) pattern described earlier. Our elaboration of Mehan's pattern is deliberate. We want to contrast two specific differences. First, whereas Mehan's construct involved having teachers initiate a question (that usually involves getting the student to offer particular bits of information), the teacher in this class elicited observations — that is, he encouraged students to look at the graph or relation of interest and observe various properties about it. The second difference between Mehan's IRE pattern and the ERE pattern we describe is that teachers falling into the IRE pattern *evaluate* the correctness of an answer without asking where or why a student arrived at the answer given. In contrast, the teacher in this case study consistently *elaborated* students' responses in an effort to encourage longer, more thoughtful discussions and to model potentially appropriate means of justification.

The PD Pattern. For their part, the students initially offered short responses and then worked toward building and describing their own imagery in conceptual terms that others could understand. As they developed more sophisticated imagery, a second pattern appeared to emerge. In this case, the teacher or a student would make a *proposition*, and others would *discuss* it. This PD (proposition-discussion) pattern was generally marked by students' efforts to either explain their own thinking or ask a classmate for clarification.

We identified four distinct ERE→PD phases over the course of the semester. Each is briefly outlined below.

Phase I. During the first phase, the teacher introduced and helped the students refine their views of what constitutes a sound mathematical explanation. In particular, he introduced the concept of a "conceptual explanation" and juxtaposed this with a "calculational" explanation. During ERE conversations, the teacher would elicit comments and observations to help the students develop vocabulary and experience recognizing conceptual explanations. As the class transitioned into the PD phase, the students and teacher entered into more animated discussions such as a debate over how a conceptual explanation differs from simply giving a formulaic answer, and why such a skill would be valuable to a mathematician or teacher. The mathematical context of these discussions involved describing graphs in terms of the phenomena they portrayed, rather than their superficial shape.[2]

Phase II. During the second phase, the teacher and students focused on developing a conceptual orientation toward learning mathematics that involved using one's own words to justify an answer, and also learning to question or follow up on others' thinking. The mathematics discussed during this phase included describing a graph as a set of points and considering translations and dilations of the functions in terms of effects on individual points rather than as stretching lines.

Phase III. During the third phase, the class expanded their views and, eventually, conceptual explanations of complex mathematical ideas, such as the Fundamental Theorem of Calculus, through the imagery of covariation of functions.

Phase IV. During the final phase the students in this class, most of whom were prospective teachers, were asked to "try on the hat of a teacher" in order to design, implement, and analyze lesson plans. To this end, each student was required to develop a lesson plan and then use it to tutor one child. The mathematical goal of the lessons they constructed was to highlight dependency relations. The pedagogical goal was to apply their conceptual orientation for learning toward an orientation for teaching.

Given space limitations, it is not possible to present examples of ERE→PD exchanges that occurred during each of the phases of the semester. In what follows, we present one brief example from Phase II during which the class developed a conceptual orientation for learning mathematics.

Example of an ERE→PD transition

Phase II began during the third week of the semester. The teacher introduced an instructional sequence in which the students were asked to describe the graphical behavior of novel functions. Examples of functions the class considered include $f(x) = x^2 \bmod (x)$ and $f(x, y) = xy$. The teacher's purpose for using these non-conventional and non-intuitive examples was to support the students' efforts to visualize relationships between changing quantities and to challenge their beliefs regarding the value of purely calculational instruction. In the following episode, the teacher's objective was to encourage students to visualize how individual points on the Cartesian graph of $y = \sin(b\theta)$ can be related to a varying arc length of measure θ on the unit circle.

---

[2] These activities were specifically designed to elicit a "Graph as Path" misconception wherein students favor graphs of functions that "look like" the situation being represented rather than a graph showing the relation between time and position or speed of a traveling object (cf., Dugdale, 1993).

Episode IIa: Refining a conceptual explanation through the use of visualization (ERE pattern). To begin this discussion, the teacher used the *Geometer's Sketchpad* software to create a unit circle with a free point on the circle that controlled two arc lengths: θ, the independent variable, and $b\theta$ (measured in radians) for some value of the parameter b. The imagery that the teacher was hoping to support required that the students focus on the covariation of two quantities: the arc length of θ and the length of $\sin(b\theta)$, which the students identified as line segment $B'F$ in Figure 1. During class, the teacher changed the value of θ and asked the students to visualize the Cartesian graph that coordinated the arc length, θ, with the length of line segment $B'F$.

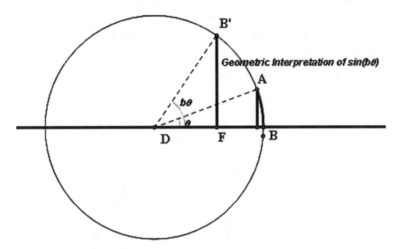

Figure 1. Teacher's GSP sketch showing θ and $\sin(b\theta)$.

Next, the teacher asked the class to consider the Cartesian graph of the function $y = \sin(30x)$. We claim that the discussion followed an ERE (Elicitation-Response-Elaboration) pattern because the teacher first elicited a number of observations from the students regarding the graph they were envisioning. After each student gave a response, the teacher engaged students in a discussion, elaborating on their observations by asking them to consider the mathematical causality underlying them.

Teacher: What will happen when I increase b?

Hugh: It will stretch it, make it higher [sic].

Teacher: Let's consider a slightly different question: What happens if we have $y = ax^2$. What would happen if we increase the value of a? Valerie?

Valerie: The graph [of the parabola] would get thinner and thinner.

Teacher: We have a tendency to talk about graphs being 'skinny' and 'stretching.' But we are not making them skinnier; we are *dilating* the graph [teacher's emphasis]. For each x, a times $f(x)$ dilates $f(x)$ by a factor of a. Now, how does this relate to the changing of b in the graph of $y = \sin(b\theta)$?

In this exchange, the teacher was questioning Hugh's imagery but instead of simply evaluating it (saying, for example, "No, that is not correct"), the teacher chose to challenge the students to consider the causality behind their observations. In particular, he was challenging them to elaborate their observations by illustrating the difference between describing a graph's visual shape (e.g., 'getting thinner' or 'going higher') and describing its point-dilations (e.g., relating it to how $f(x)$ varies as the parameter varies).

This episode illustrates another critical issue that we have found to be very important in the evolution of students' revised views of the nature of mathematics and mathematical argumentation. Although the teacher may appear to be telling the students what language to use, we argue that he was not telling them *what* to say but *what imagery to focus on* and how to describe their imagery more conceptually by focusing on the mathematical causality that supported their observations. In later debriefing sessions with the teacher, we discussed the issue of telling the students what to say versus helping them develop their imagery. He pointed out that students do not know how to talk in ways that are conceptual for the reason that they have not thought this way before. From his perspective, even if he had told the students exactly what to say, he would have expected a period of interpretation and negotiation during which the students' constructed images would be internalized. This type of social insight characterizes the elaboration stage of ERE pattern exchanges and offers a sharp contrast with the shorter, more evaluative answers that are generally found in IRE or funneling/focusing pattern exchanges.

Episode IIb: Developing ways to describe novel functions (PD Pattern). The next few activities in the sequence led to a transition from an ERE to a PD (Proposition-Discussion) pattern transition. During a class that occurred two days later, the teacher asked the students to discuss their methods for finding the equation of "mystery" polar graphs such as the one shown in Figure 2. After completing several of these tasks for homework, the students noted that the "trial and error" method was ineffective and that they had to rely on their imagery and notions of the mathematical relationships to identify the equations.

During one episode, Eric volunteered to explain his reasoning for determining the equation of the graph shown in Figure 2:

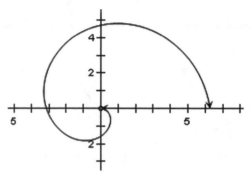

Figure 2. Graph of mystery function.

Eric:	I just broke it down into sections of how [the graph] varies from 0 to 2π. I asked myself, "What is the distance doing?" Basically, we see that when $\theta = 0$, we have distance = 6. So $f(0) = 6$. As we go around from 0 to 2π, we see distance getting smaller and smaller. Distance is decreasing at a steady rate. This suggested to me a linear relation. So I tried several linear equations with $f(0) = 6$; $f(2\pi) = 0$. I was also imagining the relation like this [draws an imaginary line in the air with his hand as if making a Cartesian graph of a line with a negative slope].
Hugh:	Is it possible for the graph to go from 0 out?
Eric:	We are starting at 0, but the point does not start at the origin! As I go up, imagine what happens to the values of the graph. It does not matter what you do to theta; it is also going to be changing linearly in relation to the trip.
Hugh:	So you imagined a line graph and then translated this back to polar coordinates?
Eric:	I just watched it in my mind and saw it was a linear function.

We classify this and the ensuing discussions (in which the class went on to discuss two other mystery graphs) as PD in nature because Eric's proposition led to a relatively long debate during which others attempted to understand Eric's method for describing the behavior of the mystery function in terms of one quantity that was varying. Eric's contribution in this excerpt can be seen as initiating a shift from the ERE to the PD patterns in that, before this, the teacher might have been trying to help Eric elaborate. Instead, Eric now *used* such an elaboration to both think about *and* explain his understanding of the mystery graph. This description of causality appeared to help other students, like Hugh, who picked up on Eric's emphasis of looking at the covariation of one quantity with another in terms of thinking about Cartesian graphs to explain polar graphs.

Focusing on Implicit Messages

As noted in the introduction to this chapter, we believe that a focus on the implicit messages that students receive as they participate in the evolution of various interaction patterns can be a key to explaining students' difficulties learning new material. When students were asked to reflect on their learning after Phase II, many indicated that they were not only learning about the material, but also learning how to learn in a new way. For example, one student wrote:

> I find it very helpful to think about graphs in this way because now I have a better understanding of the graphs I looked at. Before when I saw the graph of $\sin(x)$, I just thought, "Okay that is or is not a correct graph of $\sin(x)$," and verified it by plugging points. But then I began using the unit circle to relate to the graph to see as x varies what the sine or cosine graph looks like.

In contrast, another noted:

> My training has been biased toward a calculational approach; therefore I am unable to think of functions in this way. It confuses me. I have to keep my mind on too many things. It is easier to visualize "a sine graph" than it is to visualize angles, cosines, sines, and hypotenuses shifting and changing against one another. It provides insights to people who can think in that way, but I have difficulty thinking in those terms. Everyone approaches problems in different ways.

While this student's overall mid-semester assessment was somewhat reluctant, it was clear that he was indicating that he felt a challenge to explain in ways that he had not done before. We found it particularly gratifying to read that same student's final reflection:

> In sum, the primary idea behind the Analyzer, MacFunction, and Geometer's Sketchpad homeworks is the idea of "covariation." Students studied covariation by examining and tracing "dependency relations" (among variables or geometrical elements) and seeing how these variables or elements are tied to the others by a web of such dependencies. Thus, the student obtains a clear understanding of how every aspect of a given problem or phenomenon is related to every other aspect. This, then, is precisely the deep, full, detailed understanding of the phenomenon that "conceptually based" instruction is intended for. In contrast, an algorithmic approach would allow the student to get a correct answer (such as the optimal values in Analyzer #1) but would not force the student to understand the phenomenon or its intricacies.

Here it becomes clear that this student had developed a strong understanding of the value and structure of a conceptual *explanation*. Moreover, although he had voiced his opposition (in terms of unfamiliarity, discomfort, etc.) to PD type exchanges earlier, his final reflection indicated that he did come to develop meaningful insights and an appreciation for making meaning of mathematics. It is particularly important to note that he came to realize the value of conceptual explanations for him as both a learner and a teacher. For example, he reported that when he worked with his tutee in phase IV, he demanded conceptual explanations and that this effort opened his eyes to the superficial understandings that students generally have. The other students' write-ups also indicate, to varying degrees, a general consensus supporting the value of a conceptual orientation for both learning and teaching.

Conclusions

The goal of this chapter has been to discuss the value of having college-level mathematics teachers become aware of the ways a class may become encouraged to negotiate a productive culture with fruitful interaction patterns and to provide one rich example of an established pattern that was shown to have a positive effect on students' attitudes and understanding. Our use of the example and case study was not designed to offer a prescription or template for effective mathematics teaching. Instead, we used the data to illustrate how one classroom community developed patterns of interaction that were not characteristic of the traditional ERE or funnel patterns found in many classrooms. In so doing, we hope to point out that other patterns can emerge and be recognized once a teacher attempts to adopt a conceptual rather than calculational orientation toward teaching.

One result from the study was our somewhat surprising conclusion that two distinct patterns of interaction emerged: the ERE (Elicit observations, Respond, Elaborate) pattern, and the PD (Propose, Discuss) cycles that worked together to engage the class in the process of developing appropriate methods of justification and supporting more sophisticated means of reasoning. Of course, we do not mean that all interactions followed the same template, or that every student had developed the same sense of expectations for what types of answers were considered appropriate. However, we do claim that both the ERE and PD patterns working in a cyclic manner facilitated the robust discussions of mathematical ideas and eventually provided a social basis for changing not only the students' mathematical understandings, but their abilities to explain mathematical relations and their overall views of mathematics as well.

A second result of this work has been the development of a methodology for identifying patterns of interaction, which, we hope, might help teachers and researchers identify classroom interaction patterns. In short, we found that the patterns we identified were inclusive, descriptive, and repeated. These characteristics are consistent with the identification of practices described by Bowers (2001).

When reviewing this chapter, one mathematician told us that the most surprising conclusion to him was that interaction patterns cannot simply be implemented or dictated by the teacher alone. Our claim is based on the view that all communication practices are inherently social and therefore follow the rules of all interpersonal communication: the messages that are received are not necessarily those intended by the sender, and meanings are often implicitly negotiated between speakers (in this case, the students and the teacher). For these reasons, classroom communication patterns cannot be explicitly laid out by the teacher alone. Instead, they are negotiated through an implicit process of trial and error by which the students might offer an explanation that serves the implicit function of an opening offer: in essence, the student is asking, '*is this acceptable?*' The reaction — from both the teacher *and the students* — sends an implicit message back to the student and the rest of the class which, in turn, moves the process of negotiation forward one more round. It is not surprising, therefore, that the same teacher may see different patterns emerge in different classrooms due to the differences in student compositions.

In summary, we believe that a study of classroom interaction patterns can enable a mathematics teacher to identify the types of implicit messages that students might be forming regarding what it means to "understand" a math topic. For example, teachers can determine the degree to which they are asking calculational or conceptual questions and assess whether the patterns of interaction resemble Mehan's IRE patterns or the ERE $\rightarrow$ PD patterns identified in this study wherein students are first asked to wrestle with ideas, and then use them in a proposition-discussion mode.

If students become accustomed to engaging in conceptually-based interaction patterns (as opposed to calculationally-based IRE or funneling/focusing patterns), our research suggests that they will be able to write more effectively on exams, respond better to open-ended tasks such as those involved in technology explorations, and, if they are prospective teachers, encourage their own students to think more deeply about mathematics. In this way, we hope to make some headway in addressing Stigler and Hiebert's (2004) strong recommendation that real improvement in mathematics education comes not at the superficial level of textbooks, but at the deeper, cultural level of teaching practices.

References

Au, J. (1993). *Literacy instruction in multicultural settings*. Orlando FL: Harcourt Brace.

Ambrose, R. A. (2004). Initiating change in one elementary school teacher's orientations to mathematics teaching: A case of building on beliefs. *Journal of Mathematics Teacher Education, 7,* 91–119.

Arcavi, A., Kessel, C., Meira, L., & Smith, J. P. (1998). Teaching mathematical problem solving: An analysis of an emergent classroom community. In A. H. Schoenfeld, J. Kaput, & E. Dubinsky (Eds.), *Research in collegiate mathematics education, III* (pp. 1–70). Providence, RI: American Mathematical Society.

Ball, D. L. (1990). Breaking with experience in learning to teach mathematics: The role of a preservice methods course. *For the Learning of Mathematics, 10*(2), 10–16.

Boaler, J. (2000). Exploring situated insights into research and learning. *Journal for Research in Mathematics Education, 31,* 113–119.

Bowers, J. S., & Nickerson, S. (2001). Identifying cyclic patterns of interaction to study individual and collective learning. *Mathematical Thinking and Learning, 3*(1), 1–28.

Bowers, J. (2001). *The practice of identifying practices.* Paper presented at the 2001 Annual Conference of the American Educational Research Association, Seattle, WA.

Cazden, C. B. (1988). *Classroom discourse: The language of teaching and learning*. Portsmouth, NH: Heinemann.

Cobb, P. (1996). Where is the mind? A coordination of sociocultural and cognitive constructivist perspectives. In C. Fosnot (Ed.), *Constructivism: Theory, Perspectives, and Practice* (pp. 34–52). New York: Teachers College Press.

Cobb, P., & Bowers, J. (1999). Cognitive and situated learning perspectives in theory and practice. *Educational Researcher, 28*(2), 4–15.

Cobb, P., & Whitenack, J. (1996). A method for conducting longitudinal analyses of classroom videorecordings and transcripts. *Educational Studies in Mathematics, 30*, 213–228.

Dillon, J.T. (1990). *The practice of questioning.* New York: Routledge, Chapman and Hall, Inc.

Dugdale, S. (1993). Functions and graphs: Perspectives on student thinking. In T. A. Romberg, E. Fennema, & T. P. Carpenter (Eds.), *Integrating research on the graphical representation of functions* (pp. 101–130). Hillsdale, NJ: Lawrence Erlbaum Associates.

Lakatos, I. (1976). *Proofs and Refutations.* Princeton, NJ: Princeton University Press.

Lampert, M. (1990). When the problem is not the question and the solution is not the answer: Mathematical knowing and teaching. *American Educational Research Journal, 27*(1), 29–63.

Lave, J. (1997). The culture of acquisition and the practice of understanding. In D. Kirshner & J. Whitson (Eds.), *Situated cognition: Social, semiotic, and psychological perspectives* (pp. 17–35). Mahwah, NJ: Erlbaum.

Medina, P. (2001). The intricacies of initiate-response-evaluate in adult literacy education. Retrieved October, 2004, from http://www.edst.educ.ubc.ca/aerc/2001/2001medina.htm.

Mehan, H. (1979). *Learning lessons: Social organization in the classroom.* Cambridge, MA: Harvard University Press.

Moschkovich, J. N., & Brenner, M. E. (2000). Integrating a naturalistic paradigm into research on mathematics and science cognition and learning. In A. E. Kelly & R. A. Lesh (Eds.), *Research design in mathematics and science education* (pp. 457–486). Mahwah, NJ: Lawrence Erlbaum Associates.

Puro, P., & Bloom, D. (1987). Understanding classroom communication. *Theory into Practice, 26*, 26–31.

Richardson, V. (1996). The role of attitudes and beliefs in learning to teach. In J. Sikula (Ed.) *Handbook of research on teacher education* (pp. 102-119). New York: Macmillan.

Steinbring, H. (1989). Routine and meaning in the mathematics classroom. *For the Learning of Mathematics, 9*(1), 24–33.

Stephan, M., & Rasmussen, C. (2002). Classroom mathematical practices in differential equations. *Journal of Mathematical Behavior, 21*, 459–490.

Stigler, J. & Hiebert, J. (2004). Improving mathematics teaching. *Educational Leadership, 61*(5), 12–19.

Thompson, A. G., Philipp, R. A., Thompson, P. W., & Boyd, B. (1994). Calculational and conceptual orientations in teaching mathematics. In D. B. Aichele & A. F. Coxford (Eds.), *Professional development of teachers of mathematics* (pp. 79–92). Reston, VA: National Council of Teachers of Mathematics.

Thompson, P. W. (1994). Students, functions, and the undergraduate mathematics curriculum. In E. Dubinsky, A. H. Schoenfeld, & J. J. Kaput (Eds.), *Research in collegiate mathematics education I* (pp. 21–44). Providence, RI: American Mathematical Society.

Voigt, J. (1995). Thematic patterns of interaction and sociomathematical norms. In P. Cobb & H. Bauersfeld (Eds.) *The emergence of mathematical meaning: Interaction in classroom cultures* (pp. 163–201). Hillsdale, NJ: Erlbaum.

Vygotsky, L. S. (1978). *Mind in society: The development of higher psychological processes.* In M. Cole, V. John-Steiner, S. Scribner, & E. Souberman (Eds.), Cambridge, MA: Harvard University Press.

Wood, T. (1994). Patterns of interaction and the culture of mathematics classrooms. In S. Lerman (Ed.), *The culture of the mathematics classroom* (pp. 149–168). Dordrecht, The Netherlands: Kluwer Academic Press.

Wood, T. (1999). Creating a context for argument in mathematics class. *Journal for Research in Mathematics Education, 30*(2), 171–191.

Wood, T., & McNeal, B. (2003). Complexity in teaching and children's mathematical thinking. In N. Pateman, B. J. Dougherty, and J. T. Zilliox (Eds.), *Proceedings of the 27th Conference of the International Group for the Psychology of Mathematics Education held jointly with the 25th Conference of PME-NA* (Vol. 4, pp. 435–441). Columbus, OH: ERIC Clearinghouse.

Wood, T., & Turner-Vorbeck, J. (2001). Extending the conception of mathematics teaching. In T. Wood, B. S. Nelson, & J. Warfield (Eds.), *Beyond classical pedagogy* (pp. 185–208). Mahwah, NJ: Erlbaum.

15

Mathematics as a Constructive Activity: Exploiting Dimensions of Possible Variation

John Mason, *Open University*
Anne Watson, *Oxford University*

Introduction

Mathematics is often seen by learners as a collection of concepts and techniques for solving problems assigned as homework. Learners, especially in cognate disciplines such as engineering, computer science, geography, management, economics, and the social sciences, see mathematics as a toolbox on which they are forced to draw at times in order to pursue their own discipline. They want familiarity and fluency with necessary techniques as tools to get the answers they seek. For them, learning mathematics is seen as a matter of training behaviour sufficiently to be able to perform fluently and competently on tests, and to use mathematics as a tool when necessary.

Unfortunately this pragmatic and tool-based perspective may cut people off from the creative and constructive aspects of mathematics, making it more difficult for them to know when to use mathematics, or to be flexible in their use of it. On its own, this perspective can reinforce a cycle of de-motivation and disinclination. The result is a descending spiral of inattention, minimal investment of energy and time, and absence of appreciation and understanding, leaving learners disempowered from pursuing their discipline through the use of mathematics.

By contrast, mathematicians see mathematics as a domain of creativity and discovery in its articulation, proof, and application. Full appreciation of a mathematical topic includes the exposure of underlying structure as well as the distillation and abstraction of techniques that solve classes of problems, together with component concepts. Mathematicians construct objects as examples of concepts, as illustrative worked examples of the use of techniques, and as possible examples of or counter-examples to conjectures. As Paul Halmos put it,

> A good stock of examples, as large as possible, is indispensable for a thorough understanding of any concept, and when I want to learn something new, I make it my first job to build one. ... Counter-examples are examples too, of course, but they have a bad reputation: they accentuate the negative, they deny not affirm. ... the difference ... is more a matter of emotion. (Halmos, 1983, p. 63)

> If I had to describe my conclusion [as to a method of studying] in one word, I'd say *examples*. They are to me of paramount importance. Every time I learn a new concept I look for examples ... and non-examples. ... The examples should include wherever possible the typical ones and the extreme degenerate ones. (Halmos, 1985, p. 62)

Halmos is one of many authors who express a similar sentiment. Feynman expressed it this way: "I can't understand anything in general unless I'm carrying along in my mind a specific example and watching it go" (Feynman and Leighton, 1985, p. 244).

We take the view that mathematics can be presented and experienced as a constructive activity in which creativity and making choices are valued, not just for their own sake, but in order to stimulate learners to use their own powers to make sense of phenomena mathematically, whether in the domain of mathematics itself, or in cognate disciplines that make use of mathematical tools and mathematical thinking. By powers here we mean things like imagining and expressing, specialising and generalising, conjecturing and convincing, and so on (Mason, Burton, & Stacey, 1982; Mason, 2002; Mason & Johnston-Wilder, 2004a, 2004b).

For example, presenting learners with the function $x \to |x|$ as an example (often the only one) of a function which is differentiable everywhere on the reals except at one point, invites them to see it as a monster to be barred (Lakatos, 1976), as a pathological object to be ignored (MacHale, 1980). Yet there is an opportunity to invite learners to use it to construct whole classes of functions which have the same property at $(0, 0)$, such as $x \to \lambda|x| + xg(x)$ for any differentiable function g. They can also construct for themselves a function that is differentiable everywhere except at some other specified point, and then extend this to non-differentiability at several points (e.g., $x \to \lambda|x - a| + \mu|x - b|$ where $a \neq b$ and $\lambda\mu \neq 0$). Not only does this invite learners to make use of their control over functions through translation and scaling, but it demonstrates that any example can be expanded to whole classes of functions.

We suggest that seeing mathematics as a constructive activity requires a small but important shift in thinking that can have significant impact on the interest and commitment of learners, the way they use their acquired concepts and tools, and hence the way they use mathematics in their own discipline. After many years of teaching in high school and university contexts, often working with disaffected students, and countless experiences working with others on mathematics, we take it as axiomatic that people who are encouraged to use their powers not only experience pleasure, but sharpen and extend those powers. They can become more motivated to enquire further. Learners who are encouraged to be creative and to exercise choice respond by becoming more committed to understanding rather than merely automating behavioural practices.

Finally, we suggest that mathematics seen and presented as a creative and constructive activity can engage and motivate those who might pursue mathematics further. It can also yield insight into how learners are thinking in ways that complement what teachers discover from learners' responses to routine tests on standard exercises.

This chapter begins by showing that many, or most, tasks and exercises presented to students for pedagogic purposes can be seen as constructions, and that adopting this view can assist learners in appreciating what the problem is asking, as well as finding solutions and also motivating them. The second section shows how the notion of dimensions-of-possible-variation can be used to inform the structuring of tasks, so that learners can be induced to move from exercising in order to train their behaviour, to exploring in order to sharpen their thinking and deepen their awareness of underlying mathematical structure. The third section moves beyond exercises as tasks, and considers problems in the two forms identified by Pólya. The fourth section shows how the same principle of dimensions-of-possible-variation can inform the way learners encounter mathematical concepts as they develop rich example-spaces which will, it is hoped, come to mind when they encounter future problems. The fifth section briefly comments on the need to prompt and provoke learners to shift from the metaphor of learning mathematics as exercise, to learning mathematics as exploration.

The chapter is theoretical in nature, presenting possibilities for practice based on mathematical structures. Thus, the warrants for this approach are to be found in mathematics itself rather than in empirical research[1]. Nevertheless, the practices arise from our extensive, evaluated, self-critical experience as mathematicians, teachers and educators, and they have been used with learners ranging from adolescents to adults, in school and university. Our method of enquiry is to observe the behaviour of people as they explore and learn mathematics: our own behaviours, as well as those of our students and colleagues. We intentionally construct, and put ourselves and colleagues into, mathematical situations that appear to be analogous to what learners encounter. From these we gain insight into strategies based on mathematical thinking of which learners may be unaware. We test, in many contexts, both the strategies and ways of drawing them to the attention of learners who do not yet use them spontaneously. For example, this chapter arises from some eight years of work during which we focused on the potential contribution the construction of mathematical objects can make to an enhanced experience of mathematics. We always seek feedback from workshop and teaching sessions. A particularly powerful form of feedback comes from teachers who are energised to use our ideas in their

---

[1] Our method (Mason 2002a) relates to both action research and design research paradigms. It takes place in naturalistic settings; we evaluate and develop actions in a deliberate and cyclic manner. Ideas are not only put back into practice, but also to critical audiences of peers and learners; the path of our enquiry branches frequently; the opportunities to work on ideas arise during our normal working lives. To those who want detail about numbers of cycles, numbers of students, numbers of occasions we can only answer 'hundreds'.

next teaching sessions and report back to us of their experiences. We do not carry out quantitative controlled studies because these are not appropriate in a domain that is highly sensitive to the world views of teacher and learners, the relationship between them, and their propensity to reflect on their experience.

Pedagogic Phenomena Being Addressed

The particular pedagogic phenomena we are addressing can be summarised as the tendency for many learners to do as little as necessary in order to complete what they are asked to do, and the widespread assumption that 'doing' the tasks they are set will mean that they will learn. This is the basis for the implicit *didactic contract* between teacher and learners described and elaborated by Brousseau (1984, 1997). Too often 'doing' means 'getting answers' rather than using the task to appreciate general concepts and methods as exemplified by the particular. There is a tendency for learners to be satisfied with the particular rather than trying to see through the particular to the general (Whitehead, 1919), or, on the other hand, leaping to vast generalisations without checking particular cases. Learners are often satisfied by assenting to what they are told, by trying to work out what the teacher thinks rather than (re-) constructing for themselves, by using worked examples as templates in doing assignments without probing the underlying reasoning, and by internalising as little as possible. This response to the *didactic contract* is rarely sufficient to reach what Skemp (1976) described as *instrumental understanding* (knowing enough to succeed only at routine tasks), much less the more desirable *relational understanding* (appreciating connections and knowing enough to be able to respond flexibly in novel situations). Christiansen and Walther (1986) observed that "Even when students work on assigned tasks supported by carefully established educational contexts and by corresponding teacher-actions, learning as intended does not follow automatically from their activity on tasks" (p. 262). As human beings, learners respond well to being asked to make choices, including creating mathematical objects for themselves. It gives them a sense of involvement and it enriches the space of examples to which they have access when someone else is talking in generalities. Indeed, understanding of concepts and techniques can usefully be thought of in terms of the richness of the *example space* that comes to mind (Watson & Mason, 2002, 2005), and the complexity of transformations learners can use to modify those examples. We have found that with prolonged exposure to an atmosphere in which learners are expected to construct examples, their example spaces become both more extensive and more richly interconnected.

We have also found that learners respond well to being called upon and expected to use their own powers to specialise and to generalise (Pólya, 1945, 1962), to imagine and to express, to conjecture and to convince, to organise and to characterise. This is born out by numerous studies, such as Boaler (2002), Senk & Thompson (2003), and Watson, De Geest & Prestage (2003): when learners' powers are engaged, they display behaviour beyond what is normally expected. Thus the challenge is to promote a movement from merely *assenting* to what they are told or asked to do, to taking the initiative and *asserting* (in the form of making, testing and validating conjectures, constructing examples which illustrate conditions, and generalising particular tasks to a class of 'types' of tasks) through using and developing their natural powers (Mason & Johnston-Wilder, 2004a).

An atmosphere of competitive seeking of single correct answers does little to foster mathematical thinking. Calling upon learners to make choices, to act creatively and to use their powers is best supported in a *conjecturing atmosphere*, in which what is said by anyone is treated as a conjecture uttered with the intention of possibly modifying it according to critique and counter-examples. There are close analogies with what Legrand (1993) calls *scientific debate*. Among other things this implies that learners are constantly challenging, constantly seeking examples and counter-examples. It means that learners use examples in order to re-construct generalisations and to appreciate mathematical reasoning for themselves.

Our theoretical position can be summarised as follows:

- promoting mathematical thinking improves motivation and confidence, and hence both competence and effective use of mathematics as a tool, even among those only taking mathematics as a service subject.

- trusting learners to make choices and so to exercise creativity and to explore available freedoms enriches their mathematical understanding and appreciation of concepts, techniques, and heuristics, as well as fostering involvement in and getting pleasure from learning mathematics.

Exercises as Construction Opportunities

A set of exercises may be seen by a learner as an obstacle, a necessary hurdle to be overcome as quickly as possible, but it can also be seen as an opportunity to develop fluency and facility by reducing the amount of attention required to get solutions using a standard technique, to the point of automating that technique. A set of exercises can also be taken as an opportunity to seek generality which encompasses all of the particulars. This turns it into a construction task: to re-construct the *question space*[2] from which the particular questions have been drawn. This view of exercises as an opportunity to appreciate generality has a long history, since the earliest written records contain worked examples of mathematical calculations. Babylonian cuneiform tablets and Egyptian papyri often include statements such as 'do thou like this' and 'thus it is done' (Gillings, 1972) suggesting that the learner is intended to do more than simply 'follow' the template. The point of doing exercises is to appreciate the generality of a method, and to internalise and automate its functioning. Girolamo Cardano (1501-1576) writing in Latin in his famous work *Artis Magnae Sive de Regulis Algebraicis* includes phrases such as

> In accordance with these demonstrations, we will formulate three rules and we attach a jingle in order to help remember them; We have used a variety of examples so that you may understand that the same can be done in other cases and will be able to try them out for the two rules that follow, even though we will there be content with only two examples; It must always be observed as a general rule … ; So let this be an example to you; by this is shown the *modus operandi* …. (quoted in Cardano, 1545/1969, p. 36–41)

Thus the point of a set of exercises can be seen as a behavioural aspect of the *didactic contract*, or it can be seen as an opportunity to construct not only solutions to the particular tasks, but the general class of which these are particulars.

Take for example, the factoring of the difference of two squares. The basic idea is straightforward, but it can be masked by varying different features. Asking learners to factor the following expressions

$$x^2 - 1, \quad a^2 - b^2, \quad 4x^2 - 9y^2, \quad 64a^4 - 81b^4, \quad 4x^2a^6 - 25y^4b^{10}, \quad (2x + 3a)^2 - (3x - 2a)^2$$

could be intended to expose learners to the variety of possibilities for factoring the difference of two squares. Most texts would have several examples of each type; some would mix them up, and others would arrange similar ones together. The former are probably intending learners to detect similarities from amongst apparent differences, whereas in the latter case the intention is probably to get learners to detect what is the same about several so as to appreciate a range of possibilities. In both cases there is an implicit aim towards speed and accuracy, but to gain speed one has to recognise and exploit similarity, while to gain accuracy one has to focus on specific details. But unless learners are prompted to reflect on *what* is changing and *how*, they are unlikely to appreciate the various aspects which can change, nor what changes are permissible to maintain the underlying structure. Here, constants and multiple letters can change, but each must be raised to an even power. Perhaps it is no wonder that learners often fail to recognise the difference of two squares outside of the section of text devoted to it.

Seeing sets of exercises as construction opportunities opens the way to further meta-tasks intended to promote reflection and construal by learners. Karp (2004) provides an excellent illustration in the context of quadratic equations that have been obscured by the use of reciprocals, linear expressions, and square roots. In almost every topic, learners can be asked to construct:

- another example like the ones they have done;
- one which obscures the use of the technique as much as possible;
- one which shows that they know how to do 'questions of this type' (and they can be asked to describe how they would recognise a 'question of this type');
- one which would be a good test for learners next year;
- one which they think might challenge the teacher (or some other relative expert);
- the most general question of this type;

and so on. Focus and emphasis switch from doing particular questions to appreciating the technique as a general

---

[2] A term introduced to us by Chris Sangwin (personal communication) arising from his project to use software to generate random questions for learners from a space of possible questions but paying attention to internal structure and constraints within that space (e.g., quadratic equations with real or complex roots).

method and appreciating the whole exercise as representative of a space of possibilities. This is where Marton's notion of *dimensions of variation* is so useful.

Dimensions of Possible Variation, and Related Ranges of Permissible Change

Marton proposed that learning can be seen as extending awareness of what constitutes an example (Marton & Booth, 1997; Marton & Trigwell, 2000; Marton & Tsui, 2004, see also Runesson, 2005). He observed differences in learning according to the nature and range of variation to which learners were exposed. To capture this, he introduced the notion of *dimensions of variation* to refer to the different aspects of an object which can be varied and still that object remains an example of a specified concept. For example, in meeting limits, learners need to be aware that a limit can be approached from one or the other side only, or from both sides, depending on the context. This constitutes a dimension of variation that is vital to appreciating limits. Other more elementary dimensions include the fact that the variable expressing the limit can be something other than x or h; that the point being approached can be specific (a number) or general (a letter); that the function whose limit is sought may have other variables and parameters present; and that in any case, there may or may not be a limit. In group theory, learners need to be aware that not all groups can be displayed as the symmetries of objects in two or three dimensional space; that the same group can be generated by different actions on different objects; that a finite cyclic group can be of any finite order; and that the generators can be denoted by any symbol.

Since teachers and learners are usually aware of different dimensions, and since the same person may be aware of different dimensions of variation at different times, in Watson & Mason (2002, 2005) we extended the notion of dimensions to *dimensions of possible variation* of which someone may be aware. Thus a teacher may be aware of many more possibilities than learners, but may not be aware of this difference. By being explicitly aware of important dimensions of possible variation, a teacher can choose to direct learner attention to relevant dimensions as they develop their sense of a concept. Furthermore, within each dimension of possible variation, learners may be aware of different *ranges of permissible change*. For example, learners might have encountered $|x - a|$ for various integer values of a but it may never have occurred to them that a could be any real number. Indeed, their sense of 'any real number' may itself have a limited range of permissible variation. If learners only see systems of equations with integer coefficients, they may never think of the possibility that they could be other numbers.

The notion of dimension of possible variation is, in some sense, the dual to an important and ubiquitous theme in mathematics, that of *invariance in the midst of change*. Any theorem in mathematics can be seen as a statement of what remains (relatively) invariant when other things are permitted to change. Usually it is some relationship or property which is being preserved. However, it is not always clear that learners are aware of what it is possible to change.

Seeing generality through particular instances is basically detecting some features to keep invariant while others are permitted to change. A useful pedagogic strategy that calls upon this theme is to ask learners to look for what is the same, and what is different between two or more objects (Watson & Mason, 1998; Coles & Brown, 1999; Brown & Coles, 2000) as well as with the problem solving strategy of asking yourself 'what if … were to change' (Brown & Walter, 1983).

Applied to a set of exercises, looking for dimensions of possible variation and the associated range of permissible change within each dimension calls upon the learner to go beyond merely solving each individual exercise. It draws attention to the exercises as representative of a potential question space and invites learner re-construction of such a space. Next time learners encounter a similar question, they are much more likely to recognise the type and so have access to a solution approach or technique than learners who have contented themselves with obtaining solutions solely to the particular questions in the exercise set.

For learners, aspects of exercises that could vary include context, explicit numbers, implicit structural numbers, and choices of which elements of a task are given as data and which are to be found. This last opens the mathematical theme of *doing & undoing*, in which learners attempt to characterise all the similar questions that have the same answer, and all the possible answers to similar types of questions. A learner who has explored dimensions of possible variation is much more likely to recognise the structure of a novel task rather than being misled by superficial similarities with previously solved tasks.

For teachers, in addition to these dimensions, the order of presentation can take into account a supposed hierarchy of complexity, a supposed order of conceptual development, or significant aspects, which is essential for learners to appreciate in order to understand the topic, concept or technique (as illustrated in Karp, 2004).

Problems as Construction Opportunities

Users of mathematics are more in need of experience of using mathematics to resolve problems in their own domain of interest than they are of rehearsing to mastery a collection of techniques. Techniques can be quickly automated if there is need for repeated and efficient use. But the same is true of the mathematics student: "The mathematician's main reason for existence is to solve problems, and that, therefore, what mathematics *really* consists of is problems and solutions." (Halmos, 1980, p. 519)

What constitutes a *problem* is a non-trivial matter. What is a problem to learners is not usually problematic for the author or setter. For example, in a recent study Lara Alcock & Keith Weber asked learners in a university course to prove that if f is an increasing function on R to R, then there is no real number c that is a global maximum for f (Alcock & Weber, 2005). An experienced mathematician has an instant intuition that can be translated into a formal argument with only a minimum of refinement. A learner may not find the technical terms 'speak to them', may lack intuition or insight and so may have to explore, and is very likely to be unfamiliar with writing formal reasoning. They may experience some or all of this as 'not knowing what is expected of them'.

The 'problem' for most learners is simply 'how to get the answer as quickly and painlessly as possible', whereas the author or setter sees it as an opportunity for learners to re-construct and appreciate techniques and concepts, and to experience mathematical thinking. Learners need to be supported in moving from a 'don't – can't – won't try' frame to one of 'can take on as a challenge to use my powers of thinking' (Dweck, 1999). If the word *problem* is only applied to tasks set by a teacher, we lose the importance of the learner's perception of worthwhile challenge. It is the learner who problematises tasks, for until the learner experiences the task as a mathematical problem to understand and make sense of (rather than a problem of getting an answer), tasks are simply tasks that 'have to be done'. Tasks which bring learners into contact with significant themes and concepts without requiring teacher intervention are known as *adidactical situations* (Brousseau, 1997).

When you encounter an unfamiliar problem, or when a problem involves unfamiliar technical terms, it can be very helpful to construct a specific example. To be useful, such an example must be confidence inspiring if not familiar. The purpose of this *specialising* as Pólya called it (Pólya 1945, 1962) is to develop confidence and to try to see what is going on so that you can re-generalise for yourself (Floyd, Burton, James & Mason, 1981, Mason & Johnston-Wilder, 2004b). You don't just 'do calculations', you watch what you do and try to see what is particular and what is general. By seeing through the particular to the general (Whitehead 1919) learners come to appreciate the generality, and to solve not just the particular problem, but other problems like it. Clearly experience of and disposition towards constructing examples both help in tackling new and unfamiliar problems. George Pólya (1945) famously divided problems into two kinds: *problems to find* (something), and *problems to prove*. In the following subsections we argue that both kinds can usefully be seen as construction opportunities.

Problems to Find

The language of *find* has the potential to summon up an image of someone looking around in dusty corners for something that someone, probably someone else, has mislaid. It is of course intended to signal the creative side of mathematics, but most often such problems come across to learners as a requirement to 'find the right technique and to apply it correctly'. An alternative perspective is to see 'problems to find' as construction tasks. Asked to solve some simultaneous equations in n variables, the challenge is to construct an n-tuple which satisfies all those constraints; to solve a differential equation the challenge is to construct functions which meet the constraint; to find an extremal value of a differentiable function on an interval the challenge is to construct a point that lies on the function and is a local extremum for that function; to find the definite or indefinite integral of some function the challenge is to construct a number, but more than that, a way of looking at the function which makes it amenable to one or more integration techniques.

Seeing a 'problem to find' as a construction task, which may be facilitated by the use of some familiar, and perhaps some not so familiar tools, is completely different psychologically from trying to work out the answers at the back of the book and from worked examples, what sequences of symbols of varying degrees of meaningfulness to write down and in what order, so as to satisfy the marker. Consider for example the following apparently simple 'problem': find $\int_0^2 (x-1)dx$. This might appear at the beginning of a collection of integrals of polynomials, with the intention that learners rehearse integration and evaluation of definite integrals. It is hardly 'a problem', except to learners who perhaps have no mental images and so are unaware that geometrically the answer is clearly zero, or to learners who are not overly confident in integrating and substituting correctly. Seen as one of many hurdles to overcome in sequence, it requires getting an answer. Alternatively, the task can be seen as a challenge to construct something. Seen this way, questions begin to arise about what sort of an object is sought (a number) and what this number might mean (apart from being 'the answer'). The number being sought is not just any number, but one which meets the constraint that it measures the area calculated by the integral.

Thinking in terms of dimensions of possible variation leads to questions such as how the number depends on the various parts of the question that could be altered. What if one or both of the limits were changed? What about changing the constant in the integrand? A learner approaching the task in this vein might be encouraged to find a general class of integrals, all of which give the same answer, such as $\int_a^b \left(x - \frac{a+b}{2}\right)dx$, which highlights the role of the arithmetic mean of the limits of integration and stimulates questions such as: 'what about changing the integrand?'; 'what possibilities are there for a general quadratic with the same answer?' One could go further and try to use geometric insight to predict the shape of multiple integrals and line integrals that also give zero as the answer, and to ask oneself about the significance of the zero. It is well known for example, that learners are often perplexed by 'negative areas' (Mason 2002). Suddenly what looks like a routine question turns into a possibility for exploration, for encountering important mathematical ideas, and for clarifying the mathematical meaning of 'area'. In the process of investigation, learners are likely to do several integrations (presumably the original intention) while at the same time making choices as to what functions to integrate, and using their own powers to specialise and generalise, conjecture and convince.

Problems 'to find' may sometimes have only a single answer, but often there are several or even infinitely many possibilities, all of which constitute a potential example space. Where learners are asked to construct not just a single example, but several examples from that space, one after another, we have found that they begin to appreciate the extent of that space, so that their thinking changes from being satisfied with the first (usually simple) example that comes to mind, to looking for the scope and breadth of generality possible. As this appreciation grows, as they consider more and more extreme or 'peculiar' examples (Bills, 1996a; Bills, 1996b) they can give rein to their creativity. For example, learners exposed to $|x|$ as a continuous function differentiable everywhere except at one point often do not appreciate the class of functions in the teacher's example space. By prompting them to use that idea to construct others, such as $|x| + |x-1| + |x-2|$, inserting coefficients, and then creating other cusps for themselves from other functions, they are likely to appreciate both cognitively and affectively the plethora of examples over which they have control (as distinct from a foreboding sense of functions lurking in the shadows of which they have no idea!). Appreciating the significance of a general solution to a differential equation as producing a class of functions which (in the case of first order at least) do not intersect and which 'cover' the space, makes the selection of a particular example, according to initial conditions, much more meaningful. Comparing what is the same and what different about different members of a general class can help learners comprehend the differential equation as a property being satisfied by a class of functions that differ in other ways.

Problems 'To Prove'

Pólya's distinction between 'find' and 'prove' is not rigid, because of course you can argue that a problem 'to prove' is a problem 'to find a proof', but this playfulness fails to appreciate the significant psychological difference between trying to find something you don't know, and trying to find a chain of reasoning that justifies something you do know (or conjecture). Perhaps the clearest example of the distinction is found in induction problems. There is a huge difference between

(a) Find $\displaystyle\sum_{k=2}^{n} \frac{1}{k(k-1)}$.

(b) Prove that $\displaystyle\sum_{k=2}^{n}\frac{1}{k(k-1)}=1-\frac{1}{n}$.

The former requires an element of creativity, with access to 'partial fractions' coming to mind, or else the construction of some particular cases and the formulation of a conjecture on the basis of these. Once there is a conjecture, the problem becomes one of proving. The problem 'to prove' expects learners to have 'induction' come to mind, or else to deconstruct $1-\frac{1}{n}$ as a telescoping sum:

$$1-\frac{1}{n}=\left(1-\frac{1}{2}\right)+\left(\frac{1}{2}-\frac{1}{3}\right)+\cdots+\left(\frac{1}{n-2}-\frac{1}{n-1}\right)+\left(\frac{1}{n-1}-\frac{1}{n}\right).$$

Note that during the induction step it is often difficult to gain insight into 'why' the proof works, into the structural underpinnings that make the statement true. See Harel and Brown (this volume) for a further discussion of students' conceptions of inductive proofs.

Learners who are satisfied with getting the answers to a suite of exercises are likely to learn much less than learners who develop the confidence that they could 'do another question of this type' in the future. The former can dwell in the particular, but the latter are engaging with the general, with "what does it mean to be a problem of 'this type'?". In our view, this is what constitutes mathematical thinking, as opposed to clerical proficiency. This perspective is supported by Vadim Krutetskii's research (Krutetskii, 1976). He found that learners who were quick at mathematics tended to remember numbers and structure, whereas learners who were not so quick tended to remember contexts and other surface features of the tasks they encountered.

Constructive Approach to Solving

When asked to solve a particular problem, usually through the use of a technique recently encountered, a typical strategy for learners is to find a worked example which looks similar. As Chi & Bassock (1989), Sweller & Cooper (1985), and Anthony (1994) have pointed out, the problem with worked examples is that choices are made as to what action to take at each step, and the basis for that choice may not always be clear to learners. Consequently, using the worked example as a template may achieve a solution, perhaps involving some tinkering due to slightly different setting, but it may not shed any light on 'how the technique is used', much less on when and why it 'works'. When problem solving is seen as construction, metaphors such as *bricolage* are available, where an attempt at a solution is assembled from available bits and pieces, which might at least shed light on where the real difficulty lies, or even on how to then set about constructing solutions systematically.

Alcock (2004) found through interviewing mathematicians that learners do not use examples in the same ways as mathematicians. Whereas mathematicians use examples to illuminate through instantiation, to see through the particular to the general, and to consider possible counter-examples, many learners see them at best as templates and at worst as 'more stuff to learn'. They often select examples for reasons which do not appear to be mathematically robust, relevant or informative, and to use them as demonstrations as if this constituted a proof. She goes on to suggest ways in which learners could be supported to make more effective use of the examples they do encounter, and in particular to get learners to construct their own examples (see also Dahlberg & Housman 1997, Watson & Mason 2002, 2005).

Freedom and Constraint

A pervasive theme throughout mathematics concerns the imposition of constraints on some otherwise general object, leading to characterising the collection of objects (if any) that satisfy those constraints. For example, starting with the freedom of an arbitrary pair of numbers, you can follow Diophantos (1964) and impose constraints such as that the sum is given, or perhaps that the sum of their squares is itself a square. An additional constraint can be included, such as that the difference is given, or that the numbers are also to be in a given ratio. Instead of searching around for a template to deal with the specific task, learners can become accustomed to starting from an unconstrained generality, and then imposing the constraints sequentially, seeking as general a solution as possible at each stage. For example,

- to find a number that leaves a remainder of 1 on dividing by 2, 3, 4, and 5, it can be useful to start with n as any integer, then to impose the constraint the remainder be 1 on dividing by 2, and to construct the most general number possible, and so on.

- to construct a solution to three simultaneous linear equations in three unknowns, it can be useful to think first in terms of a general triple of numbers with no constraint, as a point roaming through three dimensions. Imposing one constraint limits the movement to a plane, and expressing one variable in terms of the others effectively 'solves' one equation. A second constraint then confines the point to the intersection of two planes (if they are not parallel), and the first solution can be substituted in the second to yield a general expression for all those points.

The effect may be the same as using a prepared technique, but the psychological experience can be quite different as the learners' powers are made use of to develop a resolution. Sometimes the heuristic of trying particular cases can be useful as a means to locate underlying structure; sometimes a single particular case can be used as a generic example, as Hilbert is reported to have done:

He [Hilbert] was a most concrete, intuitive mathematician who invented, and very consciously used, a principle; namely, if you want to solve a problem first strip the problem of everything that is not essential. Simplify it, specialize it as much as you can without sacrificing its core. Thus it becomes simple, as simple as can be made, without losing any of its punch, and then you solve it. The generalization is a triviality which you don't have to pay much attention to. This principle of Hilbert's proved extremely useful for him and also for others who learned it from him. Unfortunately, it has been forgotten. (Courant, 1981, p. 161)

By attending to the actions taken in a particular case, it is sometimes possible to create your own template that can be generalised.

Seeing a problem in terms of constructing an object that satisfies a number of constraints not only evokes a spirit of construction, but opens the way to identifying and dealing with those constraints. Starting from the most general unconstrained object can be followed by expressing the most general object satisfying one constraint, then two constraints, and so on until a solution is found or shown not to exist. This is certainly powerful in algebra, where it can help learners to locate and express constraints symbolically in order to produce a symbolic statement of the problem. Isaac Newton was one of those who worked at a time when the mathematical focus of solving problems shifted from expressing constraints in symbols, to developing techniques for solving the resulting equations and inequalities. Many learners may not realise that the techniques they encounter in textbooks and lectures are the fruits of this kind of labour. But techniques are only useful once the problem has been expressed symbolically.

Geometry is a domain in which it often helps to see construction tasks such as: 'given three distinct concurrent lines, construct all the triangles for which these are the medians, or the altitudes, or the angle bisectors'. Even the problem that stimulated much of Schoenfeld's (1985) work, 'find a circle tangent to a given pair of lines and passing through a specified point on one of those lines', can fruitfully be seen not just as a construction task, but as a task with constraints imposed on initial freedom. The problem is to construct a circle. That is easy enough, but what matters is a sense of the freedom available: choice of centre and radius. Then it must pass through a specified point. Then it must be tangent to a given line through that point. Then it must also be tangent to another given line as well. By becoming aware of the possibilities at each stage, the solver not only gets a sense of the impact of the constraints, but also, by looking for the most general class of solutions at each stage, may find access to the consequences of each constraint in turn. Choosing the constraints in a different order is sometimes more helpful, but learners need to be aware that they are dealing with a sequence of constraints before they can change the order.

Some geometry problems succumb to the removal of one constraint and the construction of a locus that captures the freedom available without that constraint. Where the locus is recognisable (a straight line, a circle, …) it may suggest a conjecture that can then be used to complete the construction (Love, 1996). The algebraic version of this is to let go of one or more constraints and express the general class of all objects before imposing further constraints.

Concept Development as Construction

Exercises are the most visible, but by no means the most significant, aspect of learners' pedagogic experience (see the chapter by Weber, Porter, and Housman in this volume for a further description of the pedagogical advantages of using examples). Every concept — indeed every idea — has behind it a culturally rich collection of images and connections that Tall and Vinner (1981) call the *concept-image* (see also the chapter by Edwards and Ward in this volume for a further description of concept images). Indeed, it seems a reasonable conjecture that every technical term in mathematics signals an important shift in the way of perceiving and thinking that someone made in the past, and has to be re-experienced by each learner. A significant component of a person's concept image is the collection of examples and non-examples to which they have access, what Watson and Mason (2005) call their *example space*. Awareness of dimensions of possible variation can inform both the teacher in preparing encounters for learners, and learners as ways to probe their understanding of concepts.

The notion of dimensions of possible variation is particularly powerful when learning is seen as appreciating variation. Only if you know what can be varied can you appreciate the delicate relationship between particular and general, between an example and that which is exemplified.

Trying to understand a mathematical concept or theorem is much like trying to make sense of a mathematical problem. You have to ground yourself in something familiar, and this usually takes the form of an example that is sufficiently familiar to enable you to proceed with confidence. By following the theoretical or the general through the particularities of an example, it is possible to get a sense of what the theorem is saying or what the concept is encapsulating.

However, when a teacher offers an example and works it through, it is the teacher's example. Learners mostly assent to what is asserted. In a textbook, the words and examples become 'yet more to be learned', or ignored: yet more to which to assent, *en route* to the tasks. When learners construct their own examples, they take a completely different stance towards the concept or theorem. They 'assert'; they actively seek to make sense of underlying relationships, properties and structure which form the substance of the theorem or concept.

Developing appreciation of a mathematical concept involves finding yourself using the concept to express what you are thinking. As has been pointed out by many people (e.g., Lakoff, 1987), concepts are inextricably entwined with examples of those concepts. The richer the range of examples and the more extensive the sense of how to construct examples, the richer the appreciation of the concept. Halmos raises the important question of sources:

> Where can we find examples, non-examples, and counter-examples? Answer: the same place where we find the definitions, theorems, proofs and all other aspects of mathematics — in the works of those who came before us, and in our own thoughts. ... we find them first, foremost, and above all, in ourselves, by creative thinking. (Halmos, 1985, p. 64)

Furthermore, awareness of, or being able to construct examples which lie 'just over the boundary' of the concept, or providing counter-examples to weakened conditions of a theorem, alerts the learner to difficulties that may arise when the theorem is applied. A task or mathematical situation initiates a *space of examples* that may be given in the text (Michener, 1978), and may be enriched through learners being stimulated to construct their own examples (Watson & Mason, 2002, 2005). The richer and more complex that space, the richer the connections and sense of appreciation of the concept.

If learners are in the habit of constructing their own examples, and if they are supported in seeking generalities that encompass a range of examples, then they are likely to feel more secure in the use of the concept, as well as having access to a range of possible examples on which to test conjectures and through which to seek structure. For example, as was pointed out in the introduction, many learners treat $f(x) = |x|$ as an isolated, even perverse, example of a function, designed just to 'give them trouble' because they want continuous functions to be differentiable. Whereas it is easy and even natural to *monster-bar* a single example, it is much harder to do this to a huge class of examples. Whenever a counter-example is offered, what matters is whether learners are aware of how the single counter-example is just one illustration of a class or space of such examples, and how those other examples could be constructed. Klymchuk (2005) provides not only a wealth of counter-examples to erroneous but common learner conjectures and assumptions in the calculus, but also points to ways of making the most of the examples.

From Exercising to Exploring

One of the features of developing facility in the use of a technique is to reduce the amount of attention required to carry out the technique. The novice requires full attention to each step, with the result that they may not have enough free attention to watch out for slips, and they may lose their sense of direction overall. By contrast, having automated a technique, the expert has free attention both to catch slips and unusual features arising, and also to maintain a sense of the overall plan. As Jerome Bruner (1986) points out, a skilled teacher can act as *consciousness for two* by retaining a sense of the overall plan while the learner engages with particular details. Developing fluency in a technique involves becoming familiar with the overall flow and procedure. Gaining facility requires practice.

Traditionally, facility is developed through the doing of a large number of repetitive exercises, literally 'exercising' the fluent use of a technique or way of thinking about a problem. The trouble is that it is possible to do repetitive tasks without actually gaining facility, indeed without even thinking very much about anything! When doing a set of exercises, learners are rarely interested in the answers, but only whether they conform with those provided at the back of the book. Learners are not often supported and provoked into treating exercises as the raw material for mathematical development.

A much more effective approach advocated by Caleb Gattegno (1987) and developed by, among others Dave Hewitt (1994), is to engage learners in an exploration which, *en passant*, involves them in constructing examples for themselves in order to try to see what is going on. For example, finding the points in the plane through which a specified number of tangents can be drawn to a given curve such as $y = x(x - 1)$ or $y = x(x - 1)(x + 1)$ invites learners to choose points, to construct lines, to make them tangent to the curves, and to try to locate some overall pattern to the results which they might then be able to justify. In the process they are likely to gain considerable practice in differentiating and constructing straight lines. There is an opportunity to appreciate the growth of functions as x gets large (positively and negatively) in relation to potential tangents. Similarly, finding the distance between pairs of straight lines in three dimensions can be tedious as a set of exercises, but finding the minimum distance between two ruled surfaces amounts to the same thing, except that the learner is the one who constructs the pairs of straight lines, and is furthermore induced to work as generally as possible rather than being content with particular distances between particular pairs of lines.

If the particular examples that learners construct for themselves require the use of a technique with which they have some fluency, then it is in their interest to 'get the answers' because it will contribute to working out what is going on in the exploration. Thus learners can be induced to rehearse a technique on examples constructed by them rather than imposed from outside, and whose solutions matter to them.

Summary

In order to appreciate a concept, it is vital to be aware of the variation it entails, and the invariance that is preserved. In order to make sense of a technique for solving a class of problems, it is vital to be aware of the range and scope of that class. Learners who focus their attention on getting through tasks as quickly and painlessly as possible do themselves an enormous disservice, for they hasten the moment when they decide that mathematics is too much for them, and they cut themselves off from the pleasure of creativity which mathematics affords, and which is necessary in order to use mathematical concepts and techniques flexibly.

In order to promote learners' encounters with mathematical creativity, teachers can look for opportunities for learners to make choices, to construct objects as examples, and to articulate generalisations of particulars. When exercises, problems, examples of concepts, and counter-examples to obvious conjectures always come from the teacher or the text, learners are cut off from access to the creative and constructive aspects of mathematics, which are the sources of pleasure and the desire to find out more. Learners are dis-empowered and it becomes even more difficult for learners to shift from mere assenting to full engagement. We have argued, above all, that by seeing and posing tasks as construction tasks, using whatever familiar mathematical objects and techniques are at hand, learners' whole attitude toward the learning of mathematics can be altered for the good.

References

Alcock, L., & Weber, K. (2005). Referential and syntactic approaches to proof: Case studies from a transition course. In H. Chick & J. Vincent (Eds.), *Proceedings of the 29th Conference of the International Group for the Psychology of Mathematics Education*, *2*, 33–40. Melbourne, Australia: PME.

Alcock, L. (2004). Uses of example objects in proving. In M. Høines & A. Fuglestad (Eds.) *Proceedings of PME 28*. Bergen, Norway: Bergen University College, 2, 17–24.

Anthony, G. (1994). The role of the worked example in learning mathematics. In A. Jones *et al.* (Eds.), *SAME papers* (pp. 129–143), Hamilton, New Zealand: University of Waikato.

Bills, L. (1996a). *Shifting sands: students' understanding of the roles of variables in 'A' level mathematics*. Unpublished doctoral dissertation, Open University, Milton Keynes, United Kingdom.

—— (1996b). The use of examples in the teaching and learning of mathematics. In L. Puig and A. Gutierrez (Eds.) *Proceedings of the 20th Conference of the International Group for the Psychology of Mathematics Education*, 2.81–2.88, Valencia, Spain: Universitat de València.

Boaler, J. (2002). *Experiencing school mathematics: Traditional and reform approaches to teaching and their impact on student learning*. Mahwah, NJ: Lawrence Erlbaum Associates.

Brousseau, G. (1984). The crucial role of the didactical contract in the analysis and construction of situations in teaching and learning mathematics. In H. Steiner (Ed.) *Theory of Mathematics Education*, Paper 54 (110–119) Bielefeld, Germany: Institut fur Didaktik der Mathematik der Universitat Bielefeld.

—— (1997). *Theory of Didactical Situations in Mathematics: Didactiques des Mathématiques, 1970–1990*, (N. Balacheff, M. Cooper, R. Sutherland, & V. Warfield, Trans.). Dordrecht, The Netherlands: Kluwer.

Brown, L., & Coles, A. (2000). Same/different: a 'natural' way of learning mathematics. In T. Nakahara & M. Koyama (Eds.) *Proceedings of the 24th Conference of the International Group for the Psychology of Mathematics Education*, Hiroshima, Japan, 2–153.

Brown, S., & Walter, M. (1983). *The Art of Problem Posing*. Philadelphia: Franklin Press.

Bruner, J. (1986) *Actual minds, possible worlds*. Cambridge, MA: Harvard University Press.

Cardano, G. (1969). The rules of algebra: Ars magna (T. Witmer, Trans.). New York: Dover. (Original work published 1545)

Chi, M., & Bassok, M. (1989). Learning from examples via self-explanation. In L. Resnick (Ed.) *Knowing, learning and instruction: essays in honour of Robert Glaser*. Hillsdale, NJ: Lawrence Erlbaum Associates.

Christiansen, B., & Walther, G. (1986). Task and activity. In B. Christiansen, G. Howson, & M. Otte (Eds.), *Perspectives in mathematics education*. (pp. 243–307). Dordrecht, The Netherlands: Reidel.

Coles, A., & Brown, L. (1999). Meta-commenting: Developing algebraic activity in a 'community of inquirers'. In L. Bills (Ed.) *Proceedings of the British Society for Research into Learning Mathematics,* (pp. 1–6), Warwick University, Coventry, United Kingdom: MERC.

Courant, R. (1981). Reminiscences from Hilbert's Göttingen. *Math Intelligencer, 3*(4) 154–164.

Dahlberg, R., & Housman, D. (1997). Facilitating learning events through example generation. *Educational Studies in Mathematics 33*, 283–299.

Diophantos, (Heath, Trans.). (1964). *Diophantus of Alexandria : a study in the history of Greek algebra*. New York: Dover.

Dweck, C. (1999). *Self-Theories: Their role in motivation, personality and development*. Philadelphia, PA: Psychology Press.

Feynman, R., & Leighton, R. (1985). *"Surely you're joking, Mr. Feynman!": Adventures of a curious character* (E. Hutchings, Ed.). New York: Norton.

Floyd, A., Burton, L., James, N., & Mason, J. (1981). *EM235: Developing Mathematical Thinking*. Open University Course. Milton Keynes, United Kingdom: Open University.

Gattegno, C. (1987). *The science of education part I: Theoretical considerations*. New York: Educational Solutions.

Gillings, R. (1972 reprinted 1982). *Mathematics in the time of the pharaohs*. New York: Dover.

Halmos, P. (1980). The heart of mathematics. *American Mathematical Monthly, 87*(7) 519–524.

—— (1983). *Selecta: Expository writing*. In D. Sarasen & L. Gillman (Eds.), New York: Springer-Verlag.

—— (1985). *I want to be a mathematician: An automathography*. New York: Springer-Verlag.

Hewitt, D. (1994). *The principle of economy in the learning and teaching of mathematics*. Unpublished doctoral dissertation, Open University, Milton Keynes, United Kingdom.

Karp, A. (2004). Examining the interactions between mathematical content and pedagogical form: notes on the structure of the lesson. *For the Learning of Mathematics, 24*(1) 40–45.

Klymchuk, S. (2005). *Counter-examples in calculus*. Auckland, New Zealand: Maths Press.

Krutetskii, V. A. (1976). *The psychology of mathematical abilities in school children* (J. Teller, Trans.). In J. Kilpatrick & I. Wirszup (Eds.). Chicago, IL: University of Chicago Press.

Lakatos, I. (1976). *Proofs and refutations: The logic of mathematical discovery*. Cambridge, United Kingdom: Cambridge University Press.

Lakoff, G. (1987). *Women, fire, and dangerous things: What categories reveal about the human mind*. Chicago, IL: Chicago University Press.

Legrand, M. (1993). *Débate scientifique en cour de mathématiques*, Repères IREM, No. 10, Topiques Edition.

Love, E. (1996). Letting go: An approach to geometric problem solving. In L. Puig & A. Gutiérrez (Eds.). *Proceedings of the 20ᵗʰ Conference of the International Group for the Psychology of Mathematics Education*. Valencia, Spain: Universitat de València.

MacHale, D. (1980). The predictability of counterexamples. *American Mathematical Monthly, 87*, 752.

Marton, F., & Booth, S. (1997). *Learning and awareness*. Mahwah, NJ: Lawrence Erlbaum Associates.

Marton, F., & Trigwell, K. (2000). Variatio est mater studiorum. *Higher Education Research & Development, 19*(3) 381–395.

Marton, F., & Tsui, A. (2004). *Classroom discourse and the space of learning*. Mahwah, NJ: Lawrence Erlbaum Associates.

Mason, J., Burton, L., & Stacey, K. (1982). *Thinking mathematically*. London, United Kingdom: Addison Wesley.

Mason, J., & Johnston-Wilder, S. (2004a). *Fundamental constructs in mathematics education*. London, United Kingdom: Routledge Falmer.

Mason, J., & Johnston-Wilder, S. (2004b). *Designing and using mathematical tasks*. Milton Keynes, United Kingdom: Open University.

Mason, J. (2002). *Mathematics teaching practice: A guide for university and college lecturers*. Chichester, Sussex, United Kingdom: Horwood Publishing.

—— (2002a). *Researching your own practice: The discipline of noticing*. London: Routledge Falmer.

Michener, E. (1978). Understanding mathematics. *Cognitive Science, 2*, 361-383.

Pólya, G. (1945). *How to solve it: A new aspect of mathematical method*. Cambridge, MA: Princeton University Press.

—— (1962). *Mathematical discovery: On understanding, learning, and teaching problem solving* (Combined edition). New York: Wiley.

Runesson, U. (2005). Beyond discourse and interaction. Variation: A critical aspect for teaching and learning mathematics. *Cambridge Journal of Education, 35*(1) 69–88.

Schoenfeld, A. (1985). *Mathematical problem solving*. New York: Academic Press.

Senk, S., & Thompson, D. (Eds.). (2003). *Standards-based school mathematics curricula: What are they? What do students learn?* Mahwah, NJ: Lawrence Erlbaum.

Skemp, R. (1976). Relational and instrumental understanding. *Mathematics Teaching, 77*, 20–26.

Sweller, J., & Cooper, G. (1985). The use of worked examples as a substitute for problem solving in learning algebra. *Cognition & Instruction, 2*, 58–89.

Tall, D., & Vinner, S. (1981). Concept image and concept definition in mathematics with particular reference to limits and continuity. *Educational Studies in Mathematics, 12*(2) 151–169.

Watson, A., de Geest, E., & Prestage, S. (2003). *Deep progress in mathematics: The improving attainment in mathematics project*. Oxford, United Kingdom: Dept. of Educational Studies, Oxford University.

Watson A., & Mason, J. (1998). *Questions and prompts for mathematical thinking.* Derby: Association of Teachers of Mathematics.

—— (2002), Student-generated examples in the learning of mathematics. *Canadian Journal of Science, Mathematics and Technology Education, 2*(2), 237–249.

—— (2005). *Mathematics as a constructive activity: Learners generating examples.* Mahwah, NJ: Lawrence Erlbaum Associates.

Whitehead, A. (1919) (12th impression, reprinted 1948). *An introduction to mathematics.* London, United Kingdom: Oxford University Press.

16

Supporting High Achievement in Introductory Mathematics Courses: What We Have Learned from 30 Years of the Emerging Scholars Program

Eric Hsu, *San Francisco State University*
Teri J. Murphy, *University of Oklahoma*
Uri Treisman, *University of Texas at Austin*

Introduction

This article is aimed toward faculty in mathematics departments who are working to increase the number of high-achieving mathematics students from racial and ethnic minorities and for researchers investigating these endeavors. The Emerging Scholars Program (ESP) is one of the most widespread models for supporting such increases. It is also one of the oldest, so there is a considerable body of research, both quantitative and qualitative, related to its impact. Whether or not one chooses to implement an ESP, this discussion of the history, philosophy, structure, impact, and future of the program will highlight important and emerging themes that any related student support efforts must confront.

Individuals, departments, and colleges are invested in issues related to diversity for a number of reasons. One reason is a belief that all students should have the opportunity to pursue a satisfying career path. While acknowledging that personality plays some role in student academic choices, research and practice point to systemic factors that also affect students' choices (some of which we elaborate in the section below, A Brief Look at the Emerging Scholars Program Model). This implies that institutions themselves have a responsibility to address such factors. Another reason for attention to issues of diversity is the increased health that disciplines experience from drawing on diversity—e.g., a variety of backgrounds, goals, and perspectives. This reason grows ever more important as minorities continue to constitute an increasing proportion of the U.S. population. The effects of diversity in the discipline reach beyond mathematics and researchers in science and engineering fields. In fact, the large majority of mathematics majors go on to jobs in industry and into teaching jobs at the K–12, community college, and university levels. The effectiveness of minority mathematics teachers at these levels has not been definitively researched. However, small-scale qualitative studies have found that

> Students of color tend to have higher academic, personal, and social performance when taught by teachers from their own ethnic groups. (However, this finding does not suggest that culturally competent teachers could not achieve similar gains with students of color from different ethnic groups.) (National Collaborative on Diversity in the Teaching Force, 2004, p. 6)

Moreover, calculus in particular is a significant filter for premedical tracks, so the effects of student success—and particularly the success of minority students—in calculus cascade into the makeup and health of the medical professions.

For all these reasons and many others, many institutions and individuals are taking active responsibility for the high achievement of minority students in mathematics. This paper offers an overview of recent related research, as well as lessons learned during thirty years of work in Emerging Scholars Programs, some of which have been documented in the literature and some of which have been passed on as program folklore. This paper also documents some of the most recent relevant changes in the social and political landscape.

The year 2004 marks the 30th anniversary of the first piece of research directly related to the ESP model. It was an ethnographic study started in 1974 by Uri Treisman (Treisman, 1985, 1992; Fullilove & Treisman, 1990; Treisman & Asera, 1990), who was then a graduate student in mathematics at the University of California at Berkeley. The timely observation, based on an analysis of academic transcripts, that even high-achieving, high-potential minority students were struggling in freshman calculus, led to the question, "what was the overall experience of African-American and Hispanic students in calculus at UC Berkeley?" (Asera, 2001, p. 9). More details about Treisman's study and the resultant ESP model are provided in the next section of this paper, A Brief Look at the ESP Model.

The last thirty years have seen many attempts across the country to adapt ESP to local conditions. These decades have also seen the birth and growth of a variety of programs. Gándara (1999) offers an instructive overview of other program models that have documented systematic analysis of impact. In this study of 20 programs (including ESP), Gándara identified five strategies that such programs typically employ: "mentoring, financial support, academic support, psychosocial support, and professional opportunities" (p. 29). In this scheme, ESP strategies were classified as academic support and psychosocial support. Other programs have used different combinations of strategies and have successfully addressed different needs, goals, and target populations (e.g., graduate students). Throughout the report, Gándara stressed the need for more programs that emphasize excellence (as opposed to non-failure).

There have also been many changes in the social and political landscape since the beginning of ESP. One of the most encouraging changes has been the increase in bachelor's degrees awarded to black and Hispanic students. According to the latest comprehensive surveys performed by the National Science Foundation, there was an impressive increase between 1990 and 2001 in the absolute number of bachelor's degrees earned by underrepresented minorities, both overall and specifically in science and engineering. The proportion of bachelor's degrees earned by blacks and Hispanics has increased noticeably as well (see Table 1). We see slower growth when we focus on mathematics degrees, but there is still an increase in the proportion of mathematics degrees awarded to blacks and Hispanics. However, despite the general improvement in numbers, the proportion of blacks and Hispanics graduating with bachelor's degrees in science and engineering still lags far behind their respective proportions of the population as a whole. Nonetheless, the efforts of individuals, departments, and colleges to increase the success of minorities in mathematics-based fields have made a noticeable difference.

We must offer one caution. The existing research on minority academic performance is in its first stages. Issues with effect size, replication, and scalability thwart attempts to be absolute about conclusions, despite calls from the mathematics community to be definitive. In short, educational research on minority advancement in higher education is not in a mature state where any definitive results can be claimed.

Table 1. Bachelor's degrees earned by Blacks and Hispanics in 1990 and 2001 (National Science Foundation, 2001, 2004; U.S. Census Bureau, 2001b)

	Blacks (1990)	Blacks (2001)	Hispanics (1990)	Hispanics (2001)
Number of Bachelor's Degrees				
All Fields	59,301	106,648	43,864	89,972
Science and Engineering	18,230	33,869	13,918	29,262
Mathematics	720	845	413	668
% of Total Bachelor's Degrees				
All Fields	5.6%	8.5%	4.1%	7.2%
Science and Engineering	5.3%	8.1%	4.0%	7.0%
Mathematics	4.9%	7.2%	2.8%	5.7%
% of Population	11.7%	12.1%	9.0%	12.5%

To complicate matters, any results and recommendations considered from research need to be analyzed and adapted to local history and resources. In fact there is often a practical disconnection between educational research and practice. The reasons for this disconnection are many and well summarized elsewhere (e.g., Burkhardt & Schoenfeld, 2003). Indeed, this disconnect has been the general experience of ESP practitioners and program designers. They are often working at the edge of their time and material resources and cannot hope to be simultaneously immersed in the sprawling mathematics education literature on minority academic performance.

Despite these cautions, educational research can strengthen and refine practice. Excellent research and theory can call attention to hidden structures and subtle phenomena that have practical implications for program design. Research can also reframe one's understanding of the practical terrain in which programs are implemented or bring to mind fruitful and essential questions. As an example of research-informed practice, this paper begins with a brief description of the ESP model and an overview of the documented effects of programs based on this model. We will then discuss three major questions that can be addressed from new research around minority academic performance of the last 15 years. We conclude with a look at future challenges both for program designers and for education researchers hoping to make more practical impact with their work.

A Brief Look at the Emerging Scholars Program Model

The original version of the ESP model, the Professional Development Program Mathematics Workshop at the University of California at Berkeley, was designed in the 1970s in response to ethnographic research at UC Berkeley that described the contrasting study habits of black students and Chinese students (Fullilove & Treisman, 1990; Treisman, 1985, 1992; Treisman & Asera, 1990). In particular, the black students, who were more likely to fail or barely pass calculus, tended to study in isolation, separating their academic lives from their social lives. In contrast, their Chinese classmates, who were more likely to achieve high grades in calculus, tended to form cohesive support groups that factored heavily in the ability of these students to navigate the system. The first workshops began in Fall 1977, and the program that developed from this research proved to be not only successful, but portable. This section summarizes some of the literature on implementations at public research universities.

The overall goal of a mathematics workshop program is to increase student achievement by creating small diverse communities of learners who work on challenging mathematics in visible and collaborative ways. A key aspect of the program is that it emphasizes honors-level mathematics, not remedial. It is also important to note that across ESPs there is little uniformity of theoretical frameworks around group work and classroom culture (cf. the excellent overview of theoretical perspectives on classroom interaction in this collection by Bowers and Nickerson). This lack of uniformity is a result of the decentralized, spontaneous, and pragmatic spread of the ESP model. Some structural features, however, are much the same from one instantiation to another (Alexander, Burda, & Millar, 1997; Asera, 2001; Moreno, Muller, Asera, Wyatt, & Epperson, 1999; Murphy, Stafford, & McCreary, 1998; Treisman & Asera, 1990). In general, students in calculus at large research universities attend a lecture along with several hundred other students. They also enroll in a recitation section. During traditional recitations, a graduate teaching assistant answers homework questions at the chalkboard. In the ESP model, however, these recitations are replaced with sessions that have come to be known as workshops. Students in these workshops continue to attend the same lectures and take the same exams as the students in the traditional recitations. However, workshops typically are not only smaller than traditional recitations (12–20 students instead of 25–30), but also meet for longer blocks of time (75–120 minutes instead of 50) and more often (2–3 times per week instead of 1–2), a commitment for which students may receive additional credit hours.

For each workshop session, the graduate teaching assistant (GTA) constructs a worksheet, sometimes with the aid of a problem database (Hsu, 1999), composed of challenging problems. Asera (2001) characterized the features of the problems used on ESP worksheets: good problems pull ideas together from across multiple chapters, fill in gaps in student preparation without resorting to remediation, and are challenging enough to incite student collaboration and to teach students to persevere. During the workshop class time, students work individually and collaboratively on these problems (not on homework), while the GTA keeps the conversations moving forward without directly answering students' questions (Alexander, Burda, & Millar, 1997; Kline, 1994). Although the ESP literature does not tend to discuss GTA training and development in detail, many programs have in-house training sessions (e.g., Hsu, 1996) or send their GTAs to an annual session at the University of Texas at Austin (Epperson, 2003).

Several journal articles provide evidence of the ESP model's success and identify several kinds of effects. For example, ESP participants tend to achieve higher grades in calculus than underrepresented students who do not participate in the program—and often ESP participants receive higher grades than their (nonparticipant) white and Asian classmates (Alexander, Burda, & Millar, 1997; Bonsangue, 1994; Fullilove & Treisman, 1990; Moreno, Muller, Asera, Wyatt, & Epperson, 1999; Murphy, Stafford, & McCreary, 1998). ESP participants are more likely to complete the calculus sequence than nonparticipants (Alexander, Burda, & Millar, 1997; Murphy, Stafford, & McCreary, 1998). Furthermore, underrepresented minority students who participated in an ESP are more likely to persist in a calculus-based major (Bonsangue, 1994; Murphy, Stafford, & McCreary, 1998). The studies have had issues with defining appropriate control groups, if for no other reason than not being able to control for self-selection bias, i.e., the fact that ESP students choose to enter the program. Nevertheless, for the most part, the outcomes cited in these studies cannot be explained by preexisting differences in admissions criteria (e.g., SAT scores) between the ESP participants and the control groups. While there is scant research examining which program features contribute most to which outcomes and for what reasons, Herzig and Kung (2003) attempted to isolate the effects of cooperative learning and length of time in class. They concluded that "it is likely that the success of ESP results from some combination of the various aspects of the program, including length of time, use of group learning, types of students, and community-building activities" (p. 46).

Another question that arises regarding participant selection is the question of race identification. On the one hand, in the current post–affirmative action political climate, identifying potential ESP participants can be nontrivial. On the other hand, ESP has always emphasized diversity, recruiting students from all ethnic and racial backgrounds. For the most part (with the exception of Fullilove and Treisman [1990], which is outdated), details related to recruiting participants have been left to oral tradition and have not been systematically documented in the publicly available literature. It is believed that personal interactions with other units on campus and with high school teachers can facilitate the participant selection process. Some ESPs target specific high schools or geographic regions. Some work closely with units on campus that serve incoming freshmen. Several of the existing programs contact potential participants by letter. Then during summer freshmen advising time, ESP personnel meet with these students to explain the structure and philosophy of the program and, in some cases, ask students to pledge participation in the program.

In many cases, ESPs make use of both graduate and undergraduate teaching assistants. Although, again, this has not been the subject of research in the context of ESP, there is a belief among some program personnel that employing former participants as undergraduate teaching assistants has benefits at multiple levels. It is likely that the current ESP participants benefit from interacting with an upperclassman role model. It is also likely that the undergraduate teaching assistants themselves benefit both by mastering the course content at a new level and by experiencing teaching as a profession. This structure, where newcomers participate in increasingly legitimate and central ways, parallels that of the successful sites of stable communities of practice discussed in varied settings by Lave and Wenger (1991) and in mathematics teacher settings by Hsu (2004).

The process used to select graduate teaching assistants (GTAs) seems to be even more varied. For example, some coordinators identify potential graduate teaching assistants through informal conversations with department staff who interact with the GTAs and can discuss personalities and interests. Other departments rotate the teaching assignment through the GTAs so that it is not seen as a special teaching assignment (which might marginalize the program). The impact of these various processes has not been systematically investigated. However, teaching an ESP workshop is considerably more difficult than teaching in an average section, and we believe that ESP GTAs need to be carefully chosen for skill and diligence. Analogous to the belief that undergraduate teaching assistants themselves benefit, there is a belief that ESP experience can positively affect the teaching philosophy that graduate students take into their positions as teachers (Kung, in press).

In her milestone report, Asera (2001) was one of the first researchers to document the role of the ESP coordinator. She noted that, behind the scenes, the ESP coordinator is responsible for the program environment. In practice, this means that the coordinator is responsible for identifying and recruiting students, and perhaps GTAs, as well as monitoring the structures that enable a community to form. Other coordinator responsibilities include student advising and organizing social events. In the ESP context, there is a need for research related to each of these aspects, including the role of the coordinator and the relative importance of program features such as use of challenging problems, use of group learning, length of class time, designated space, student advising, and social events. Some of these features have been studied in other contexts, including college in general (Light, 2001; Tinto, 1993), but the relative importance of each to the success of ESP is as yet undetermined. Nevertheless, twenty years after the initial ESP workshops, Seymour and Hewitt (1997) published work in which minority participants characterized programs that work best: "well-advertised, departmentally-based, field-specific, open to all students and accessible" (p. 389). Although this description was not specifically referring to ESP, the ESP model certainly emphasizes all of those features.

Essential Questions from Recent Research

As noted above, the initial research base for ESP began in 1974. More than three decades have passed since then, with several important key issues emerging from careful research and the hard-earned wisdom of practitioners. Anyone who hopes to support high minority achievement in collegiate mathematics can expect to grapple with these issues or risk repeating the mistakes of the past. This section is devoted to considering these questions in detail, pointing to relevant research and examining the ways ESP has wrestled with these emerging issues.

Question 1: Do Your Introductory Courses Introduce? Do Your Preparatory Courses Actually Prepare?

The original Emerging Scholars Programs, located at research institutions, focused on calculus courses as being the most important and tractable. The calculus courses' importance stemmed from their being the primary filter course for mathematics majors, with research revealing that African-Americans and Hispanics were failing calculus courses at a higher rate than their Caucasian and Asian counterparts despite strong preparation and high test scores. The calculus courses were tractable in the sense that they were viewed by the mathematics departments as important enough to the major that resources could be directed to building a community around complementary workshops. Many ESPs found success focusing on first-time calculus takers, mainly first and second-year students. The rationale was that newer students were vulnerable due to lack of experience, but also not yet scarred by failing calculus, so one could still build around them a community with a challenging honors environment.

While ESPs have historically focused on calculus, research has increasingly called attention to the role of the preparatory mathematics courses in determining a student's future trajectory in the field. It is worth noting that what counts as preparatory varies from one institution to another. In some institutions—for instance, research universities— the preparatory course might be precalculus; at other institutions—for instance, community colleges—it might be college algebra or below. In any case, minority students tend to begin their coursework one course before the official start of the mathematics major (Ruddock, 1996). In fact, national data reveal that underrepresented minority undergraduates disproportionately place into remedial mathematics courses in college (see Table 2). This is symptomatic of a national phenomenon, as reported in a study by the National Science Foundation (1999):

> The mathematics course-taking patterns of black students have an effect on their participation in other science, mathematics, and engineering fields. Participation rates are high at both the precollege level and at the levels of college algebra and precalculus. The participation rate is lower in calculus and lower than it should be in finite/discrete mathematics given the proportion of black students who major in computer science. (p. 44)

Certainly not all students who enroll in mathematics courses at the developmental, algebra, trigonometry, and precalculus levels intend to continue into calculus. However, these courses serve many students who do intend to pursue a calculus-based major (e.g., about 69% in Bergthold and Ng, 2004), and these courses are disproportionately populated with minority students. Thus, the experiences of minority students taking preparatory classes should be rethought. This may mean redistributing department attention and energy. Often mathematics departments value, in decreasing order, graduate courses, upper-division courses, lower-division courses, introductory major courses, and preparatory courses. That is, departmental investment tails off as one looks earlier into the major sequence—from courses that are seen to produce majors to courses that are not seen as feeding the mathematics-major pipeline. To effectively support the success of minority students, however, we need to work on the courses in which these students actually begin.

If one frames the obstacle to building diversity in mathematics as high failure rates, a tempting solution is to achieve low failure rates by helping students not fail, as opposed to the ESP philosophy of pushing students to excel. Ruddock (1996) reframes the problem in terms of the course trajectories of students, showing that one must worry not just about how students do in preparatory courses, but about how they do in subsequent courses. Ruddock studied

Table 2. Percentage of first- and second-year undergraduates who reported (1999–2000) ever taking remedial mathematics courses (National Center for Education Statistics, 2002, p.132)

Ethnic Group	% Taking Remedial Mathematics
White	13.3%
Black	19.5%
Hispanic (any race)	20.4%
Asian/Pacific Islander	12.6%
All Undergraduates	15.0%

the success of students at the University of Texas at Austin who began their mathematics careers in precalculus compared to those who began in calculus. The course title *precalculus* emphasizes its role not as a proper subject of study, but as preparation for success in calculus. Thus, Ruddock tracked the success of groups of *precalculus-first* and *calculus-first* students. Defining *success* as receiving an A or B in a course, she found the precalculus group had a significantly lower percentage of success in both first-semester and second-semester calculus. This result held even when she restricted the study population to students scoring between 500 and 600 on the SAT Mathematics test to control for mathematics aptitude. The result also held when disaggregated by ethnic group. One particularly intriguing result was that students who scored a B in first-semester calculus were less likely to succeed in second-semester calculus if they were precalculus-first. She also examined mathematics major graduates to see what their beginning course was. Overwhelmingly, they began in calculus. Ruddock found similar results from the available data at the University of California at Berkeley. Schattschneider (2006), in a study of two four-year colleges and two two-year colleges, also found that between one-half and two-thirds of all students who passed precalculus did not pass first-semester calculus.

Attempts to adapt the ESP model to address underpreparation have met with varying degrees of success. For example, at the University of California at Santa Barbara, the initial ESP calculus results were so encouraging that they began including participants with lower and lower levels of preparation. The effect was negative both for the students and for the workshops as a whole. On the one hand, some students were put into a situation that was outside their capacity at that time; on the other hand, the workshops suffered under the strain of trying to support these students. The guideline seems to be that students in a calculus workshop need to be within one standard deviation of the mean in terms of their mathematics background. In a more positive example, the University of Kentucky successfully took advantage of local history and resources to focus on precalculus as well as calculus. This work continues there as the successful MathExcel program. The strength of ESP lies in emphasizing excellence, but the details of how to implement this philosophy successfully and broadly in courses that precede calculus have yet to be pinned down.

Question 2: How Salient Should One Make a Student's Minority Status?

Minority academic performance is typically framed as an issue requiring the institution to serve its minority students differently, with solutions then framed as the creation of special programs for minority students (e.g., Minority Engineering Programs). Often, students are recruited for their minority status, and this status is highlighted throughout the proceedings. Over the last three decades, this has been a pragmatic, simplified approach to the complicated issue of ethnic identity. Yet this approach is really too simplistic. In a seminal study, Seymour and Hewitt (1997) interviewed science, mathematics, and engineering (S.M.E.) undergraduate majors at seven institutions (none of which had an ESP). Among the 335 participants were 88 students of color—black (27), Hispanic (20), Asian-American (35), and Native American (6). From these interviews, Seymour and Hewitt identified areas that specifically affect students from these populations: "differences in ethnic cultural values and socialization; internalization of stereotypes; ethnic isolation and perceptions of racism; and inadequate program support" (p. 329). They also caution, however, that despite research efforts to characterize issues common to all students, and additional issues common to minority students, it is naive to think of minority students, or any group of minority students, as a uniform population. Seymour and Hewitt emphasized that

> differences among and within different racial and ethnic groups have greater significance for the chances of success than had previously been assumed. ... Failure to take such differences into account may, in and

of itself, explain why programs intended to recruit or support "minority students" have not improved their chances of survival in S.M.E. majors. ... [Indeed], any statement purporting to summarize the experience and attrition risks of all non-white S.M.E. students tends to distort and mislead. "Minority programs" based on presumption of needs common to all "minorities" tend to founder, quite largely, because they do not address the needs of specific racial and ethnic groups. (p.322)

Race and ethnicity have always been much more complicated than can be captured by the simple categories used to track students. But race is more complicated today than it was even a decade ago. Indeed, the ethnic landscape has been changing. Racial integration of neighborhoods has been steadily increasing, and the number of multiracial families is growing rapidly. Furthermore, recent research calls attention to the idea that the act of highlighting people's identity (gender or ethnicity) can have unexpected and unintended negative consequences. Also, one cannot ignore the significant shifts in the political landscape with respect to affirmative action, which have large implications for many race-targeted programs. In the following sections, we present four themes from research related to the question of minority status: (1) Stereotype threat, (2) Multiethnic and multicultural racial identities, (3) Legal and political challenges to race-targeted programs, and (4) The changing landscape of race.

Stereotype threat. A fascinating series of experiments show that by making students' minority status salient, one can influence their test performance measurably. The original series of studies by Claude Steele, Joshua Aronson, and Steven Spencer used difficult questions from a GRE verbal test (Steele & Aronson, 1995). Two groups of students were given the same test. Each group had both white and black sophomore college students of equal academic qualifications randomly assigned. However, one mixed-race group was told the test measured their ability, and the other group was told it was not a test of ability but was instead designed to discover how students approached these problems. The performances of all the white participants and the "non-ability-measured" African-American students were similar, but the "ability-measured" group of African-American participants did measurably worse.

The theoretical explanation is that African-American students in the "ability-measured" group felt significant pressure not to confirm the negative stereotype about African-Americans having low intellectual abilities. This pressure, dubbed *stereotype threat*, caused them to underperform on the test. Neither the "non-ability-measured" group nor the white students experienced stereotype threat, and their performance was as expected. To show that students were actually experiencing a heightened awareness of negative stereotypes, Steele and Aronson repeated the original experiments with two intriguing pre-tests. First, participants were given a series of words with some letters known and some unknown, and were asked to complete the word by finding the missing letters. The twist was that in some cases there were multiple correct responses, some of which were related to stereotypes and some not. They found African-American participants in the "ability-measured" group answered with significantly more words relating to racial stereotypes than the "non-ability-measured" African-American participants or any of the white students. Students were also given a sports and music preference survey right before the test, and African-American students in the "ability-measured" group reported considerably less interest in stereotypically African-American preferences like basketball and hip-hop compared to the "non-ability-measured" group. This last result was interpreted to mean that students about to take an "ability-measured" test were unusually intent on not being stereotyped by their race.

One question left open by the original work was whether stereotype threat was the result of internalized self-doubt within African-American students or whether stereotype threat was triggered externally by making negative stereotypes salient. Later experiments extending the original work seem to have settled the question in favor of external triggers (Steele, 1997; Aronson, Quinn, & Spencer, 1998). For instance, white male students (who, as a group, are supposed to not have the same internalized self-doubt) had their mathematics test performance depressed when they experienced external triggers. That is, when they were told that Asian students outperformed whites, they experienced an externally induced stereotype threat (Aronson et al., 1999). Stereotype threat performance depression has been replicated for high-achieving female mathematics students (Spencer, Steele, & Quinn., 1999) and Hispanic students (Aronson & Salinas, 1997) as well. One intriguing small-scale study by Inzlicht and Ben-Zeev (2000) showed that female mathematics test performance decreased in proportion to the number of males in the room, even though verbal test performance was unaffected. In a fascinating experiment with Asian-American women, participants were given one of three questionnaires that induced external stereotype threats: Questionnaire 1 made their female identity salient by asking questions about their sex and gender identity, Questionnaire 2 made their Asian identity salient, and Questionnaire 3 (the control condition) asked questions unrelated to sex or race. Once these questionnaires were completed, all of the groups were given a quantitative test. The results revealed that the "female" group (those receiving Questionnaire 1) performed worse than the control group, while the "Asian" group (those receiving Questionnaire 2) outperformed the control group (Shih, Pittinsky, & Ambady, 1999).

There is some evidence that telling participants explicitly that stereotype threat could affect them can eliminate its effects. For instance, Johns, Schmader, and Martens (2005) performed a test comparing performance on a mathematics test by women in a "non-ability-measured" group, the usual threatened "ability-measured" group, and a threatened "ability-measured" group that was then informed about stereotype threat and its likely effects on performance. While the second group performed worse than the first and the third, the "informed" group performed as well as the first. Another study (McIntyre, Paulson, & Lord, 2003) showed that reading about successful women helped alleviate the effects of stereotype threat.

Multiethnic and multicultural racial identities. A good deal of research and practice makes a simplifying assumption that when we assign students to an ethnic category, they themselves identify with that categorization. This can be a helpful working assumption, and indeed some ESPs make this working assumption, but research has shown that it is becoming increasingly problematic. In what follows, we present four recent trends that complicate issues of race.

First, this generation of students entering college, regardless of ethnicity, is more likely than past generations to have grown up in a racially diverse community. To be certain, racial integration is not occurring at a constant level and is not ubiquitous. According to the Lewis Mumford Center's analysis (2001) of census data, both urban and suburban neighborhoods still tend to be segregated by race. However, there was a decrease in segregation in the 1980s, and a smaller, but still significant, decrease in the 1990s. Though the Mumford Center expresses worries about the slowing rate of diversification of neighborhoods, the fact remains that segregation continues to decrease, and the effects have been cumulative. Hence, our students may have very complicated ethnic identities compared to those who grew up with greater segregation a generation ago.

Second, this generation of students is more likely to identify with multiple or mixed ethnicities. Since 1960, rates of intermarriage between ethnicities have increased in every ethnic category. In 1960, 0.4% of all marriages were interracial; in 1980, 2% were interracial, and by 1992, 2.2%. In 2000, the census began recording a wider range of multiethnic identifications, and interracial marriages (including situations with two partners of mixed race) were recorded as 7.4% of total marriages. Interracial cohabitation made up 15% of the cohabitation responses (U.S. Census Bureau, 2003a). These rates are higher when restricted to the college-educated population (Qian, 1997, and Kalmijn, 1998), which makes this effect even more significant for college settings, as college-educated families are more likely to send children to college.

Third, what constitutes "black" identity continues to evolve. While a number of programs focus on urban poor African-Americans, there continues to be a growing African-American middle class. Pattillo-McCoy (1999) surveys the current research on the black middle class, and finds that the proportion of blacks with white-collar jobs has been increasing since World War II. At first the growth was explosive, triggered by the post-war economic expansion and the Civil Rights movement; growth slowed in the recession of the 1970s and has not to date recovered its rapid pace. Nonetheless, the percentage of African-Americans in middle class occupations rose from 39.6% (1980) to 44.9% (1990) to 49.8% (1995). Alba, Logan, and Stults (2000) argue that middle-class African-Americans live in neighborhoods that are significantly more integrated than the neighborhoods of inner-city blacks. To be sure, there is still significant racial inequality between the black and white middle classes: the black-white income gap remains, black middle class jobs are lower paying, and the black middle class lives on average in less safe neighborhoods. By almost every economic and social measure, the black middle class lags behind its white counterpart. Nonetheless, the black middle class continues to grow, is socially distinct from the urban black lower class, and is overrepresented in the college population. Sacks (2003, p. B7) reports that among black students at elite colleges, "60 percent of their fathers and more than half of their mothers were college graduates. One-third of their fathers had advanced degrees" (citing Massey, Charles, Lundy, & Fischer, 2002) and "nearly nine of 10 African-American students admitted to the most competitive colleges had come from families in the top two tiers of the social and economic ladder" (citing Bowen & Bok, 1998).

Another recent phenomenon affecting "black" identity is the new attention to subgroups of African-Americans: blacks whose ancestors were forcibly taken to America for slavery, immigrants from the West Indies, and immigrants from Africa. Students from each of these different subgroups are treated the same as African-Americans, despite the subgroups having completely different social histories. In fact, tensions between American blacks and West Indian blacks over jobs and culture date back to the early 20th century (Woodbury, 1993). In a provocative article, Phelps, Taylor, and Gerard (2001) claim that descendents of slaves have increased levels of cultural mistrust and attachment to their ethnicity compared to West Indian and African immigrants. These issues will only intensify in the coming years, as U.S. Census Bureau data (see Table 3) show that the percentage of the U.S. foreign-born population born in Africa has approximately doubled every decade. This immigrant group is more educated (88% with a high school education) and more affluent (more than a third more per-capita income) than native-born Americans and Asian immigrants (Speer, 1994). Half of the African-born immigrants are black.

The growth in the numbers of blacks who are voluntary immigrants from the West Indies and Africa has led to increasing debate over what constitutes black identity and the purposes of affirmative action programs (Johnson, 2005). Rimer and Arenson (2004) report from Harvard that Professors Lani Guinier and Henry Louis Gates caused an uproar when announcing "the majority of [the 530 black Harvard undergraduates] — perhaps as many as two-thirds — were West Indian and African immigrants or their children, or to a lesser extent, children of biracial couples." Furthermore, Rimer and Arenson cite a study of selective universities which showed "41 percent of the black students identified themselves as immigrants, as children of immigrants or as mixed race." Careful studies of this phenomenon at any one site are made difficult by the general aggregation of all these groups in campus statistics as "black".

Table 3. African-Born Living in the U.S. (U.S. Census Bureau, 1999, Table 2; 2003b)

Year	1960	1970	1980	1990	2000
Number of African-born living in U.S.	35,355	80,143	199,723	363,819	~870,000

Fourth, the composition of the Hispanic population is changing rapidly. It is well-known that the Hispanic population is the fastest growing minority group in America. The Hispanic population in the United States grew 57.9% between 1990 and 2000, increasing from 22.4 million to 35.3 million (U.S. Census, 2001a). One striking finding from the census was that the national makeup of the Hispanic population is changing. Historically, most Hispanics have been of Mexican, Puerto Rican, or Cuban descent. However, in the last decade, Hispanics from other countries nearly doubled from 5.1 million to 10.0 million, and the proportion of all Hispanics has increased from 22.7% to 28.3%. Prominent subgroups were Salvadorans, Guatemalans, Hondurans, Colombians, Ecuadorians, and Peruvians. One should not underestimate the effects of identification with distinct subgroups of the Hispanic population. Students may resent being lumped in with other subgroups with whom they feel no identification, or worse, rivalry. For instance, developers of bilingual educational software for the Los Angeles public schools reported friction from test groups over voiceovers in multimedia. The conflict was over the perception that the choice of regional Spanish accent showed bias towards either students of Mexican descent or students from other Hispanic groups, who considered themselves distinct and rival subgroups.

We only have space to mention briefly that an analogous issue affects the lumping together of students as "Asian". This paper itself is an example of this error. We have concentrated mainly on issues affecting black and Hispanic students, ignoring the fact that there are wide differences in academic performance among Asian subgroups. These subgroups have differences in socioeconomic status, national identity and immigrant experiences. Even within these subgroups, students will have had widely varying amounts of time since their family's immigration to the U.S.

Legal and political challenges to race-targeted programs. The *Hopwood v. Texas* decision made a splash nationally upon its issuance in 1996, and again in 2001, when the University of Texas declared an end to its legal appeals. The decision itself was fairly limited: it forbade the use of race-based admissions criteria in higher education institutions in the 5th Circuit of Appeals (Texas, Louisiana, and Mississippi). Furthermore, the Supreme Court refused to hear an appeal on the technical basis that the University of Texas Law School had changed the admissions policy in question, leaving open the possibility of revisiting the question of the merits of the appeal in the future. Nonetheless, the *Hopwood* case signaled a broader shift in the political climate against affirmative action. For instance, the University of California system ceased the use of race for admissions decisions even before Proposition 209 was passed in 1996.

Colleges and universities are increasingly choosing to stop the use of racial preferences, even for admissions to internal programs, in response to the political atmosphere rather than out of any specific legal obligation. The Supreme Court did revisit the issues of racial preferences in the twin 2003 decisions on *Gratz v. Bollinger* and *Grutter v. Bollinger*, and the Court issued a subtle (and puzzling to many) dual decision which allowed administrations to use race as a factor in admissions, but not in a mechanical way. This decision gives some legal maneuvering room for race-based programs, but the practical climate remains negative, and as a trend, campuses are putting pressure on programs to justify targeting students by race, which is why ESP has had an advantage over most other interventions aimed at supporting minority success in that the program has always recruited students from all ethnicities.

ESP and the changing landscape of race. One of the essential features of ESP is the need for a truly diverse classroom. Part of the effectiveness of the workshops comes from having students seeing other students from a wide range of ethnicities struggling and then succeeding with the same mathematics. For example, deep stereotypes can be broken when an African-American sees a Chinese student struggling and gets a more realistic measure of the work of mathematics. It is possible that this aspect of ESP helps to defuse the effects of stereotype threat. It certainly facilitates exposure to a variety of perspectives, approaches, attitudes, and values.

To these ends, ESP workshops have always recruited students of all races, and aimed for a mix of students. Because of this, at the University of Texas, ESP suffered less than most other programs aimed at supporting minorities in the post-*Hopwood* era. The University of Texas ESP does recruit black and Hispanic students more heavily, but it also targets rural whites, another "at-risk" group, and aims for an "ideal" calculus classroom in which minorities are more represented than they are usually, but are by no means the only students. Undoubtedly, ESP will need to evolve in response to the increasing mixing of ethnicities and increasing complexity of black and Hispanic identity. However, in principle, the ESP workshops have already been constructed to bring together a rich mix of students from many ethnic backgrounds, and it seems possible that this mix will only grow richer as student ethnicity evolves.

Question 3: How Can One Design a Program That Lasts?

In many ways, this is the most important practical question to be answered. ESP is an interesting program because many people in many different settings have attempted to adapt ESP to their local setting. Hence, the ESP community has gained experience with a wide array of locales. While we will focus on the experience of Emerging Scholars Programs, the structural analysis will be relevant for any campus efforts to increase student achievement in specific disciplines.

One might think that with all the experience of the years of work on ESP, there might be some kind of algorithm for optimally designing a local ESP. But the reality is that the reasons for success and failure at different sites are still mysterious. There are, of course, some principles and patterns that have emerged, but this is not the same as a systematic exploration of the factors that contribute to lasting effectiveness of a program. The most definitive work on the factors contributing to ESP's effectiveness is by Asera (2001). However, Asera cautions:

> When [ESP] staff described the program in comprehensive detail, other campuses tended to reproduce those details as exactly as possible, which was usually inappropriate. But when the staff resorted to describing instead the program's driving principles, that strategy, too, proved problematic, since the principles without the weight of specific examples were far too easily misunderstood. (p. 29)

The first step in designing a successful program is understanding local conditions. We present here a framework the authors have found useful for this purpose. One often refers to a system's potential to change as its *capacity*. Such a notion is more productive when broken into the components of *human capital, social capital, material resources,* and the *structural support* of the institution. These components change with time; it is crucial to understand them not just as resources at hand, but ones with a local history and a future trajectory.

Human capital refers simply to the knowledge, skills, and goals of the people available to do the work of a project. In the case of ESP, this means the project staff, the campus faculty and administrators that support and guide the project, and the graduate instructors who teach in the program. *Social capital* refers to the bonds of trust and reciprocity residing in a social network—e.g., the extent to which the members of a department cooperate internally, with the administration, and with other departments and student support units on campus. *Material resources* refers to funding, physical space, equipment, and other tangible resources. *Structural support* refers to the institutional arrangement of hierarchical relations, obligations, and incentives—e.g., policies encouraging joint work between academic departments and student support units, the process of reviewing and altering the sequence and the content of courses, and the college incentives for doing service or building innovative programs.

Starting a program requires a different configuration of resources than maintaining and developing a program. The experience of ESP is that there needs to be at least one driving advocate, and that advocate should be a senior faculty member. Such a person will have a better understanding of the structural supports available and, through reputation and history, wield more social capital in the department and campus social network. It is important to gain the help of people who have experience and access to the higher campus administration and national networks of scholars.

Compatibility with a college's structures of support is a crucial consideration. A program can exist briefly on short-term resources, such as a heroic individual leader (human capital), outside grants and other soft money (material resources), or the piecemeal support of sympathetic faculty and staff (social capital). However, for a program to last, it must ensure a future flow of each of these resources. Good structural support ensures the future resources that keep a program thriving and growing. What this means will change depending on the campus. But there are several themes that have emerged across many ESP projects.

First, as quickly as possible, the initial efforts should take the form of departmental committees given the power to develop the project. Committee work is a framework that can attract further human capital and invoke a broader sweep of a department's social capital. Perhaps most importantly, departmental committee work fits into the usual work and reward structure of a college institution. Furthermore, a committee institutionalizes a project as a departmental priority and helps a program survive turnover in faculty leadership.

Similarly, it is important for ESP workshops to be given for mathematics department course credit (as opposed to no credit or general education credit) as a lower-division seminar attached to a calculus course. Giving course credit structurally commits the mathematics department to devoting ongoing and future resources to the program in a way that is impossible if the program is housed in a non-departmental student support unit or in some other administrative unit. It commits the department to assigning teachers and staff (human capital) and classrooms and materials (material resources) into the future.

Some ESPs have designated office and class space on campus, and anecdotal evidence shows it is a great advantage to have this. It certainly adds an air of stability and legitimacy to an effort. Students and staff come to consider the physical location a "home" and it gives people a common site to establish continuity across different generations of staff and students. Structurally speaking, once an academic program has a physical location, it takes more effort for an institution to move it or eliminate it. If a program is lucky enough to get a physical foothold someplace, it will be under continual pressure from different sides to justify the use of the space. Nonetheless, having a site is a powerful way to ensure future access to space and equipment material resources.

One common structural obstacle is institutional discouragement of cooperation between departmental efforts and student support units. In principle, the student support units are potential resources in any effort to support minority students. In practice, they exist in a historical context. Often, support programs are underfunded and live on the edge of being merged or eliminated. It is not uncommon for such programs to develop an adversarial attitude toward both the administration and the academic departments. Historically, departmental initiatives may have proceeded without input from the support unit and served to sap both resources and students from the support units. In these cases, it will take work and real collaboration to avoid "turf wars". Unfortunately, many colleges don't give incentives or support for such difficult efforts, often using the excuse that such partnerships already ought to exist. If members of the department have good relations with the support units, that would be powerful social capital to draw upon.

One enduring challenge to ESP is dealing with turnover in human capital. For instance, ESP workshops are run by graduate instructors, and thus are guaranteed turnover in the classroom leadership. Naturally, different ESPs address this issue differently. The ESP at the University of Texas at Austin holds multiday teaching assistant orientations run by veteran workshop leaders. Workshop mathematics problems are passed from generation to generation in paper copies and in Internet-accessible searchable databases (Hsu, 1999). However, as ESP has become institutionalized in certain locations, it has been a challenge to maintain the highest standards for workshop leaders. In the early stages of such a program, often the first instructors are highly committed educators, sometimes drawn from the mathematics education group on campus. As the program becomes less novel and more routine, there is a danger the mathematics department will see the program as just another piece of its academic program and either make little effort to recruit excellent instructors for the program, despite the increased difficulty of the teaching task, or, in the worst case, intentionally randomly distribute graduate instructors out of a misguided sense of fairness.

The recurring, underlying caution is that every project must suit local institutional goals and culture. Indeed, it is essential that early in the process the leadership for a new ESP investigate the department and college's true needs. This means investigating the backgrounds and aspirations of the students one means to support and getting a true picture, both historical and projected, of the landscape of the different components of capacity mentioned above. What courses are students taking? What preparation do they have? Are there enough students to fill a calculus workshop, or are the students placing into precalculus or directly out of first-year calculus with Advanced Placement credit? Are majors getting weeded out in a course later than calculus? Because of the diversity of local situations, there is no simple formula for designing a program that lasts. The one thing that is certain is the better one understands institutional history, resources, and needs, the better one's chances at making an enduring difference.

Conclusion: Thirty Years of Research to Practice

ESP was created in response to a practical crisis at the University of California at Berkeley of overwhelming failure at the calculus level of black and Hispanic students, but was rooted in ethnographic research on black and Chinese communities of mathematics students. The last three decades have seen significant progress for minorities in mathematics, but there is still farther to go, and the issues are evolving with time and new knowledge. Research has highlighted the importance of different hidden structures in people's relationships to their ethnicity and academic environments, the landscape of race relations has been changing over time, and the political landscape has shifted as well.

Although ESP and its adaptations have evolved, the underlying philosophy remains much the same: "the philosophical stance that informs all the essential elements of the ESP model is that its purpose is not to 'fix the students,' but rather to change at least a small part of the university environment, by making it more welcoming, both socially and academically" (Asera, 2001, p. 19). In particular, the model continues to emphasize excellence, diversity, and community.

Along with the general trends and issues identified in this paper, three other major challenges are emerging specific to ESP. First is the issue of creation of mathematics majors. ESP has documented great success at having its students succeed in calculus and go on to science and engineering majors. However, in practice, it takes a significant mentoring structure to encourage ESP students to major in mathematics. In fact, this issue afflicts mathematics as a college discipline nationally. From 1993 to 2001, the absolute number of mathematics bachelor's degrees awarded has decreased every single year, from 14,870 in 1993 to 11,748 in 2001 (NSF, 2004), to the point where mathematics constitutes a mere 2.8% of all science and engineering bachelor's degrees. ESP has in fact produced a number of mathematics majors, but mainly in situations where extraordinary opportunities were created for students to engage in the work of mathematics, such as independent study and summer internships.

A second challenge is one of cooperation across ESP sites. One of the strengths of ESP has been its adaptability across diverse institutions. However, this has led to a lack of systematic and sustained cooperation across ESP sites. There has been cooperation to be sure, but it has tended to be ad hoc, e.g., the gathering of ESP veterans from the University of Wisconsin, the University of California at Berkeley, and the University of Texas at Austin to categorize and annotate the database of worksheet problems described in Hsu (1999). Other efforts have been informal, e.g., the sending of graduate teaching assistants to the new-GTA training at the University of Texas at Austin ESP. There is a growing movement by ESP sites to share experiences and resources as well as to document collectively the findings of practice.

A final challenge is the issue of leadership transition. The most successful programs are starting to encounter the issue of generational change in the program leadership. The University of California at Berkeley, the University of Kentucky, and the University of Illinois at Urbana-Champaign have seen the recent departure of the original torch-bearing faculty who established the local ESPs. Time will tell how these ESPs and others evolve and how well they continue to show resilience as a second generation of leadership takes charge, in a climate that is socially and politically different than it was thirty years ago.

Acknowledgments. This article was improved dramatically due to the editing efforts of Rachel Jenkins and due to suggestions made by Cynthia Foor. This material is based in part upon work supported by the National Science Foundation under Grant No. 0347784. Any opinions, findings, and conclusions or recommendations expressed in this material are those of the authors and do not necessarily reflect the views of the National Science Foundation.

References

Alba, R. D., Logan, J. R., & Stults, B. J. (2000). How segregated are middle-class African Americans? *Social Problems, 47*(4), 543–558.

Alexander, B. B., Burda, A. C., & Millar, S. B. (1997). A community approach to learning calculus: Fostering success for underrepresented ethnic minorities in an Emerging Scholars Program. *Journal of Women and Minorities in Science and Engineering, 3*(3), 145–159.

Aronson, J., Lustina, M. J., Good, C., Keough, K., Steele, C. M., & Brown, J. (1999). When white men can't do math: Necessary and sufficient factors in stereotype threat. *Journal of Experimental Social Psychology, 35*, 29–46.

Aronson, J., Quinn, D. M., & Spencer, S. J. (1998). Stereotype threat and the academic underperformance of minorities and women. In J. K. Swim & C. Stango (Eds.), *Prejudice: The target's perspective* (pp. 83–103). San Diego, CA: Academic Press.

Aronson, J., & Salinas, M. F. (1997). *Stereotype threat, attributional ambiguity, and Latino underperformance.* Unpublished manuscript, University of Texas at Austin.

Asera, R. (2001). *Calculus and community: A history of the Emerging Scholars Program.* New York: The College Board.

Bergthold, T., & Ng, H. K. (2004). *Precalculus in transition: A preliminary report,* Retrieved August 23, 2005, from http://www.maa.org/SAUM/new_cases/new_case_07_04/SJSU_case_study.html

Bonsangue, M. V. (1994). An efficacy study of the calculus workshop model. In E. Dubinsky, A. H. Schoenfeld, & J. Kaput (Eds.), *CBMS issues in mathematics education Vol. 4: Research in collegiate mathematics education I* (pp. 117–137). Providence, RI: American Mathematical Society.

Bowen, W. G., & Bok, D. (1998). *The shape of the river: Long-term consequences of considering race in college and university admissions.* Princeton, NJ: Princeton University Press.

Burkhardt, H., & Schoenfeld, A. H. (2003). Improving educational research: Toward a more useful, more influential, and better-funded enterprise. *Educational Researcher, 32*(9), 3–14.

Epperson, J. (2003). *The Emerging Scholars Program Instructor Workshop handbook.* (Available from The University of Texas at Austin, College of Natural Sciences–Office of the Dean, 1 University Station, Austin, TX 78712-0549).

Fullilove, R. E., & Treisman, P. U. (1990). Mathematics achievement among African American undergraduates at the University of California, Berkeley: An evaluation of the Mathematics Workshop Program. *Journal of Negro Education, 59*(3), 463–478.

Gándara, P. (1999). *Priming the pump: Strategies for increasing the achievement of underrepresented minority undergraduates.* New York: The College Board.

Herzig, A., & Kung, D. T. (2003). Cooperative learning in calculus reform: What have we learned? In A. Selden, E. Dubinsky, G. Harel, & F. Hitt (Eds.), *CBMS issues in mathematics education 12: Research in collegiate mathematics education, V* (pp. 30–55). Providence, RI: American Mathematical Society in cooperation with Mathematical Association of America.

Hsu, E. (1996). *The Professional Development Program teaching assistant teaching manual.* Berkeley, CA: Professional Development Program. Retrieved August 23, 2004, from http://betterfilecabinet.com/pdp (Also available from PDP Director, 230B Stephens Hall, University of California at Berkeley, Berkeley, CA 94720-1500).

—— (1999). A web-based database of problems and practices and real communities of ESP teachers and students. In F. Hitt & M. Santos (Eds.), *Proceedings of the twenty first annual meeting of the North American chapter of the International Group for the Psychology of Mathematics Education* (p. 824). Columbus, OH: ERIC Clearinghouse for Science, Mathematics, and Environmental Education. SE 062 752. [Actual database may be found at betterfilecabinet.com]

—— (2004). Re-considering on-line and live communities of practice. In *Proceedings of the Society For Information Technology & Teacher Education 15th international conference*. Retrieved on September 2nd, 2005 from www.editlib.org/index.cfm?fuseaction=Reader.ViewAbstract&paper_id=13114

Inzlicht, M., & Ben-Zeev, T. (2000). A threatening intellectual environment: Why females are susceptible to experiencing problem-solving deficits in the presence of males. *Psychological Science, 11*(5), 365–371.

Johnson, J. B. (2005, February 22). Shades of gray in black enrollment: immigrants' rising numbers a concern to some activists. *San Francisco Chronicle*. Retrieved September 2nd, 2005 from www.sfgate.com/cgi-bin/article.cgi?file=/c/a/2005/02/22/MNGIJBF3LP1.DTL

Johns, M., Schmader, T., & Martens, A. (2005). Knowing is half the battle: Teaching stereotype threat as a means of improving women's math performance. *Psychological Science, 16*(3), 175–179.

Kalmijn, M. (1998). Intermarriage and homogamy: Causes, patterns, trends. *Annual Review of Sociology, 24*(1), 95–421.

Kline, B. (1994). Cooperative learning and the role of the TA. *The Journal of Graduate Teaching Assistant Development, 2*(2), 53–61.

Kung, D. T. (in press). Teaching assistants learning how students think. To appear in *Research on Collegiate Mathematics Education, Volume 7*. Washington, DC: Conference Board of the Mathematical Sciences.

Lave, J., & Wenger, E. (1991). *Situated learning: Legitimate peripheral practice*. New York: Cambridge University Press.

Lewis Mumford Center for Comparative Urban and Regional Research. (2001). *Ethnic diversity grows, neighborhood integration lags behind*. Albany, NY: University at Albany.

Light, R. (2001). *Making the most of college: Students speak their minds*. Boston: Harvard University Press.

Massey, D. S., Charles, C. Z., Lundy, G., & Fischer, M. J. (2002). *The source of the river: The social origins of freshmen at America's selective colleges and universities*. Princeton, NJ: Princeton University Press.

McIntyre, R. B., Paulson, R. M., & Lord, C. G. (2003). Alleviating women's mathematics stereotype threat through salience of group achievements. *Journal of Experimental Social Psychology, 39*(1), 83–90.

Moreno, S. E., Muller, C., Asera, R., Wyatt, L., & Epperson, J. (1999). Supporting minority mathematics achievement: The Emerging Scholars Program at the University of Texas at Austin. *Journal of Women and Minorities in Science and Engineering, 5*(1), 53–66.

Murphy, T. J., Stafford, K. L., & McCreary, P. (1998). Subsequent course and degree paths of students in a Treisman-style workshop calculus program. *Journal of Women and Minorities in Science and Engineering, 4*(4), 381–396.

National Center for Education Statistics. (2002). *National study of profile of undergraduates in U.S. postsecondary institutions: 1999–2000* (Report No. NCES 2002–168 by Laura Horn, Katharin Peter, and Kathryn Rooney. Project Officer: Andrew G. Malizio). Washington, DC: Author.

National Collaborative on Diversity in the Teaching Force. (2004). *Assessment of diversity in America's teaching force: A call to action*. Washington, DC: Author.

National Science Foundation. (1999). *Women, minorities, and persons with disabilities in science and engineering: 1998* (Report No. 99-338). Arlington, VA: Author.

National Science Foundation: Division of Science Resources Statistics. (2001). *Science and engineering degrees, by race/ethnicity of recipients: 1990–98, Author, Susan T. Hill* (Report No. 01-327). Arlington, VA: National Science Foundation.

—— (2004). *Science and engineering degrees, by race/ethnicity of recipients: 1992–2001, Project Officers, Susan T. Hill and Jean M. Johnson* (Report No. 04-318). Arlington, VA: National Science Foundation.

Pattillo-McCoy, M. (1999). Middle class, yet black: A review essay. *African American Research Perspectives, 5*(1), 25–38.

Phelps, R. E., Taylor, J. D., & Gerard, P. A. (2001). Cultural mistrust, ethnic identity, racial identity, and self-esteem among ethnically diverse black university students. *Journal of Counseling and Development, 79*(2), 209–216.

Qian, Z. (1997). Breaking the racial barriers: Variations in interracial marriage between 1980 and 1990. *Demography, 34*(2), 263–276.

Rimer, S., & Arenson, K. W. (2004, June 24). Top colleges take more blacks, but which ones? *New York Times*. Retrieved September 2, 2005 from query.nytimes.com/gst/abstract.html?res=F00917FC355D0C778EDDAF0894DC404482

Ruddock, M. S. (1996). *The efficacy of prerequisite courses in mathematics for minority and white college students.* Unpublished doctoral dissertation, University of Texas at Austin.

Sacks, P. (2003, July 25). Class rules: The fiction of egalitarian higher education. *Chronicle of Higher Education*, p. B7.

Schattschneider, D. (2006). College precalculus can be a barrier to calculus; integration of precalculus with calculus can achieve success. In N. Baxter-Hastings (Ed.), *A fresh start for collegiate mathematics: Rethinking the courses below calculus* (MAA Notes Vol. 69, 285-294). Washington, DC: The Mathematical Association of America.

Seymour, E., & Hewitt, N. M. (1997). *Talking about leaving: Why undergraduates leave the sciences.* Boulder, CO: Westview Press.

Shih, M., Pittinsky, T. L., & Ambady, N. (1999). Stereotype susceptibility: Identity salience and shifts in quantitative performance. *Psychological Science, 10*, 80–83.

Speer, T. (1994). The newest African Americans aren't black. *American Demographics, 16*(1), 9–10.

Spencer, S. J., Steele, C. M., & Quinn, D. (1999). Stereotype threat and women's math performance. *Journal of Experimental Social Psychology, 35*, 4–28.

Steele, C. M. (1997). A threat in the air. How stereotypes shape intellectual identity and performance. *American Psychologist, 52*, 613–629.

Steele, C. M., & Aronson, J. (1995). Stereotype threat and the intellectual test performance of African Americans. *Journal of Personality and Social Psychology, 69*, 797–811.

Tinto, V. (1993). *Leaving college: Rethinking the causes and cures of student attrition.* Chicago: University of Chicago Press.

Treisman, P. M. U. (1985). *A study of the mathematics performance of black students at the University of California, Berkeley.* Unpublished doctoral dissertation, University of California, Berkeley.

Treisman, U. (1992). Studying students studying calculus: A look at the lives of minority mathematics students in college. *College Mathematics Journal, 23*(5), 362–372.

Treisman, U., & Asera, R. (1990). Teaching mathematics to a changing population: The Professional Development Program at the University of California, Berkeley. In N. Fisher, H. Keynes, & P. Wagreich (Eds.), *Issues in mathematics education: mathematicians and education reform, proceedings of the July 6–8, 1998 workshop* (pp. 31–62). Providence, RI: American Mathematical Society in cooperation with Mathematical Association of America.

U.S. Census Bureau. (1999). *Historical census statistics on the foreign-born population of the United States: 1850–1990* (Population Division Working Paper No. 29). Washington, DC: Author.

—— (2001a). *The Hispanic population: 2000* (Report No. C2KBR/01-3). Washington DC: Author.

—— (2001b). *Population by race and Hispanic or Latino origin for the United States: 1990 and 2000* (Report No. PHC-T-1). Retrieved August 23, 2005 from www.census.gov/population/www/cen2000/phc-t1.html

—— (2003a). *Census 2000* (Report No. PHC-T-19). *Hispanic origin and race of coupled households: 2000.* Special tabulation. Washington DC: Author.

—— (2003b). *The foreign-born population: 2000* (Report No. C2KBR-34). Washington DC: Author.

Woodbury, S. A. (1993). *Culture, human capital, and the earnings of West Indian blacks* (Staff Working Paper 93-20). Kalamazoo, MI: W.E. Upjohn Institute for Employment Research.

Part II
Cross-Cutting Themes
b. Using Definitions, Examples, and Technology

17

The Role of Mathematical Definitions in Mathematics and in Undergraduate Mathematics Courses

Barbara Edwards, *Oregon State University*
Michael B. Ward, *Western Oregon University*

Introduction

One of the earliest subjects of undergraduate mathematics education research was students' difficulties in writing formal mathematical proofs. Some research focused on the heuristics involved in proof-writing, but early attempts to show that the teaching of heuristics and strategies benefited students' proof-writing skills (Bittinger, 1968; Goldberg, 1973) failed to produce statistically significant results. Other difficulties have been identified, including students' weak understanding of logic and/or mathematical concepts and their definitions (cf. Hart, 1986; Moore, 1994). Several recent studies have looked further at students' proof-writing skills (cf. Dreyfus, 1999, Harel & Sowder, 1998; Selden & Selden, 2003); this topic is also addressed in this volume in chapters by Selden & Selden, Harel & Brown, and Zazkis. The purpose of this chapter is to look closely at one topic that arose from research on proof-writing, mathematical definitions, and most importantly, the role that these definitions play in the mathematical enterprise as well as in the teaching of undergraduate mathematics courses.

Mathematical definitions are of fundamental importance in the axiomatic structure that characterizes mathematics. The enculturation of college mathematics students into the field of mathematics includes their acceptance and understanding of the role of mathematical definitions, that the words of the formal definition embody the essence of and completely specify the concept being defined. But definitions also play a role in the students' experiences in mathematics courses themselves, in the sense that definitions are often used as a vehicle toward a more robust understanding of a given concept.

In this chapter we first discuss a framework for thinking about mathematical definitions derived from literature in the fields of mathematics, mathematics education, philosophy and lexicography. Next, we discuss research on student understanding and use of definition and on the role of definitions in the teaching of mathematics. Finally, we discuss the implications of this research and important pedagogical decisions that should govern the use of mathematical definitions in the teaching of mathematics.

Definitions

Definitions play a key role in mathematics, but their creation and use differs from those of "everyday language" definitions. This distinction is outlined by philosopher Richard Robinson (1962) and lexicographer Sidney Landau (2001), and from their work we derive the terms we use in our work. We distinguish between *extracted* definitions and *stipulated* definitions. According to Landau, extracted definitions are "definitions that are based on examples of actual usage, definitions *extracted* from a body of evidence" (2001, p. 165). Robinson describes extracted definitions (which he refers to as *lexical* definitions) as "that sort of word-thing definition in which we are explaining the actual way in which some actual word has been used by some actual person" (1962, p. 135).

In contrast, *stipulated* definitions are an "explicit and self-conscious setting up of the meaning-relation between some word and some object, the act of assigning an object to a name (or a name to an object)" (1962, p. 59). Their chief advantage is "the improvement of concepts or the creation of new concepts, which is the key to one of the two or three locks on the door of successful science" (1962, p. 68). Landau says such definitions "are imposed on the basis of expert advice" with the goal of "ease and accuracy of communication between those versed in the language of science" (2001, p. 165).

Thus, as we observed in Edwards and Ward (2004), *extracted* definitions report usage, while *stipulated* definitions create usage, indeed create concepts, by decree. Moreover, when a term is defined by stipulation, it is to be free from connotation, that is, free from all the associations the term may have acquired in its non-technical use. Finally, stipulated definitions have no truth value. Extracted definitions have a truth value. They either accurately report usage or they do not. Along with Robinson and Landau, we think of mathematical definitions as stipulated, whereas most, "everyday language" definitions are extracted.

It is important to note here that mathematical definitions frequently *do* have a history — that is, they can and do evolve. The definition we use for function, for instance, may not be the one that mathematicians favored two-hundred years ago; the concept of connectedness has two definitions, path-connected and set-theoretically connected. However, in formal mathematics we do not leave the meaning of a term to contextual interpretation; we *declare* our definition and expect there to be no variance in its interpretation in that particular work.

Mathematical definitions have many features, some critical to their nature and others, while not necessary to the categorization of mathematical definitions, preferred by the mathematics community. Van Dormolen and Zaslavsky (2003) outline the features of a good mathematical definition as follows.

Necessary Features

- Criterion of hierarchy: According to Aristotle, any new concept must be described as a special case of a more general concept — a square is a quadrilateral (general concept) with four congruent sides and one right angle (special case).

- Criterion of existence: Also required by Aristotle this criterion demands proof that at least one instance of the newly defined concept exists.

- Criterion of equivalence: If one gives more than one definition for the same concept, one must prove that they are equivalent.

- Criterion of acclimatization: A definition must fit into and be part of a deductive system.

Frequently Preferred Features

- Criterion of minimality: Only the minimal number of properties necessary to "reconstruct" the concept should be mentioned. Thus the definition of a square requires one right angle, not four.

- Criterion of elegance: When choosing between two equivalent definitions we want the one that uses fewer words and symbols, or the one that "looks" nicer.

- Criterion for degenerations: Sometimes the consequences of a definition allow degenerate cases that one may wish to exclude (or not).

The role and use of mathematical definitions is deeply imbedded in the culture of working mathematicians. However, to think about the use of definitions by novices in the field, a well-known framework that describes how mathematical knowledge in the form of conceptual ideas and their definitions is "stored" and used is helpful. This framework, known as *Concept Image/Concept Definition* (Tall, 1992; Vinner, 1991), describes the interplay of the individual's understanding of a particular mathematical concept and its formal definition. The concept image is the set of all the mental pictures associated in one's mind with the name of a particular concept as well as all the properties that characterize them. The concept image may be incomplete or mathematically incorrect, and can include naïve, non-mathematical associations with the concept name. For example, the notion of speed limit on a highway may be associated with the concept image of a mathematical limit.

On the other hand, the concept definition is the mathematical definition of the given concept. This may be known to the individual, he/she may be able to repeat a correctly memorized definition for a concept, or he/she may "remember" an incomplete or incorrect version of the definition. When faced with a task involving a given concept, rigorous mathematics demands that students base their solutions on the concept definition. Vinner (1991) postulated that students often rely instead upon the concept image. In the section that follows we discuss research on student understanding of the role of definitions, including some that have shown that even post-calculus mathematics majors may rely more on their concept images than on the concept definition when doing mathematical tasks.

Student Understanding of the Role of Definitions

It seems to be common knowledge in mathematics departments that many students do not "know" the definitions they need to know in order to perform mathematical tasks such as proving theorems. Often, in an attempt to solve this problem students are asked to memorize the pertinent definitions in the course and sometimes they are given credit in examinations for repeating those definitions. However, there has been research that shows that just knowing the definition may not be enough. Rasslan and Vinner (1998) studied 180 Israeli Arab high school students and their concept definitions and concept images of the increasing/decreasing function concept. One result of this study was that although 68 percent of the students could state the definition, only 36 percent of the students applied the definition successfully and well. Another 28 percent of the students applied the definition with varying levels of success.

The notion of "operable definition" is discussed by Bills and Tall (1998) in a report on their study of five students (three mathematics majors and two physics majors) in a twenty-week real analysis course that included work with the least upper bound property. The authors define operable in the following way: "A (mathematical) definition or theorem is said to be *formally operable* for a given individual if that individual is able to use it in creating or (meaningfully) reproducing a formal argument" (p. 104). Bills and Tall found that forming operable definitions is a task beyond that of just knowing the words of a definition and that "some students meet concepts at a stage when the cognitive demands are too great for them to succeed, others never have operable definitions, relying only on earlier experiences and inoperable concept images" (p.104).

It was necessary for the students in both of these studies to work with a concept definition that they could evoke from their own understanding. Their "success" (or lack thereof) in the given tasks thus depended at least in part upon their ability to remember and apply the appropriate definition. What if students could be relieved of that cognitive load and asked to work with definitions that were available to them throughout the tasks that they were asked to do? Edwards (1997a, 1997b) postulated that if students had mathematically correct definitions in front of them at all times during interviews and written tasks, it would be possible to see evidence of their understanding of *how mathematical definitions should be used,* unencumbered by worry about the actual wording of a particular definition.

Edwards' original study involved eight mathematics majors enrolled in an introductory real analysis course that had as one of its goals helping students learn to write proofs. Specifically, the course was described as an introduction to rigorous analytic proofs in the context of the properties of real numbers, continuity, differentiation, integration and infinite sequences and series. Although the nature of mathematical definitions was implicit in the delivery of the course in which the eight students were enrolled, the teacher did not explicitly draw attention to that issue. We will discuss the methodology and results of this study later, but briefly, Edwards found that even with the definitions in front of them, many of the undergraduate mathematics majors of her study had some difficulty using mathematical definitions in a mathematically appropriate way. The question arose, however, whether this result could have been influenced by the fact that many of the definitions in Edwards' study were of concepts that students had encountered in elementary calculus or were in some way related to those concepts (e.g., absolute continuity). Would this familiarity cause students to misread definitions and cause some uncertainty in the results? Although there was no evidence of this in the original study, these questions led to the second study conducted by Edwards and Ward (2004).

The Edwards and Ward study involved eight mathematics majors enrolled in an abstract algebra course taught by one of the researchers and observed by the other. In this study, the definitions that were chosen for the task-based interviews conducted by the observer were designed to minimize the chance of previous mathematical connections, in other words, students would be *forced* to work from the definitions that were provided for them because they would supposedly have no other mathematical "memories" to use in performing the tasks.

We discuss the methodologies and results of the two studies together since the findings from the second study corroborated those of the first. In our description and discussion of the two studies we will henceforth refer to the Edwards study as the analysis study and the Edwards and Ward study as the algebra study.

Design of the Studies

The purpose of both studies was to look beneath students' understandings of the *content* of mathematical definitions to discern their understandings of the *role* played by formal definitions in mathematics. This is somewhat tricky since it is possible that a student might apply a definition in a mathematically incorrect way for at least two reasons.

- A student could have an incomplete or faulty understanding of the content of a particular definition; or
- A student could have a mathematically incorrect understanding of the role or nature of mathematical definitions in general.

For example, a student may decide that $f(x) = 3$ is not a function because the symbolic form "has no x in it" (a faulty understanding of the function definition itself); or he may decide that the particular example is not a function even after reading the definition because the requirement of having an x is something we just know and it does not need to be mentioned in the definition (a faulty understanding of the role and character of mathematical definitions in general).

A further difficulty influencing the design of these studies arose from the possibility that, if asked directly, students might profess a seemingly adequate understanding of the role of formal definitions in mathematics without really understanding this role. It is not uncommon for students (or people in general) to repeat something they have heard without full understanding. For instance students may say, perhaps to please their teachers, that "mathematics is necessary in all walks of life," without being able to cite even one non-trivial example beyond the day-to-day interactions involved in commerce.

Both studies employed similar research methods, which we now briefly outline. Participants in both studies were volunteers from an upper division mathematics course, introductory real analysis or abstract algebra. Participants in both studies were in the last two years of an undergraduate major in mathematics. Each participated in task-based interviews spaced through the course. The intent of these interviews was to investigate each student's understanding and strategies in dealing with definitions including ones that the students had previously encountered, ones that students were currently encountering in their course, and ones that students had not encountered before.

The format of the interviews depended upon the student's familiarity with a given definition. For instance, if the definition had been encountered before but had not been discussed in the course, the students were asked first to explain in their own words their understanding of the associated concept and then to provide a definition for it if they could do so. The students were then given a copy of the stipulated definition and were asked to explain its meaning and to discuss how their previous explanation agreed or did not agree with their understanding of the given formal definition. Such was the approach used with the definitions of continuity and infinite decimal in the analysis study, for example. When students had no familiarity with a concept, the interview began with the student considering some stipulated formal definition. The definitions of "group" and "coset multiplication" from the algebra study fall into that category.

In all cases, following the introduction of the stipulated definition, students were given tasks to complete which required use of the definition. Two of the tasks were determining if a given function was continuous and determining if the set of cosets of a given subgroup formed a group under the operation of coset multiplication. In the latter case, definitions and tasks were selected because the researchers' experience and that of others suggested they would be difficult for the students (Asiala, Dubinsky, Mathews, Morics, & Oktac, 1997; Brown, DeVries, Dubinsky, & Tomas, K, 1997). In all the tasks, the goal was to observe in what ways the students used, or did not use, the definitions to complete the task and to overcome their difficulties. Students had access to the written definitions at all times during the interview; thus, the researchers hoped that inaccurate memories of the mathematically correct definitions would not compromise the students' ability to do the given task in a way that was not consistent with the goals of the study.

In addition, in one of the interviews, all students were asked: "What is mathematics?" which was an indirect way for the interviewer to probe students' understandings of the role of mathematical definitions. The interviews were audio-taped and video-taped. Verbatim transcripts of the interviews were then made and analyzed by the researchers.[1]

---

[1] For a description of the analysis process for the analysis study see (Edwards, 1997a or 1997b).

Results

The results of both studies are discussed in detail in Edwards & Ward (2004). Briefly, both studies showed that many undergraduate mathematics majors do not categorize mathematical definitions as stipulated, and that they may, under some circumstances, defer to their image of a concept rather than the definition if the two do not agree. Further, the algebra study showed that some students do not use definitions the way mathematicians do, even in the apparent absence of any other course of action.

Participants in both studies were analyzed in light of what they said about the role played by definitions in mathematics as well as how each one used the definitions to complete the tasks involved in the interviews. Students were not necessarily consistent in their views and actions. For instance some students would seem to understand the role of mathematical definitions, but later on in the interview would use or not use a definition in a way that was mathematically inappropriate.

It is important to note that the students in both studies had all successfully completed at least one advanced mathematics course prior to their involvement in one of our studies and that all of the students comfortably passed the courses in which they were enrolled during the two studies. Thus our research indicates that some undergraduates with advanced mathematical training and decent, sometimes excellent, grades do not completely understand the nature and role of mathematical definitions.

We will illustrate the results of our studies with one example from each study, beginning with Jesse in the analysis study and his work with the definition for a point-wise continuous function. Before seeing the formal definition Jesse said he remembered a definition from high school for continuity, that a function was continuous at a point "if the limit at the point equals the actual value of the point." After some discussion Jesse and the interviewer formalized his (mathematically acceptable) definition and wrote it out in the following way.

A function f is continuous at a point $x = a$ if $f(a)$ exists and if $\lim_{x \to a} f$ exists and if $\lim_{x \to a} f = f(a)$.

He was then given the standard stipulated definition which was also the same definition that soon would be discussed in his introductory real analysis course.

Definition: Let f be a real-valued function whose domain is a subset of R. Then f is continuous at $x_0 \in$ dom(f) if for each $\varepsilon > 0$ there exists $\delta > 0$ such that $x_0 \in$ dom (f) and $|x - x_0| < \delta$ imply $|f(x) - f(x_0)| < \varepsilon$.

When Jesse read the standard stipulated definition he remembered seeing it at one time and he was able to discuss the first two function tasks of the interview using both forms of the definition (although he preferred his version). The third task was to state whether or not the following function was continuous at $x = 0$.

$$f(x) = |x|$$

Jesse considered his definition and the "epsilon/delta" definition, and said,

Jesse: Cusps, or there were a whole bunch of things that were not continuous. And, I think this is one of them. Although, it looks pretty continuous…. I'm pretty sure I remember that this is not continuous and my definition isn't cutting it, so I'm looking at the, at the real one. [Jesse pointed to the "epsilon/delta" definition. His definition wasn't "cutting it" because it was telling him that the given function was, in fact, continuous.]

For several minutes he went back and forth between "knowing" that this function was not continuous and seeing that both definitions before him indicated that the function was continuous. Finally, he said,

Jesse: But that, but I know cusps, and sharp peaks are not, but from the definition, if we're saying that the limit of these two equals that, and f of a equals that, then that would be continuous. (Short pause.) But it's not.

Clearly, although Jesse seemed to understand the definition (especially his own version), was able to refer to it during the interview at any time, and had used it in previous tasks, he based his mathematical decision on (misleading) memories from his elementary calculus class.[2] It is interesting to note a comment that Jesse made about definitions toward the end of this interview — that they were a "lot of jargon." During the second interview Jesse had indicated, "After about the first day in calculus, we didn't care about this [formal definition] … if you had the concept right,

---

[2] The following day he reported that he had remembered that the absolute value function *was* in fact continuous, it just was not differentiable at $x = 0$. The fact remained however that it was not the definition that convinced him.

not really the definition, that was all that really mattered." Although several current popular calculus texts and many calculus teachers (including the authors of this chapter) do not focus on formal definitions in elementary calculus, one must consider the impact of these pedagogical decisions and perhaps do something to mitigate against later misunderstandings by the student.[3]

Heidi, a student from the algebra study, represents another example of a student who seemed not to categorize mathematical definitions as stipulated. (See Edwards & Ward, 2004, for more details.) However, analyzing her understanding was more difficult. Sometimes she seemed to be exhibiting a mathematically correct view of definitions, as when she talked about a hierarchy of definitions and axioms being used to prove theorems. But a few lines later she said, "You have to make the definitions from what something actually is," which would seem to indicate a view of definitions as extracted. While doing a task involving cosets in the second interview she pointedly avoided using the definitions that were available to her. She tried instead to do the task by remembering how she had done similar tasks before.

Heidi seemed to teeter on the cusp of understanding, however, and we see her case as evidence that a student's understanding of the categorization of mathematical definitions is not necessarily clear cut. Students do not fit nicely into one group (those who understand) or the other (those who do not understand). This notion is consistent with what Burger and Shaughnessy (1986) describe as transitional stages of students' understanding between van Hiele levels in geometry. According to Burger and Shaughnessy, it is possible that students may exhibit different levels of understanding on different tasks and some may even oscillate between levels of understanding on the same task.

For us, Heidi's case is evidence that the notion of mathematical definitions may be a "teachable" concept. According to Vygotsky (1978) each individual's understanding of a given concept resides in a zone that reaches somewhat beyond his or her understanding. With the help of a teacher, that individual's understanding can grow within that zone. We interpret Heidi's understanding to be in the zone of a mathematically correct understanding of the role of definitions in mathematics, one that would categorize mathematical definitions as stipulated. In the next section we will discuss the role of definitions in mathematics courses as a pedagogical question.

The Role of Definitions in Undergraduate Mathematics Courses

So far we have focused on how students perceive the way mathematical definitions are used in advanced mathematics. However, there is also the issue of the role mathematical definitions play in the teaching of mathematics. This is a separate issue, and it is possible that in some ways these two roles may conflict. There is mathematics education literature that focuses on the role of mathematical definitions in the teaching of mathematics. Some researchers have addressed the issue in the context of K–12 students, often in the teaching of geometry (de Villiers, 1998; Mariotti, & Fishbein, 1997; Van Dormolen & Zaslavsky, 2003). Others have focused on college classrooms, especially in courses populated by prospective teachers (Hershkowitz, Bruckheimer, & Vinner, 1987; Rasmussen, & Zandieh, 2000; Harel & Brown, this volume; Zazkis, this volume). One prominent notion in this literature is that students should have experiences creating their own definitions. De Villiers (1998) writes about Felix Klein's notion of the *bio-genetic* principle as a way to employ definitions to enhance students' understanding of mathematical concepts. In Klein's view, mathematics topics should not be presented to students as completed axiomatic-deductive systems. Rather students should retrace (to some extent) the path of the original thinking about, or discovery/invention of the concept. In this paper, de Villiers (1998) describes a study that focused on developing students' abilities to construct formal definitions for geometric concepts. This study involved tenth grade students in 19 schools and showed that students who were given defining activities in the course of learning geometry had much more success when asked to complete tasks involving writing correct, economical definitions of geometric concepts.

Like Klein, Freudenthal (1973) strongly criticized the traditional pedagogical practice of providing for students extant definitions of geometric concepts. He believed that since definitions were not preconceived by the students themselves, but were the final touch of a mathematical activity, mathematics instructors were denying students the chance to participate in the entire activity by merely giving them the final product. In his view students should be allowed to participate in the mathematical enterprise from the beginning, including creating definitions.

---

[3] In the authors' view, this would *not* include returning to a "formal definitions" approach in the teaching of elementary calculus, but it could include discussions and activities focusing on the role of definitions in mathematics. Some of these are described later in this chapter.

Definition activities in mathematics courses can have several pedagogical objectives, some of which could be

- promoting deeper conceptual understanding of the mathematics involved,
- promoting an understanding of the nature or the characteristics of mathematical definitions, and/or
- promoting an understanding of the role of definitions in mathematics.

Definitions are frequently and most obviously used to promote the first objective, deeper conceptual understanding of mathematics. Indeed the traditional method of communicating mathematics between professional mathematicians begins with a statement of the pertinent definition or definitions. Activities that involve studying a mathematical definition carefully and deciding from a collection of items which are and which are not examples of the defined concept are plentiful. Some of the tasks of the interviews for both the algebra study and the analysis study are examples of such activities. These activities also indirectly address the role of mathematical definitions. For a related discussion on this topic see Wilson (1990).

Activities that address the second objective are common also, although we feel that these alone might also not be sufficient for encouraging proper use of definitions in formal mathematics. An activity used by both of the authors is to ask students working in groups to define a given concept, for example, prime number. Each group agrees upon a definition for prime number and all definitions are then displayed for the whole class to compare and then choose the "best" definition. Of course, what it means to be "best" is also discussed. Students often mention criteria from Van Dormolen and Zaslavsky (2003) and when necessary we guide the discussion toward considering the entire list. We also encourage discussions about the consequences of various conditions included in the definition for prime. For instance, do we want 1 to be prime? Further discussion of the second objective and some additional activities for undergraduate students can be found in Hershkowitz, Bruckheimer & Vinner (1987); Winicki-Landman, & Leikin (2000); and Leikin, & Winicki-Landman (2000).

It seems that the key issue for many of the students from the analysis and algebra studies, however, was understanding that mathematical definitions are stipulated and thus different from everyday definitions. To develop this understanding requires treating mathematical definition as a concept in its own right by promoting an understanding of the role of definitions in mathematics, our third objective.

In Edwards and Ward (2004), some activities are given that are designed to promote this third pedagogical objective in undergraduate mathematics courses. These activities include directly addressing the topic of mathematical definitions including discussing the notion of Concept Image/Concept Definition with students, or exploring the dictionary definitions of such words as *radical*, which has both stipulated and extracted definitions.

Probably most important, however, is providing activities for students that involve them in the process of creating their own definitions in authentic ways as practicing mathematicians might do. For instance, how can *triangle* on the sphere or a hyperbolic surface be defined so that the congruence theorems in Euclidean geometry will hold on these non-Euclidean surfaces? Or, how can *continuous function* be defined so that it can only describe a function that can be drawn "without lifting one's pencil"? It seems that authentic experiences in defining might help students understand that although they have a certain amount of freedom in the creation of the definitions, the purpose for having a definition in the end is that there will be no misunderstanding about the exact nature of the concept being defined.

It is, however, in activities involving the creation of definitions where we see the potential for conflicting messages if the goal of the activity is that students understand a given concept more deeply. It is our view that implementing the ideas of Klein and Freudenthal, using definitions to increase students' understanding of mathematical concepts must be done with caution and potentially can be problematic. If students are asked to create definitions in such a way that their task is actually one of discovering the *"correct"* definition for a concept, it may give them the impression that definitions can be right or wrong and that they are extracted rather than stipulated. In the analysis study, one student, Stephanie, decided that a given definition for infinite decimal was wrong in the case of the decimal .999.... (Edwards, 1997b). Stephanie's work with definitions indicated a subconscious belief that one's concept image should rule in the case of a conflict between concept image and concept definition. Some defining activities could inadvertently reinforce this notion.

A central focus in the definition work of Klein and Freudenthal was that students will understand mathematics better if they are involved in its development from start to finish. This is indeed an excellent way to learn mathematics especially because the student can have a better notion of what mathematics is (cf. Maher & Martino, 1996). Too often, however, defining activities are like games that are won if the student can guess what the teacher is thinking. We believe that if students create their own definitions for concepts with existing definitions, they should first be allowed to carry the activity through to its logical conclusion, including realizing unintended consequences. Later students can compare their definitions to the definitions that were created before them, probably with a much greater understanding of the defining process. This is possibly a more difficult and longer process, but it is more authentic.

Conclusion

In this paper we have outlined a framework that looks at everyday definitions as extracted and mathematical definitions as stipulated. We have also discussed the notion of Concept Image/Concept Definition as a way of describing how students work with definitions in mathematical tasks and we have described research on students' understanding of the role of definitions in mathematics as well as research on the role of definitions in mathematics education. We believe there are still questions regarding the interplay between students' understanding of the nature and role of mathematical definitions and their experiences with definitions while learning mathematics. Research is needed to determine appropriate pedagogical approaches that encourage the development of the concept of definition throughout a student's experience in mathematics classes. Further, research is needed to assess the effectiveness of activities designed to help students create more robust understandings of the concept of definition.

Definitions are essential to the mathematical enterprise, thus it seems that focus on the role of definition should be central to the education of mathematics majors. However, according to Vinner (1991),

> The role of definition in mathematical thinking is somehow neglected in official contexts.... We are not sure whether this is because it is taken for granted or because it is overlooked. It is obligatory to remember that there are some contexts in which referring to the formal definition is critical for a correct performance on a given task (p. 80).

Although many students will eventually "figure out" how to use formal definitions in a mathematically correct way, it seems that it is important not to leave this to chance. This is especially true for students who will become K–12 mathematics teachers for it is these students who will have the greatest impact on our future students. Our research has shown that there exist mathematics majors who have been successful in advanced courses in mathematics, at least from the standpoint of the grades they earn, but who really do not understand the role of mathematical definitions in a mathematically acceptable way. This concept, like any other mathematical concept, can and should be addressed in undergraduate mathematics classrooms.

Acknowledgements. Support for the algebra study research and the writing of this chapter was provided through the Oregon Collaborative for Excellence in the Preparation of Teachers (OCEPT) funded by National Science Foundation Grants DUE-9653250 and DUE-0222552.

References

Asiala, M., Dubinsky, E., Mathews, D., Morics, S., & Oktac, A. (1997). Student understanding of cosets, normality and quotient groups. *Journal of Mathematical Behavior, 16*, 241–309.

Bills, L. & Tall, D. (1998). Operable definitions in advanced mathematics: The case of the least upper bound. In A. Olivier & K. Newstead (Eds.), *Proceedings of the Annual Meeting of the International Group for the Psychology of Mathematics Education, 2*, 104–111.

Bittinger, M.L. (1968). The effect of a unit in mathematical proof on the performance of college mathematics majors in future mathematics courses. Unpublished doctoral dissertation, Purdue University.

Brown, A., DeVries, D., Dubinsky, E., & Tomas, K. (1997). Learning binary operations, groups and subgroups. *Journal of Mathematical Behavior, 16*, 187–289.

Burger, W. & Shaughnessy, J.M. (1986). Characterizing the van Hiele levels of development in geometry. *Journal for Research in Mathematics Education, 16*, 31–48.

DeVilliers, M. (1998). To teach definitions in geometry or teach to define? In A. Olivier & K. Newstead, *Proceedings of the Annual Meeting of the International Group for the Psychology of Mathematics Education, 2*, 248–255.

Dreyfus, T. (1999). Why Johnny can't prove (with apologies to Morris Kline). *Educational Studies in Mathematics, 38*, 85–109.

Edwards, B. (1997a). Undergraduate mathematics majors/understanding and use of formal definitions in real analysis. Unpublished doctoral dissertation, The Pennsylvania State University.

—— (1997b). An undergraduate student's understanding and use of mathematical definitions in real analysis. In J. Dossey, J. Swafford, M. Parmantie, A. Dossey (Eds.), *Proceedings of the Nineteenth Annual Meeting of the North American Chapter of the International Group for the Psychology of Mathematics Education, 1*, 17–22.

Edwards, B.S., & Ward, M.B. (2004). Surprises from mathematics education research: Student (mis)use of mathematical definitions. *The American Mathematical Monthly, 111*(5), 411–424.

Freudenthal, H. (1973). *Mathematics as an educational task*. Dordrecht: Kluwer Academic Press.

Goldberg, D. J. (1973). The effects of training in heuristic methods on the ability to write proofs in number theory. Unpublished doctoral dissertation, Columbia University.

Hart, E.W. (1986). An exploratory study of the proof-writing performance of college students in elementary group theory. Unpublished doctoral dissertation, University of Iowa.

Harel, G., & Sowder L. (1998). Students' proof schemes: Results from exploratory studies. In Schoenfeld, A.H., J. Kaput, E. Dubinsky (Eds.), *Issues in Mathematics Education 7, Research in Collegiate Mathematics Education III* (234–284). Providence, RI: American Mathematical Society.

Hershkowitz, R., Bruckheimer, M., & Vinner, S. (1987). Activities with teachers based on cognitive research. In M.Linquist & A. Schulte (Eds.) *Learning and teaching geometry, K–12 1987 Yearbook* (222–235). Reston VA: The National Council of Teachers of Mathematics.

Landau, S. I. (2001). *Dictionaries: The art and craft of lexicography, 2nd edition*. Cambridge: Cambridge University Press.

Leikin, R., & Winicki-Landman, G. (2000). On equivalent and non-equivalent definitions: Part 2. *For the Learning of Mathematics, 20*(2), 24–29.

Maher, C. A. & Martino, A. M. (1996). The development of the idea of mathematical proof: A 5-year case study. *Journal for Research in Mathematics Education, 27*(2), 194–214.

Mariotti, M.A., & Fischbein, E. (1997). Defining in classroom activities. *Educational Studies in Mathematics, 34*, 219–248.

Moore, R.C. (1994). Making the transition to formal proof. *Educational Studies in Mathematics, 27*, 249–266.

Rasmussen, C.L., & Zandieh, M. (2000). Defining as a mathematical activity: A realistic mathematics analysis. In M. Fernández, (Ed.) *Proceedings of the 22nd Annual Meeting of the North American Chapter of the International Group for the Psychology of Mathematics Education, 1, 301–305*.

Rasslan, S., & Vinner, S. (1998). Images and definitions for the concept of increasing/decreasing function. In *Proceedings of the Annual Meeting of the International Group for the Psychology of Mathematics Education, 4*, 33–40.

Robinson, R. (1962). *Definitions*. London: Oxford University Press. (Reprinted by Frome, 1954, United Kingdom: D.R. Hillman & Sons)

Selden, A., & Selden, J. (2003). Validations of proofs considered as texts: Can undergraduates tell whether an argument proves a theorem? *Journal for Research in Mathematics Education, 34*, 4–36.

Tall, D. (1992). The transition to advanced mathematical thinking: Functions, limits, infinity, and proof. In D. Grouws (Ed.) *Handbook of research on mathematics teaching and learning*. New York: Macmillan Publishing Company.

Van Dormolen, J., & Zaslavsky, O. (2003). The many facets of a definition: The case of periodicity. *Journal of Mathematical Behavior, 22*, 91–196.

Vinner, S. (1991). The role of definitions in the teaching and learning of mathematics. In D. Tall (Ed.), *Advanced mathematical thinking* (pp. 65–80). Dordrecht: Kluwer Academic Press.

Vygotsky, L.S. (1978). *Mind in society*. Cambridge, MA: Harvard University Press.

Wilson, P. (1990). Inconsistent ideas related to definitions and examples. *Focus on Learning Problems in Mathematics, 12*(3, 4), 31–47.

Winicki-Landman, G., & Leikin, R. (2000). On equivalent and non-equivalent definitions: Part 1. *For the Learning of Mathematics, 20*(1), 17–21.

18

Computer-Based Technologies and Plausible Reasoning

Nathalie Sinclair, *Simon Fraser University*

The purpose of this chapter is to describe how computer-based tools can help students in the doing and learning of mathematics, and to provide specific examples that illustrate the way in which well-designed technologies can support mathematical discovery and understanding. I begin with an example.

Task. Take the three vertices of a triangle ABC and reflect them each across the opposite side of the triangle to obtain a new "reflex" triangle DEF (convince yourself you always do indeed get a new triangle!). Repeat the process. Most people who have seen this problem conjecture that the reflex triangle, after several iterations, converges to being equilateral. But I thought it was a perfect problem for *The Geometer's Sketchpad*[1], which effortlessly allows an arbitrary triangle to be iteratively "reflexed" and its measurements to be computed. I dragged vertex A after producing DEF and quickly realized that the equilateral conjecture was false, for I could produce a DEF that was a straight line! And more: DEF seemed to change in a very chaotic way as I continuously dragged vertex A. When, if ever, would the figure become equilateral? When would it not? How did the "function" behave?

At this point, I realized that I needed some kind of measure of the degree to which the triangle had become equilateral, especially since the reflex triangles were getting increasingly large—exploding off the screen—as the iterations increased. After several false starts, I chose to use perimeter squared over area, a measure that achieves a minimum for equilateral triangles (as simple algebra will show), and that is simple for Sketchpad to compute, no matter how big the reflex triangle gets. By dragging A, I could see the measurements changing, achieving a minimum for "nice" initial triangles, but growing wildly for not so "nice" ones. However, it was hard to keep track of the "nice" and "not nice" positions of A: I needed an overall picture of the changing measurement. So, I fixed the base of the triangle BC, chose another point A in the plane to form the triangle ABC, performed the reflection twice to produce the second reflex triangle and then colored the point A according to how close that reflex triangle was to being equilateral (in Figure 1, vertex A is blue if the 2nd reflex triangle is close to being equilateral and white if not—since the measure of an equilateral triangle is 1, "closeness" here corresponds to the measure being between 1 and 1.5). Using Sketchpad's iteration and parametric coloring facilities, I could effortlessly repeat the procedure for all positions of point A in a given region of the plane. What a beautiful, surprising, symmetric picture emerged! And more, the picture provided an immediate guide to where I should look for equilaterality-breaking triangles (initial triangles that iterate to triangles that are scalene), as well as an immediate confirmation of the chaotic behavior I had intuited by dragging A. Figure 1a shows the coloring for the 2nd iteration while Figure 1b shows the 4th iteration; note how each branch in (a) has been divided into three parts in (b). The symmetry indicates that only 1/4 of the plane needs to be analyzed, while the blue swath above BC indicates that isosceles and near-isosceles triangles behave well, in the sense that they produce reflex equilateral triangles. The splitting of the "branches" confirms that minute changes in the position of point A can give rise to radically different reflex triangles.

---

[1] *The Geometer's Sketchpad*, similar to *Cabri-Géomètre* and *Cinderella*, is a software environment in which geometric constructions (and other forms of mathematical illustrations such as diagrams, figures, and graphs) can be manipulated interactively with the computer mouse.

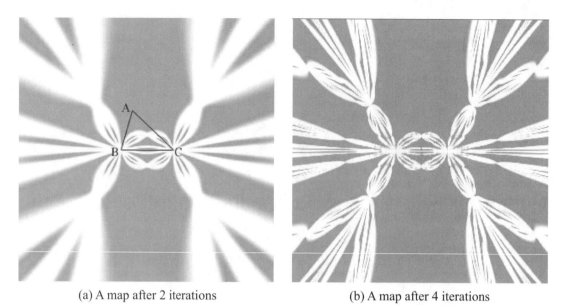

(a) A map after 2 iterations   (b) A map after 4 iterations

Figure 1. Iterations of the reflex triangle problem.

In addition to offering an attractive mathematical image, I wanted to use this example to illustrate a simple, but palpable example in which the computer provided insight and intuition, its computations provided quick and reassuring conjecture-testing (and falsifying), and the visual images yielded new insights to verify and provided avenues to explore. These characteristics, as well as others, of mathematical inquiry in a computer environment are well described by Borwein and Bailey (2004) in their book on experimental mathematics. In the following sections of this chapter, I detail the characteristics of mathematical thinking that are supported by computer-based technologies, as outlined by Borwein and Bailey, and then provide illustrative examples of student learners (middle school students as well as pre-service teachers) as experimental mathematicians.

An Updated Framework for Mathematical Reasoning

Plausible reasoning in the 21st century, the subtitle of a new book by mathematicians Jonathan Borwein and David Bailey (2004), recalls rather brazenly George Pólya's well-known two-volume book *Mathematics and Plausible Reasoning*. Written in 1954, Pólya's book continues to be a landmark contribution to mathematics—as well as mathematics education—exploring and exemplifying as it does the role that guessing, inductive reasoning, and reasoning by analogy play in the most rigorous of deductive disciplines. Borwein and Bailey's book argues, using many examples from contemporary mathematics, that computers offer a whole new form of plausible reasoning, one that may very well surpass the depth and power of Pólya's. In fact, there is a compelling sense in which Borwein and Bailey posit the computer as a tool as fundamental to mathematics as the compass and straightedge tools were to Euclidean geometry—tools that defined the very nature of the geometry that could (and could not) be done by defining the problems and the solutions, and, more importantly, the way that mathematicians could think about problems and solutions. This way of viewing the computer has profound implications for mathematics education, especially at the undergraduate level, when students—whether they are studying to be mathematicians or teachers—are developing a sense of what mathematics 'really' is and how mathematics is 'really' done. For many students, this is the first time they are interacting with mathematicians in a mathematics department.

Historically, the mathematics community held the view that 'real mathematicians don't compute' (see Borwein & Bailey, 2004). Such a feeling has led to, among other things, an absence of computers in most university mathematics courses, including those taken by pre-service teachers. If 'real mathematicians don't compute' then perhaps 'real mathematics isn't done with computers,' and therefore, good school mathematics should not be done with computers. This stance would dismiss the computer as superfluous, optional to mathematics, and to mathematics learning.

However, things may be changing in the mathematics community. Indeed, part of what Borwein and Bailey do in their book is to reveal the many ways in which experimental mathematics has been integral to the development of mathematics for centuries (even before we had digital computers). And if computers really do offer a new form of plausible reasoning, then mathematics departments might be able to help students use this new tool fluently and productively, whether these students are studying to be teachers or mathematicians.

New Kinds of Plausible Reasoning

Typically, when writing on technology in mathematics education, researchers focus either on specific content areas (such as geometry, calculus, elementary number theory) or on more general pedagogical affordances (such as immediate feedback, shift of authority from teacher to student and the development of personal agency). In this chapter, I take a more mathematical approach by focusing on the ways in which current computer-based tools support mathematical discovery and understanding, proving, and problem solving across content areas. As Borwein and Bailey outline, these ways include using the computer

1. to gain insight and intuition,
2. to produce graphical displays that can suggest underlying mathematical patterns,
3. to discover new patterns and relationships,
4. to test and especially falsify conjectures,
5. to explore a possible result to see if it is worth formal proof,
6. to suggest approaches for formal proof,
7. to replace lengthy hand derivations with computer-based derivations, and,
8. to confirm analytically-derived results.

The list is long but—especially for mathematician readers—it is worth considering how each item fits into the process of mathematical inquiry and whether explicit non-computer-based mechanisms or tools exist to support it. For most mathematicians, each item will surely represent a familiar part of their own teaching practice; however, because of the pedagogical focus of this paper, our attention will be on making these items explicitly and productively available to students.

Based on examples from research, I will illustrate how several[2] of these forms of plausible reasoning can be supported by specific computer technologies and then reflect on the implications for the community of researchers in undergraduate mathematics education. Given my own research and involvement in pre-service teacher education, the research examples will focus more on content in undergraduate mathematics curricula, as will the specific computer technologies described. Despite my use of specific digital technologies in specific settings, by highlighting the forms of plausible reasoning supported by these technologies and the characteristics of the technologies that make that possible, my specific contexts might well inform contexts I will not include, such as the use of CAS in undergraduate calculus courses.

As Borwein and Bailey point out, many prominent mathematicians throughout history have engaged in experimental mathematics, including Carl Friedrich Gauss, Leonhard Euler, Jacques Hadamard, and John Milnor. In fact, most mathematicians may ignore the way in which experimental methods—even those without the computer— contribute to their own processes of mathematical inquiry, even after having glanced at the list above. However, I hope that the reflex triangle example given at the beginning of the chapter resonated with the experiences of many readers, and drew attention to the ways in which the computer directly supports and amplifies the components of the experimental methodology. I use it because it has helped me appreciate the ways in which the different components of the experimental methodology act together to fuel inquiry, to support learning and to solve problems.

---

2 Based on the non proof-oriented nature of the student examples I will be analysing, I will not discuss the two following characteristics: to explore a possible result to see if it is worth formal proof, and to suggest approaches for formal proof. The brief Sketchpad example I related above, however, shows interesting connections between these two proof-related characteristics and what Simon (1996) calls "transformational reasoning." Simon defines "transformational reasoning" as "the physical or mental enactment of an operation or set of operations on an object or set of objects that allows one to envision the transformations that these objects undergo and the set of results of these operations." Experimental evidence has shown the importance of transformational reasoning in proving (see Arzarello, Micheletti, Olivero, & Robutti, 1998; Boero, Garuti, & Mariotti, 1996; Simon, 1996; Harel and Sowder, 1998). Simon emphasizes the central role of dynamism in transformational reasoning, and the ability to consider dynamic processes, as opposed to static states. The graphical displays I created in Sketchpad certainly led me to establish some results, which were compelling enough to pursue more formal proofs. However, by physically manipulating the dynamic processes, I could grasp some of their underlying structure —which certainly suggested approaches for a formal proof.

How Computers Complement Mathematical Reasoning

What is it that makes computers such powerful aids to mathematical thinking and problem-solving? One response that quickly comes to mind has to do with its sheer calculating power. However, computers do not only produce numerical answers. They transform calculations—converting the digits of π into melodies—as well as represent calculations— generate graphs of logarithmic functions. In other words, new technologies can actually *do* mathematics, not just record results.

Numbers are not the only thing that computers are good at processing. Indeed, only a cursory familiarity with fractal geometry is needed to see that computers are good at creating and manipulating visual representations of data. There is a story told of the mathematician Claude Chevalley, who, as a true Bourbaki, was extremely opposed to the use of images in geometric reasoning. He is said to have been giving a very abstract and algebraic lecture when he got stuck. After a moment of pondering, he turned to the blackboard, and, trying to hide what he was doing, drew a little diagram, looked at it for a moment, then quickly erased it, and turned back to the audience and proceeded with the lecture. It is perhaps an apocryphal story, but it illustrates the necessary role of images and diagrams in mathematical reasoning—even for the most devoted anti-imagers. The computer offers those less expert, and less stubborn than Chevalley, access to the kinds of images that could only be imagined in the heads of the most gifted mathematicians, images that can be colored, moved and otherwise manipulated in all sorts of ways (see, for example, Hadamard, 1945).

Students as Experimental Mathematicians

By definition now, computers are quick, they can handle lots of data at once, and they can easily display this data in various graphical formats. In this section, I will illustrate how these three characteristics can support mathematical problem solving in a unique way by helping people become aware of patterns and relationships, giving them ways to independently test ideas and hypotheses and enabling them to confirm results obtained through non-experimental means. I will begin by drawing on research[3] conducted with Rina Zazkis and Peter Liljedahl using two internet-based applets focusing on a wide range of concepts and problems in elementary number theory (Zazkis, Sinclair and Liljedahl, 2006; Sinclair, Zazkis and Liljedahl, 2004)—an area of mathematics that is taught from kindergarten to undergraduate mathematics, and one that poses well-documented and numerous challenges to learners.

Each of the two environments I describe here was designed as a visual representation and experimental interface to help students learn and understand a wide range of number theory-related ideas. After describing each microworld briefly, I highlight a few of the common characteristics that research has shown to be supportive of student learning and problem solving.

Number Worlds

Figure 2 shows a snapshot of the Number Worlds applet[4], which allows users to highlight different types of numbers (such as even or prime) on a grid of whole numbers. The centre grid contains a two-dimensional array of clickable cells. The numbers shown in the cells depend on the 'world' that has been chosen. Although the basic objects of Number Worlds are the positive integers, or natural numbers forming the Natural World, the user can choose among other sets: the **Natural World, Whole World, Even World, Odd World**, and **Prime World**[5]. It is also possible to change the numbers shown in the cells. The user can increase or decrease the start number by 'one row,' that is, by the value of the grid width, or simply by one 'cell.'

The appearance of the grid can be affected by changing the value of the **Grid width** menu. By selecting values from one to twelve, the user can change the number of columns displayed, and thus the total number of cells. There are always exactly ten rows.

Within each world, the user can highlight certain types of numbers: **Squares, Evens, Odds, Primes, Factors**, and **Multiples**. Further, the multiples that are chosen can be shifted by an integer in order to create any arithmetic sequence.

---

[3] The research results that will be presented were based on approximately 4-6 hour sessions, including open lab time and guided interviews, with 90 elementary pre-service teachers.

[4] The Number Worlds microworld was written in Java and is available on-line as an applet at www.math.msu.edu/~nathsinc/java/NumberWorlds/

[5] This approach was inspired by Brown (1978), who uses similar ways of re-examining ideas in elementary number theory by shifting the focus from natural numbers to other domains.

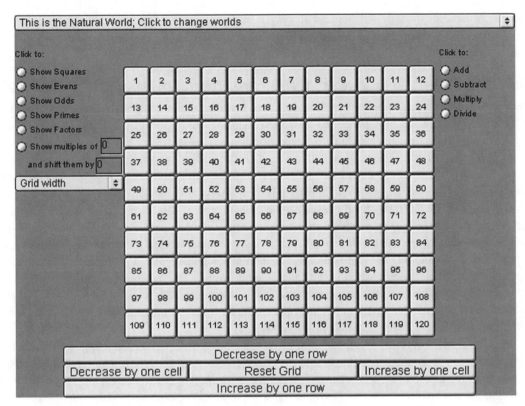

Figure 2. The Number Worlds microworld.

The Color Calculator

The Colour Calculator (CC)[6] is an internet-based calculator that provides numerical results, but that also offers its results in a color-coded table. Each digit of the result corresponds to one of ten distinctly colored swatches in the table. The calculator operates at a maximum precision of 100 decimal digits, and thus, it represents each result simultaneously by a (long) decimal string and a table, or grid of color swatches. It is possible to change the dimension, or the width, of a color table to values between one and thirty. Figure 3 shows the result of typing 1/7 into the calculator with the grid width set at ten, thus generating the associated table (which, for the purposes of this chapter, has been converted to grayscale).

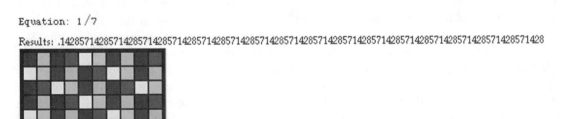

Figure 3. The Colour Calculator showing 1/7.

---

[6] Also available on-line; please click the Alive Maths link from www.math.msu.edu/~nathsinc/, then choose your preferred language, and then select the second of the two activities.

Of course, because of the way numbers are displayed in the CC—only the digits after the decimal point are represented in the table—the division operation produces the most interesting results, particularly when the rational quotient has a repeating pattern.

Common Features Relevant To Mathematics Learning and Problem Solving

The power of both microworlds, Number worlds and the CC, is rooted in three important features that are not found either on the handheld calculator or in paper-and-pencil environments: color, speed, and size. The size feature relates to the number of digits that are calculated and displayed (for the CC: 100 instead of the typical handheld calculator's eight; for Number worlds: a very large number of prime numbers can be displayed, as well as any number of multiples). With only eight digits, the handheld calculator can barely display the repeating pattern found in a fraction with a denominator as little as seven (since its repeating pattern is of length six, the rounded eighth digit can easily obfuscate the pattern). In contrast, the CC easily handles much larger units of repeat. This means that repeating decimals repeat transparently, and therefore stand in greater contrast to non-repeating decimals. Research with college-level pre-service teachers showed that the number of digits displayed helps learners create a perception that infinitely repeating numbers really *do* go on and on, and that non-repeating infinite decimals really do *not* repeat (Sinclair, Zazkis & Liljedahl, 2004). Similarly, students working with the Number Worlds applet reported coming to realize that the multiples also really do go on and on (one student even stated "I mean in high school they just said, that's a never-ending decimal, 1/7 is a never ending decimal. That's what we were told. Just put three dots beside it and don't worry about it. Well, no more little dots for students of mine, no way"!).

The speed feature relates to the computer's ability to calculate and display quickly the factors, multiples, and decimal expansions of a fraction; this contrasts with paper-and-pencil environments where, for example, the process of conversion from fraction to decimal is frequently long, tedious and error-prone. Speed thus allows learners to see a much wider range and greater number of "factor families" or decimal expansions (it makes no distinction between 'hard' fractions like 3/41 and 'easy' fractions such as 563/4), and to test conjectures during exploration and problem solving more quickly. In fact, the microworlds draw attention away from the procedure of factoring or of converting to the properties and relationships that exist between numbers and their factors or between common fractions and their decimal representations. This allows learners to work with the results of the factoring or of conversion and to treat these results as objects rather than processes. Researchers such as Sfard (1991) have discussed the great difficulty that students face in making this transition. However, as Dubinsky and McDonald (2001) show, the computer can help learners reify (or encapsulate) processes by providing a means through which actions can be applied to these processes. This is exactly what occurs in the research described with the Colour Calculator and Number Worlds; by applying actions to processes the students were able to reify mathematical processes and thus work with the more abstract concepts in elementary number theory (cf., Sinclair, Zazkis & Liljedalh, 2004; Sinclair, Zazkis & Liljedahl, 2006)

Finally, the color feature, which is most significant in the CC, is responsible for translating strings of digits into a format where patterns are more easily discernible. Since the colors are displayed within a manipulable grid, they can be seen all-at-once (compare the tabular representation with the one-dimensional array) and can be flexibly arranged to reveal certain patterns. Many of these patterns are recognisable and attractive (stripes, diagonals, and checkerboards) and can thus become motivational objects: Can you create a grid of stripes? Can you create an "all-red" fraction? These kinds of question are qualitatively different than: Can you find a fraction that has a repeating decimal representation? Color is used less extensively in Number Worlds, with different numbers toggling between only two colors. However, this seemed enough to draw students' attention to interesting mathematical patterns.

I have singled out these three factors in order to draw similarities between the two microworlds presented in this chapter, but also to achieve some greater generality beyond these two specific examples. Many of the digital tools used by researchers at the Center for Experimental and Constructive Mathematics (see http://www.cecm.sfu.ca/interfaces/), for example, have similar features and, as I will argue in the next section, play a significant role in mathematical problem solving for mathematicians and novices alike.

Research findings with Number Worlds and the Colour Calculator

I now discuss five components of the experimental methodology described by Borwein and Bailey in turn, placing special emphasis on those that provide a compelling way to generate understanding and insight; generate and confirm or confront conjectures; and make mathematics more tangible, lively and appealing.

To gain insight and intuition. Gaining insight and intuition may sound like a mystical process that leads to "aha" moments or momentous breakthroughs, but for many students the process may be much more modest, and lead to deepened learning rather than problem solving breakthroughs. Insight and intuition often involve qualitative understandings that are not well captured by propositional statements or definitions but that inevitably support and strengthen them. The first example I discuss will strike many readers as slightly simplistic, but it highlights the forms of insight often lacking in students and the ways in which images and large example spaces (see Mason and Watson, this volume) can help students define and distinguish mathematical concepts.

Previous research has shown that students often confuse factors and multiples, both conceptually and linguistically (Zazkis, 2000; this volume). They will characterize arithmetic sequences of non-multiples, that is, multiples shifted, as being "sporadic" whereas sequences of multiples will be described as "orderly" (Zazkis & Liljedahl, 2002). After spending 1–3 hours using Number Worlds, research participants (college level pre-service teachers) in the study conducted by Sinclair, Zazkis and Liljedahl (2004) consistently and overwhelmingly described all arithmetic sequences (whether of multiples or non-multiples) using adjectives previously reserved for sequences of multiples. In fact, the following rich adjectives were used by the participants: "constant," "equally spaced," "continuously repeating," "distinctively patterned," "continuous," "sequential," "repeating," "predictable," "extending infinitely," "uniform," "regulated," "symmetrical," "systematic" and "orderly." In contrast, the participants described factors as being: "scattered," "sporadic," "non-patterned," "inconsistent," "not symmetrical," "random," "chaotic," "limited," "discordant," and "unorganized." In short, they constructed very distinct visual images of factors and multiples, as evidenced in the rich vocabulary of adjectives.

It could be that by working with multiples and factors at the same time—and having access to visual representations of both—the participants were able to see that sequences of multiples are more like arithmetic sequences of non-multiples, that is, multiples shifted, than they are like factors. In most classroom settings, without such a strongly contrasting notion, students might focus more on the differences between sequences of multiples and arithmetic sequences of non-multiples than on their similarities. We see here the power of students creating images of mathematical concepts, something that requires both the image-making capacities of the computer, and the opportunity to become familiar with a wide example space.

A similar phenomenon occurred with the CC. Whereas many students possess a small and weakly characterized example space of fractions (such as 1/2, 1/4, 3/4, 1/3, 2/3 and perhaps other unit fractions with denominators less than 10), the research participants who have worked with the CC developed strong images and even personalities for a much larger set of fractions. In addition to being able to call upon 'basic' fractions such as 1/3, 1/4, 6/10, 2/5, the participants knew what fractions having denominators such as 7, 13, 17, or 47 would look like (one participant commented that "Well I know, I know I can now say well 7 is actually a routine example with a length of 6"), and they could describe some relationships between the denominators and the decimal expansions (particularly for prime number denominators and denominators of the form 10^n-1). This kind of relational, instead of algorithmic, understanding might account for Danielle's claim "I now have a better understanding of how certain fractions create different types of decimals, such as finite or infinitely repeating." She may not have had a better understanding of how to convert fractions to their decimal representations, but she had a larger set of concrete (and colorful) examples corresponding to different types of decimal representations. Color seemed to play a role in vivifying the participants' understanding of various properties and relationships. It is difficult to tell whether she could have described 1/3 prior to using the CC, but Kimberley now says that a "denominator 3 will always give me a block of mono-colour in my mind." She adds "I'll forever see fractions and decimals in color."

Students' development of strong personalities for numbers resonates well with Keith Devlin's (2000) definition of a mathematician as someone for whom mathematics is a soap opera:

> The characters in the mathematical soap opera are not people but mathematical objects—numbers, geometric figures, groups, topological spaces, and so forth. The facts and relationships that are the focus of attention are not births and deaths, marriages, love affairs, and business relationships, but mathematical facts and relationships about mathematical objects. (p. 262)

In a similar vein, the calculating wizard Wim Klein remarked "Numbers are friends to me." Taking 3,844 as an example, he said, "For you it's just a three and an eight and a four and a four. But I say 'Hi, 62 squared!'". It seems quite plausible that "coloring" numbers is part of developing personalities for them.

Many students used Number Worlds to gain insight and intuition in the actual course of problem solving. For example, in trying to determine what kind of values produces the image in Figure 4, one participant decided to show the prime numbers, not in order to verify his conjecture, but in order to get a sense of what prime numbers look like on the grid. He knew that the image did not necessarily start at 1, and was thus aware that Number Worlds might not give him the answer. Instead, he knew that he could use the visual patterns displayed by the microworld to gain insight and intuition.

To Produce Graphical Displays That Suggest Underlying Mathematical Patterns. The way in which Number Worlds displayed multiples provided yet another visual support for students' learning of multiples: Number Worlds emphasizes the space in between subsequent highlighted numbers—and this space is constant for multiples. The participants seemed to form an image of multiples as possessing the property of 'constant spread.' One of the interview questions consisted in asking students to identify four different images, each representing, in order from left to right, factors, multiples, primes and squares (see Figure 4). Every interviewed participant immediately recognized the image of multiples, with most appealing to the visually discernible spacing between consecutive highlighted cells.

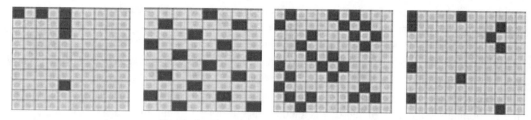

Figure 4. Images taken from Number Worlds.

The notion of spread may seem obvious to the mathematically inclined, but Andrew—one of the strongest students in the class—commented on how it was the microworld that helped draw his attention to it. "Until you see it," he explained, you don't know "there's always going to be the same amount of space" (in Sinclair, Zazkis and Liljedahl, 2004). Students also showed strong evidence of an if-and-only-if understanding of the 'constant spread' idea in several cases. Every interviewed participant was able to deduce that the three other images presented in the interview (factors, square numbers, and primes) were *not* multiples because they did not have the same amount of space between highlighted numbers.

The 'constant spread' feature of multiples is a visual perception of the property 'every nth number is a multiple of n.' This property—which is referred to as the 'every nth' property—is usually not capitalized upon by students (Zazkis & Campbell, 1996; Ferrari, 2002). In fact, an interview question—Is there a multiple of 7 in the 9th row of your grid? What about 23rd row? (grid width = 10)—was designed to probe participants' understanding of this property explicitly. Previous research indicates that students would have difficulty solving this question without appealing to an algorithm: they would *not* simply use the 'every nth' property of multiples. However, ten out of the fifteen participants in this study, using the graphical display, who were asked this question effectively used the 'constant spread' property of multiples. Based on the participants' success with this problem, it appears that the visually-based 'constant spread' property of multiples is more effective in problem-solving contexts for students than the equivalent 'every nth' property without visual representation.

The visual representation of multiples highlights their 'constant spread' property, that is, that there is an equal number of cells between each pair of subsequent multiples. Student understanding of 'constant spread' does not follow from an additive understanding of multiples, as prior research shows (Zazkis & Liljedahl, 2002). Perhaps the visual representation helps students move from an additive understanding of multiples to one which features the 'every nth' property. That is precisely what Michelle's comment suggests: "multiples follow a pattern in that every 3rd number is highlighted if we want multiples of 3, it gives us an actual image, not just words to describe it."

With regard to the visual representation afforded by the CC, Kevin articulated the increase in pattern possibilities now available: "patterns can be seen as you move left to right, up and down, and even left and up, or right and down. Patterns are more visible and meaningful in this sense than when written down on paper in the standard 0.blah blah blah blah blah blah way." Some participants talked about the way in which the colorful patterns attracted and held their attention, in a way that numbers might not. For example, Kyle noted that "without colors to represent numbers, patterns are much more difficult to discern, and can impact highly on an individual's ability to focus. Similarly, Kelly explained "the colors were important in not only clarification purposes, but were also important in keeping me attentive." The color also seems to make patterns more accessible than they are on handheld calculators, as Dianna insisted: "Being able to see the numbers represented as color helps the patterns to become more pronounced for me. Normally, on the regular calculator, you cannot see that there are sets of repeating numbers—they usually just look like a jumble with no rhyme or reason."

To discover new patterns and relationships. During their university course, the research participants had been exposed to the algorithm for converting decimal numbers into fractions. They had completed an assignment which required the use of this algorithm for a range of decimal numbers. Their work with the CC allowed them, in a sense, to "re-discover" a new pattern and relationship that led to both using the algorithm more flexibly and to understanding why the algorithm works. One task asked the students to investigate the several fractions whose denominators are of the form $10^n - 1$ (such as 9, 99, 999, etc.). The participants were very surprised to find that, for example, 24/99 yields a repeating decimal with unit of repeat 24 and that 123/999 yields a repeating decimal with unit of repeat 123. However, they easily worked through the questions in this task, generalizing the "magic 9's" pattern and noting some "exceptions" such as 24/999. It is when they began encountering these exceptions that they started to connect the phenomenon of the magic 9's to the algorithm learned in class (exemplified below), which, for several groups we interacted with, "explained" the phenomenon.

Let $n = 0.238\_\_\_$ Then $1000n = 238.238\_\_\_$

So $1000n - n = 999n = 238$

And $n = 238/999$

Though the algorithm always features nines (when the repetition starts immediately after the decimal point; see the third and fourth steps above), those nines seem to remain somewhat opaque for students, who focus on the algebraic manipulations involved. Many groups of students wrote out the algorithm on a piece of paper, accompanied by exclamations such as "Oh! Now I see what's going on." Andrew in particular commented on how working with the 9's in the CC made "theories that were out there become related." In an interview, Blake also commented on the relationship between what he learned in class and what he learned while working with the CC: "I think that really like using the numbers over 9, that really like related that, because it's, you never really work backwards with that formula, or whatever, but in this case you are forced to go the other way, which really makes it make sense. You know, it just gives you a better understanding of that, that it really works."

Perhaps students tend to view the algorithm as a process—or even a trick—that allows them to turn decimal numbers into fractions rather than as a general relationship between fractions and repeating decimals. The patterned display of the CC helped that relationship become both more apparent and interesting for our participants. It gave the participants an invitation to "play," as Blake described later in an interview: "No, I didn't know that at all, because, well like I, you know, I kind of knew them but I just, I didn't relate the two, because I hadn't seen it like that, we didn't play with any numbers like that, so I just didn't make the connection." The participants also gained an appreciation of how the "magic 9's" phenomenon yielded a process that allowed them to generate any repeating decimal they wanted, thus reversing the direction of the algorithm.

To test and especially falsify conjectures. Many students used Number Worlds to test and falsify conjectures. For example, when asked how to create the pattern shown in Figure 4b, one participant began by reasoning that the values of the multiples (= 5) would have to remain the same. She then decided to investigate what the effect of changing the value of the shift would be. By observing the effect of several different values she was able to determine empirically, and then verify, the solution.

There were two specific concepts that these experimental approaches enabled the participants to encounter. Prior research has discussed a belief found among students that large numbers have more factors than small numbers (Zazkis, 1999). It was described as belonging to the family of intuitive beliefs "the more of A, the more of B," discussed in detail by Stavy and Tirosh (2000). It appears that experimentation in Number Worlds helped the participants directly confront their misleading belief in relation to number of factors. Kelly provides another explanation for the role of the experimental aspect of the microworld in helping her understanding of factors:

However, the number of factors for each number depends on several aspects. Looking at only a few numbers is difficult to show such patterns; therefore, I have learned that it is important to look deeper and find more proof to reasoning. By finding similarities between several small numbers with four factors we were able to make generalizations about all numbers with four factors.

By experimenting with many different examples, Kelly shows that it is possible to locate patterns which can displace intuitively acquired ones such as "more of A, more of B."

Previous research has also shown that students do not immediately recognise a number represented as a product as a composite number (Zazkis & Liljedahl, 2004). In particular, they do not recognise that a product of two prime numbers cannot itself be a prime number. During the interviews, participants in this study were asked whether the prime numbers were closed under multiplication. Every single participant unequivocally provided the correct answer.

Some students spoke about the importance of experimenting and making predictions, particularly in contrast to just being told about a certain property or relationship. For example, Amber stressed how being able to work with the ideas gives her a sense of agency:

> …when you're actually working with it and it's not somebody just telling you how it is and you're actually seeing it for yourself, so if you get to try things out and you get to see what your results are and you get to make um, make predictions and then see if they're right or wrong, I think that helps.

For Jake, the ability to not only test theories but also, to test theories quickly was important: "The program allowed me to test potential theories as they entered the mind, and quickly enough so that the thought was not lost; the program did the time-consuming work." Jake's comment captures the way in which the microworld may support qualitative reasoning, i.e., by allowing students to test ideas that are not fully formed. Kyle commented on how the ease of experimentation helped make concrete his ideas-in-formation: "being able to change variables and viewing the outcomes can clarify and solidify ideas which are more often than not just floating around in a person's head." Since it was easy to perform any operation, Katherine found she was less concerned about making mistakes, and more inclined to just try things out:

> …you could just quickly click and say this, and then reset it and try something different if it didn't work, it was pretty quick that way like you could just do it and the picture was there in front of you, right or wrong, okay it's wrong, so let's just erase it and try something new.

The ease with which participants could experiment with a wide range of numbers may have helped them move beyond the kind of pattern spotting that frequently characterizes the approaches of students—where a few small-number examples are seen as sufficient to determine a rule or pattern.

To confirm analytically-derived results. The research participants spent most of their time actually working with the microworlds, solving problems designed to take advantage of the microworld's strengths. This meant that few problems required the authentic kind of back and forth, between pencil-and-paper and computer, that most mathematicians adopt in their own work. However, a few participants used Number Worlds to check results they had derived non-empirically. These participants would reason through a problem without using the microworld and then, once they had a solution, try it with Number Worlds to verify their solution. Since the students could have just as easily asked the interviewers about the correctness of these solutions, but instead chose to use the computer, suggests that the participants had developed a certain reliance on—and perhaps trust in—the visual feedback of Number Worlds. In addition to establishing correctness, these participants knew that the visual feedback might also provide additional guidance in case their solution was incorrect.

The aesthetics of experimental mathematics

I offer a comment on another aspect of working with computer-based technologies that Borwein and Bailey mention only briefly, that of aesthetics. Most mathematicians have a highly developed aesthetic vocabulary, and frequently comment on the beauty and elegance of certain mathematical entities such as theorems, proofs and definitions. And although aesthetic judgments are most frequently heard as evaluative judgments of these entities, mathematicians also rely on aesthetic feelings to motivate their choice of problem and to guide their problem-solving process (Hadamard, 1945; Poincaré, 1908/1956; Sinclair, 2004). Aesthetic choices and sensibilities are not just frivolous "extras" for mathematicians; they play a fundamental role in the creation and development of mathematics. However, educators have found it difficult to evoke and nurture them in the classroom.

Papert (1978) was perhaps the first to argue that computers are especially well-suited to developing and using students' aesthetic appreciation, and others have followed suit (see Goldenberg, 1989; Sinclair, 2001). Its visual capabilities, which have been exemplified frequently in this paper, along with its expressive capabilities, which help students feel that they are producing and not just consuming mathematics, are both seminal components of this aesthetic potential. For example, after her use of the CC, one student described fractions with the aesthetic language that mathematicians have often used to describe favorite entities and ideas:

> The repeating fractions have a flow that I find comforting. As a child I disliked fractions that did not terminate, but now I see them in a light of beauty. I find that the decimals which terminate sad. They are unable to touch the fingertips of forever, like the repeating ones can.

Similarly, Mary reflected on the emotional experience she felt working with Number Worlds:

> The repeating, infinite pattern is calming. I can see with my eyes open the pattern on the grid right in front of me. They all seem to be linked together, no matter how far away from the next multiple they are, they are all connected to each other.

For some of the research participants, the supposed beauty and elegance of number theory was made manifest in the pleasing patterns they could create and manipulate on the grid. More importantly, the general appeal of the patterns encouraged most participants to simply 'play around' and explore. In fact, the aesthetic dimension of their experiences was intimately related to learning and problem-solving, and not just tricks to make mathematics more palatable or "fun."

The experiences of these research participants resonate with many reports in the literature of the new relationships to mathematics that computer-based environments help nurture in students. Often, researchers seem almost apologetic in discussing this affective and aesthetic dimension of student learning. However, as almost any autobiography or personal commentary will show, such experiences are what keep mathematicians 'in the game.' Future research on the use of technology in mathematics learning should uncover the ways in which the aesthetic, affective and cognitive dimensions of mathematical activity interrelate.

Discussion

Several researchers have argued that the inductive patterning activities found in many reform-based curricula—whether computer-based (see Chazan, 1993) or paper-and-pencil based (see Hewitt, 1992)—may hinder the development of deductive reasoning. Several of the characteristics of experimental mathematical reasoning tend to be inductive in nature, including: 'to gain insight or intuition,' 'to discover new patterns or relationships' and 'to produce graphical displays that can suggest underlying mathematical patterns.' While other characteristics such as 'to test and especially falsify conjectures,' 'to suggest approaches for formal proof' and 'to explore a possible result to see if it is worth formal proof' call upon aspects of mathematical activity that are neither inductive nor deductive. They do, however, directly support deductive reasoning in the course of establishing a proof—they highlight aspects of proving that blur the long-standing divide between induction and deduction. It may be that the tension teachers and students experience between inductive exploration and deductive proof can be allayed by the experimental approach discussed in this chapter. By explicitly drawing attention to and supporting the use of the characteristics identified and illustrated above, teachers may be able to help students effectively engage in both inductive and deductive forms of reasoning.

References

Arzarello, F., Micheletti, C., Olivero, F., & Robutti, O. (1998). A model for analyzing the transition to formal proof in geometry, *Proceedings of PME-XXII*, Stellenbosch, vol. 2, pp. 24–31.

Boero, P. Garuti, R. & Mariotti, M.A. (1996). Some dynamic mental processes underlying producing and proving conjectures, *Proceedings of PME-XX, Valencia, 2*, 121–128.

Borwein, J., & Bailey, D. (2004). *Mathematics by experiment: Plausible reasoning in the 21st century*. Natick, MA: A. K. Peters.

Brown, S. (1978). *Some "prime" comparisons*. Reston, Virginia; National Council of Teachers of Mathematics.

Chazan, D. (1993). High school geometry students' justifications for their views of empirical evidence and mathematical proof. *Educational Studies in Mathematics, 24*(4), 359–387.

Devlin, K. (2000). *The math gene: How mathematical thinking evolved and why numbers are like gossip.* New York: Basic Books.

Dubinsky, E., & McDonald, M.A. (2001). APOS: A constructivist theory of learning in undergraduate mathematics education research. In D. Holton et al. (Eds.), *The teaching and learning of mathematics at university level: An ICMI study* (pp. 273–280). Netherlands: Kluwer.

Ferrari, P. (2002). Understanding elementary number theory at the undergraduate level: a semiotic approach. In S. Campbell and R. Zazkis (Eds.), *Learning and teaching number theory: Research in cognition and instruction* (pp. 97–116). Westport, CT: Ablex Publishing.

Goldenberg, P. (1989). Seeing beauty in mathematics: Using fractal geometry to build a spirit of mathematical inquiry. *Journal of Mathematical Behavior, 8,* 169–204.

Hadamard, J. (1945). *The Psychology of Invention in the Mathematical Field.* Princeton, NJ: Princeton University Press.

Harel, G., & Sowder, L. (1998). Students' proof schemes: Results from exploratory studies. *Research in Collegiate Mathematics Education, 3,* 234–283.

Hewitt, D. (1992). Train spotters' paradise. *Mathematics Teaching, 140,* 6–8.

Papert, S. (1978). The mathematical unconscious. In J. Wechsler (Ed.), *On aesthetics and science.* Boston: Birkhäuser.

Poincaré, H. (1908/1956). Mathematical creation. In J. Newman (ed.), *The world of mathematics, Vol. 4* (pp. 2041–2050). New York: Simon and Schuster.

Pólya, G. (1954). *Mathematics and plausible reasoning.* Princeton, NJ: Princeton University Press.

Sfard, A. (1991). On the dual nature of mathematical conceptions: reflections on processes and objects as different sides of the same coin. *Educational Studies in Mathematics, 22,* 1–36.

Simon, M. (1996). Beyond inductive and deductive reasoning: the search for a sense of knowing. *Educational Studies in Mathematics, 30,* 197-210.

Sinclair, N. (2001). The aesthetics *is* relevant. *For the Learning of Mathematics, 21*(2), 25–32.

—— (2004). The role of the aesthetic in mathematical inquiry. *Mathematical Thinking and Learning, 6*(3), 261–284.

Sinclair, N., Zazkis, R., & Liljedahl, P. (2004). Number worlds: Visual and experimental access to elementary number theory concepts. *International Journal of Computers for Mathematics Learning, 8*(3), 235–263.

—— (2006). A coloured window on pre-service teachers' conceptions of rational numbers. *International Journal of Computers for Mathematics Learning, 11*(2), 177–203.

Stavy, R., & Tirosh, D. (2000). *How students (mis-)understand science and mathematics: intuitive rules.* New York: Teachers College Press.

Zazkis, R. (1999). Intuitive rules in number theory: Example of "the more of A, the more of B" rule implementation. *Educational Studies in Mathematics, 40*(2), 197–209.

—— (2000). Factors, divisors and multiples: Exploring the web of students' connections. *Research in Collegiate Mathematics Education, 4,* 210–238.

Zazkis, R., & Campbell, S. (1996). Divisibility and multiplicative structure of natural numbers: Preservice teachers' understanding. *Journal for Research in Mathematics Education, 27*(5), 540–563.

Zazkis, R., & Liljedahl, P. (2002). Arithmetic sequence as a bridge among conceptual fields. *Canadian Journal of Science, Mathematics and Technology Education, 2*(1), 93–120.

—— (2004). Understanding primes: The role of representation. *Journal for Research in Mathematics Education, 35*(3), 164–186.

Zazkis, R., Sinclair, N., & Liljedahl, P. (2006). Conjecturing in a computer microworld: Zooming out and zooming in. *Focus on Learning Problems in Mathematics, 28*(2), 1–19.

19

Worked Examples and Concept Example Usage in Understanding Mathematical Concepts and Proofs

Keith Weber, *Rutgers University*

Mary Porter, *Saint Mary's College*

David Housman, *Goshen College*

Elsewhere in this volume, Watson and Mason discuss example generation from the students' perspective by highlighting some of the ways that example generation can be used to increase students' understanding of mathematics and improve their attitudes toward mathematics. This chapter complements this work by describing ways that teachers and textbooks might use examples to help undergraduates better understand mathematics. We distinguish between using worked examples to solve exercises and problems and using examples to help promote students' understanding of mathematical concepts and proofs. We begin with worked examples provided by the teacher or textbook. We then discuss the role of examples in building an understanding of a mathematical concept. Next we discuss how examples can be useful in understanding mathematical proofs. In each of these sections, we present specific suggestions that teachers might use in their own mathematics classrooms and we cite research studies that motivate and support these suggestions.

Worked Examples

The term "example" has multiple uses in mathematics education (cf., Watson & Mason, 2002). In some contexts, the word "example" refers to an illustration of a technique used to complete a certain type of mathematical task. For instance, a written solution to the question "Find all local minima and maxima of the function $f(x)=x^3+5x^2-8$" might be regarded as an example of how to solve minimum/maximum problems in an introductory calculus course. This is the way that the word example is often used in undergraduate textbooks, in which individual sections of the book frequently introduce a technique and then provide a series of examples in which the technique is applied. In this paper, we refer to such examples as "worked examples". In other cases, the use of the word example is meant as a particular instance of a mathematical concept (e.g., 6 is an example of an even number). Consideration of these types of examples can be used to improve students' understanding of a concept. Pedagogical uses of examples of concepts will be discussed later in this chapter.

A series of studies by Lithner (2000, 2003, 2004) suggest that undergraduates in procedure-oriented mathematics courses like calculus complete homework exercises predominantly by first locating similar worked examples in the textbook and then using them as a basis for formulating a solution to the exercise. (Note that this might also be the case in proof-oriented courses. Recent studies suggest that some undergraduates in proof-oriented courses may construct proofs via the use of worked examples (Fukawa-Connelly, 2005; Weber, 2004, 2005a, 2005b)). In one study, Lithner (2003) observed undergraduates completing their homework problems in a calculus course. He found that the students in his study almost always used worked examples to complete their homework. This strategy was employed by both weak and strong students, and was used in cases in which the undergraduates had the background knowledge to complete the problems without the use of this strategy (Lithner, 2000, 2003). In another study, Lithner (2004) analyzed the exercises in calculus textbooks to determine what proportion of exercises could be solved by the use of worked examples. He found that, for 90% of the exercises in these textbooks, there was an analogous worked example presented earlier in the section. With minor modifications, these worked examples could be transformed into solutions to the exercises. In these cases, the undergraduates could solve the problem without reasoning about the concepts in the section. Only 10% of the problems could not be solved in this way and required the students to reason about the properties of the concepts that they were ostensibly studying. Lithner (2003, 2004) expressed concern about these results, fearing that many students are completing their homework in ways that are not conducive to building their conceptual understanding or problem-solving strategies.

In spite of Lithner's reservations, many cognitive psychologists have stressed the benefits of having students use worked examples to solve problems (e.g., Zhu & Simon, 1987; Atkinson, Derry, Rankl, & Worthman, 2000). In particular, Atkinson et al. argue that worked examples "provide an expert's problem-solving model for the learner to study and emulate" (p. 181–182). Students who try to solve problems without examples typically develop, practice, and reinforce "novice strategies"—that is, strategies for solving problems that both ignore the deep structure of the problems being solved and are generally ineffective. In contrast, students who use appropriately chosen worked examples as a guide for solving problems are more likely to focus on the deep structure of the problem that they are solving and use more sophisticated strategies for solving it (Atkinson et. al., 2000). Tarmizi and Sweller (1988) analyzed what types of worked examples are most effective for helping students learn how to solve problems. They found that worked examples which reduced students' *cognitive load*—that is, solutions that would not require a student to expend a great deal of mental effort to understand—proved to be more beneficial to students than more complicated worked examples that required greater mental effort to comprehend. For instance, worked examples in geometry that relied on a single mode of representation (e.g., only analytical or only diagrammatic) helped students more than worked examples that combined two representations (e.g., a solution in which analytic reasoning interacted with a diagram), since the latter required students to expend cognitive effort to understand the links between the analytic and diagrammatic portions of the solution.

The preceding summaries serve as a basis for two pedagogical suggestions. First, when worked examples are presented to students, it is important to include examples that are simple and easy to follow. For instance, when presenting a solution to a min/max problem in calculus, it is advisable to have some examples that do not use sophisticated algebraic manipulations, the use of trigonometric identities, or other techniques that an undergraduate might not easily follow. Such examples will cause students to focus more on the details of the solution, rather than its deeper structure. Second, it might be worthwhile to ask students to complete some exercises that cannot be solved solely via the consideration of a worked example, but requires the student to think about relevant properties and concepts (Lithner, 2003). Such experiences will provide students with the opportunity to develop their understanding of the mathematics being studied and mitigate chances that they will develop the unproductive belief that mathematics consists of learning a series of procedures.

Using Examples to Build Concept Images

What does it mean to understand a mathematical concept? An undergraduate's understanding of a mathematical concept should include his or her ability to state and reason from the definition of that concept. However, mathematics educators argue that one's understanding of a mathematical concept involves much more than this. Tall and Vinner (1981) distinguished between a student's knowledge of a concept's definition and that student's *concept image*—i.e., her/his total cognitive structure, including all examples, nonexamples, facts, properties, relationships, diagrams, images, and visualizations, associated with that concept. Students' images of concepts have a significant influence on how they reason about a concept. Many students have images of concepts that are at variance with the concept's definition. For instance, Tall and Vinner (1981) found that many students claimed that functions whose graphs have cusps are not continuous, even if they could state the definition of continuity. Students with poor images of concepts often experience difficulty applying the concept definitions and writing proofs about those concepts (e.g., Moore, 1994; Weber & Alcock, 2004; see also Oehrtman, Selden & Selden, and Harel & Brown, this volume). On a more positive note, students with rich and accurate concept images are often able to reason productively about these concepts and use their intuitive reasoning as a basis for constructing formal proofs (Weber & Alcock, 2004).

These findings suggest that an important goal of mathematics education is to provide students with the opportunity to build strong concept images. In this subsection, we will address the question of how this goal might be achieved. We first present the results of two research studies demonstrating that students can build their understanding of a concept by considering and generating examples of that concept. We then discuss actions that a teacher might take in his or her classroom to lead students to generate or consider a variety of examples and describe research that supports these suggestions.

Several studies show that one way that students can develop a strong concept image is by generating examples of that concept. Dahlberg and Housman (1997) and Housman and Porter (2003) investigated the strategies undergraduate students use to learn a new mathematical concept. In both of these studies, students were given the following formal definition:

A function is called *fine* if it has a root (zero) at each integer.

At first students were given no guidance and were simply asked to come to an understanding of the definition. Students used a variety of learning strategies at this point, including generating examples, reformulating the definition in their own words, memorizing, and recalling definitions of the base concepts *function*, *root*, *at each*, and *integer*. Students were then asked to carry out a number of tasks that both measured and helped to develop their understanding of the new concept: Students were asked to give an example of a fine function, give an example of a function that was not fine (a "nonexample"), provide an explanation in the student's own words and/or pictures of what a fine function is, verify whether six functions were or were not fine, and determine whether four conjectures were true or false. In Dahlberg and Housman's study, the students who used example generation (producing one or more examples related to the concept) and concept reformulation (expressing the concept using pictures, symbols, or words different from the definition) were the ones best able to develop an accurate and useful understanding of the concept. In addition, the students who used example generation were the ones who were best able to identify the correctness of conjectures and provide explanations. The students who primarily reformulated concepts without generating examples were also able to determine whether a given object was an example of the mathematical concept, but these students were more easily convinced of the validity of a false conjecture. Although example generation and concept reformulation were the most beneficial learning strategies for these students, example usage – the use of researcher-provided examples — was also somewhat effective in helping students learn about the concept. In Housman and Porter's study, students who wrote and were convinced by deductive arguments in a separate task-based interview were successful in reformulating concepts, using examples, and generating examples when asked to do so or when it was necessary to disprove a conjecture.

The results of these two studies suggest that the consideration of examples of concepts can help students understand concepts better. In addition, many mathematicians find it useful to consider carefully chosen examples to understand concepts and their definitions (e.g., Alcock, 2004). As Paul Halmos remarked, "A good stock of examples, as large as possible, is indispensable for a thorough understanding of any concept, and when I want to learn something new, I make it my first job to build one." (Halmos, 1983, p. 63). Unfortunately, the data from these two studies also demonstrate that some students do not spontaneously consider examples when presented with a new concept. In the rest of this sub-section, we consider three ways that teachers might lead students to consider examples: (1) by presenting examples, (2) by helping students generate examples, and (3) by asking students to reason about given examples.

The most straightforward suggestion is for teachers to simply provide examples and counterexamples when they introduce a new concept. However, there are two reasons why a teacher should exercise caution in choosing which examples to present. First, students tend to overgeneralize, believing that irrelevant properties held by an example of a concept are shared by all members of the concept. Second, as Watson and Mason point out in this volume, many students will treat counterexamples as isolated cases, anomalies that can be ignored. A few researchers have suggested guidelines for example presentation that can potentially alleviate the negative effects of these student tendencies. We will discuss these guidelines below and then illustrate how they can be applied in the case of convergent sequences.

- First, teachers can present a wide range of examples that do not all share an irrelevant characteristic (Sowder, 1980).

- Second, it is often useful to pair an example and a counterexample that differ in only one characteristic, allowing students to focus their attention on relevant aspects of the concept (Sowder, 1980).

- Third, teachers can not only describe why the examples are, or are not, members of the relevant concept, but also describe ways that students can produce similar examples or counterexamples (Peled & Zaslavsky, 1997).

- Finally, consistent with Watson and Mason's chapter in this volume, after presenting one type of example of a concept, teachers could then ask students to construct similar examples or even describe how a class of such examples could be constructed (see also Watson & Mason, 2002).

To illustrate these guidelines in a concrete setting, we discuss the concept of convergent sequences. It is natural to exemplify this concept with a prototypical convergent sequence, such as $(1/n)$. The danger with only introducing this one example, or very similar examples such as $(1/n^2)$, is that students may focus on features of this sequence that do not guarantee convergence. For instance, students may infer that convergent sequences are monotonic, never attain their limit, or that each term must be closer to the limit than the last. In fact, an extensive body of research shows that many undergraduates hold these beliefs (Cornu, 1991). For this reason, it is better to present students with a range of examples, perhaps including an alternating sequence converging to zero (illustrating a non-monotonic convergent sequence), a constant sequence (showing that sequences can attain their limits), and non-prototypical convergent sequences such as (1, 2, 3, 4, 5, 6, 1, 1, 1, 1, 1…). Likewise, students should be asked to consider a wide range of counterexamples, including sequences that diverge to infinity and negative infinity, other unbounded sequences, and sequences with multiple cluster points. Each of these counterexamples could be compared to a specific convergent sequence, similar in most respects, but differing in an important respect that causes one to diverge and the other to converge. For instance, comparing the sequences

$$a_n = \begin{cases} 1 - \dfrac{1}{n} & \text{if } n \text{ is odd} \\ 1 + \dfrac{1}{n} & \text{if } n \text{ is even} \end{cases} \quad \text{and} \quad b_n = \begin{cases} 2 - \dfrac{1}{n} & \text{if } n \text{ is odd} \\ 1 + \dfrac{1}{n} & \text{if } n \text{ is even} \end{cases}$$

can illustrate that sequences defined by subsequences can converge, but only if the two subsequences converge to the same number. For examples such as (a_n) and (b_n), teachers can describe the reasoning they used to produce these examples and describe how other examples of that type could be produced. For each presented example, students can be asked to generate another sequence or a family of sequences that have the same features as the example under consideration.

Teachers can also have students play a more active role in their mathematical learning by having them generate examples of concepts themselves. When discussing convergent sequences, the teacher could ask students to generate a *particular* sequence that is convergent. Responses might include $(1/n)$. Some students may not have come up with any examples on their own, but after seeing their classmates' examples, they might contribute to the next task. The teacher could then ask students to give an example of a sequence that is *peculiar* in some way. Responses might include some of the previously reported examples together with the reasons why they are peculiar. For instance, the sequence

$$a_n = \begin{cases} 1 - \dfrac{1}{n} & \text{if } n \text{ is odd} \\ 1 + \dfrac{1}{n} & \text{if } n \text{ is even} \end{cases}$$

is given as two formulae instead of one, and the sequence $b_n = 0$ for all natural numbers n is constant. Asking students to give examples of convergent sequences in alternative representations, such as a graph of a convergent sequence or expressing a convergent sequence as a recurrence relation, can also be encouraged. Discussion of such examples can lead the class towards a *general* characterization of a convergent sequence (See also Marrongelle & Rasmussen, this volume, for a further description of proactive teacher moves).

Students can also be asked to generate examples in order to explore boundaries and extend the range of possibilities. For example, students who are studying convergent sequences could be asked to find, in this order, each of the following:

1. a convergent sequence,
2. a convergent sequence that is not monotonic,

3. a convergent sequence that does not get strictly closer to its limit with each term,

4. a convergent sequence that achieves its limit, and

5. a convergent sequence whose formula, treated as a real-valued function, would not be continuous.

In doing this task, ask students to make sure that an example given for any one item should *not* satisfy the next item. Thus, the first example would be a sequence that converges, but is monotonic. The second example would be a non-monotonic convergent sequence, but one that does become strictly closer to its limit with each term (such as $(-1)^n/n$), and so forth. When each additional constraint narrows the range of possibilities, students are induced to think more broadly to successfully complete the exercise.

Alcock (2004) suggests a third way teachers can use examples to enrich students' concept images. Students can be given definitions for a collection of concepts. Then students can be presented with worksheets with a number of objects and be asked to determine what properties each object has. To illustrate, after students are introduced to sequences, they can be given the definitions for convergent, bounded, and monotonic sequences. They can then be given a collection of sequences and asked to determine if the sequences are convergent, bounded, and/or monotonic. Conversations between students and between student and teacher can enable students to understand what the definitions of each of these concepts are asserting and to build their concept images of the properties. Further, these activities can address potential misconceptions that students might have or develop (examining sequences such as $(-1)^n/n$ will help students realize that a sequence does not have to be monotonic to converge) and to form mathematical conjectures in response to their own questions (e.g., are all convergent sequences bounded? Do all monotonic bounded sequences converge?). Such a treatment does not have all the benefits of example generation that Watson and Mason discuss in this volume; for instance, the affective benefits of example generation may not be realized here. However, Alcock's suggestion may be more efficient in terms of time, and it does ensure that students will consider the classes of examples that teachers believe are important.

Using Examples in the Form of Generic Proofs

There are many purposes of presenting proofs in university classrooms. However, mathematics educators argue that two of the most important purposes of proof are *convincing*—i.e., removing all doubt that a theorem is true—and *explaining*—i.e., providing students with insight as to *why* a theorem is true (e.g., Hanna, 1990; Hersh, 1993). Hersh (1993) argues that the formal proofs that we present to our students often fail to achieve both of these goals. First, many students obtain conviction of general assertions not by reading formal proofs, but by checking whether that assertion holds in several individual cases.[1] Formal proofs seem superfluous to these students—they feel they can find out whether an assertion is true or not just by checking a few examples themselves. Further, due both to their weak understanding of formal proofs and the highly formal way that proofs are traditionally presented, undergraduates often do not find proofs to be explanatory (Hersh, 1993).

Rowland (2002) suggests an alternative to formal proofs in number theory. When discussing a general theorem that applies to a class of objects, choose an arbitrary object from among that class. Demonstrate that the theorem holds for that particular object, but make sure that the demonstration relies in no way upon properties of the specific object under consideration that are not shared by all objects in the class to which the theorem applies. Mason and Pimm (1984) call such a demonstration a *generic proof* of the theorem. Rowland advocates presenting generic proofs of theorems prior to, or in lieu of, formal proofs of theorems. He argues that students gain more conviction and understanding from generic proofs than from formal proofs, and students' comprehension of formal proofs will improve if the presentation of a generic proof precedes the presentation of a formal one.

Rowland provides a concrete instance of a generic proof by discussing his treatment of Wilson's theorem. Wilson's theorem asserts:

$$(p-1)! \equiv -1 \ (\mathrm{mod} \ p) \text{ for all primes } p.$$

When teaching students about Wilson's theorem, Rowland justifies the theorem using the following generic approach. He first looks at the statement for the particular prime 19, although 13 and 17 would work equally well. He lists the integers between 1 and 18 (inclusively), the reduced set of integers modulo 19. He then draws lines connecting each

---

[1] In the language of Harel and Sowder (1998), we might say these students hold an *empirical proof scheme*, but not a *deductive proof scheme*.

element in this list to its multiplicative inverse modulo 19, linking 2 with 10, 3 with 13, and so on. Of course, every listed element will have an inverse that is another element in the list, with the exception of 1 and 18, which are their own inverses. Now 18! can be rewritten by lining up each integer with its multiplicative inverse modulo 19. After doing this, Rowland shows how the product

$$\prod_{i=1}^{18} i \equiv 1 \cdot 1^8 \cdot 18 \pmod{19}.$$

There are several aspects of this presentation that made this a good generic proof. The first was Rowland's choice of 19. A prime such as 2, 3, or 5 would not have enough reduced residue classes to see the general structure of Rowland's arguments. A larger prime like 37 would have so many residue classes that the argument would become more difficult to follow; the students may lose the structure of the argument in the arduous calculations of finding the multiplicative inverse of each integer modulo 37. Further, 19 appeared to be an "arbitrary prime"—i.e., it did not have any noticeable distinguishing properties not shared by other primes. Another reason 2 would be a poor prime to inspect was because it was the only even prime.[2] Second, Rowland did not make use of any special properties of the number 19 .The central reasoning in Rowland's argument was that every element except 1 and 18 (which is –1 modulo 19) is not its own multiplicative inverse modulo 19. Rowland's demonstration could easily be used to verify Wilson's theorem for any other prime. Third, all constructive aspects of the proof were identified and verified. For instance, the claim that 2 had a multiplicative inverse modulo 19 was not only justified by a theorem. The inverse of 2 was also explicitly found in the number 10, and the student could verify that 2 and 10 were in fact inverses. Finally, the reasoning was presented in such a way that it could easily be abstracted into a more general formal proof. Based on questionnaires and interview data from his own classrooms, Rowland reports that students who see this type of presentation can describe why this general assertion can be applied to any prime number and gain a strong conviction that the theorem holds for all prime numbers.

Hazzan and Zazkis (2003), following Rowland, describe another way that examples could be used to construct proofs. Many proofs have *constructive components*—they show how certain elements with desired properties can be created, but do not explicitly state what these objects are. Hazzan and Zazkis advocate having students perform the constructions themselves in particular instances before observing the proof that employs these constructions In support of this recommendation, Hazzan and Zazkis argue, "The human mind is not satisfied with the knowledge that some objects exist. There is a desire to point out *exactly* what these objects are. Similarly, unraveling a construction process with an example helps us understand *exactly* how the construction works" (italics are the authors').

Consider the proof that there are infinitely many primes. A standard proof of this proposition is given below.

Theorem: *There are infinitely many primes.*

Proof (by contradiction): Suppose there are not infinitely many primes. Then we can enumerate the primes p_1, $p_2,..., p_n$. Let $N = p_1 \cdot p_2 \cdot \cdots \cdot p_n + 1$. For all i such that $1 \le i \le n$, p_i divides $p_1 \cdot p_2 \cdot \cdots \cdot p_n$ but does not divide 1, so p_i does not divide N. Hence, no prime divides N. This contradicts the fact that every integer greater than 1 must be divisible by at least one prime.

This proof has a constructive aspect in that it describes how a number N can be constructed, but does not explicitly state what the number N is. The exact value of N, of course, depends on what numbers are elements of the hypothetical finite set of primes (Leron, 1985). Students often have trouble following this proof. However, if students were asked to construct the N in the proof for particular sets of primes, their understanding might improve. For instance, students could verify that:

2 and 3 do not divide $N = 2 \cdot 3 + 1$,

2, 3, and 5 do not divide $N = 2 \cdot 3 \cdot 5 + 1$,

2, 3, 5, and 7 do not divide $N = 2 \cdot 3 \cdot 5 \cdot 7 + 1$, and so on.

Students might also want to examine $N = 2 \cdot 3 \cdot 5 \cdot 7 \cdot 11 \cdot 13 + 1 = 30031$. Here 30031 is a composite number ($30031 = 59 \cdot 509$). Inspecting this example can make students aware that the product of the first n primes plus 1 does not always yield a prime number, only a number whose prime factors are not included in the presumably finite set of

---

[2] The lack of special properties is easier to see if we move beyond looking at primes. For instance, in a generic proof about the natural numbers, one should choose numbers that are neither prime nor perfect squares.

primes. After exploring these examples, the formal proof that there are infinitely many primes will be more accessible to students. They will have a greater appreciation for how the variable N in the proof is being defined and why none of the enumerated primes will divide it.

Hazzan and Zazkis further illustrate how these techniques can be used to enhance students learning of other proofs with constructive components. One proof they looked at was a standard proof of the Basis Theorem in linear algebra. The Basis Theorem asserts that, in a finite-dimensional space, all bases have the same cardinality. Standard proofs of the Basis Theorem often rely on the Replacement Lemma, which asserts: "Let B be a set of linearly independent vectors in the spanning space of a set of vectors A. For all subsets $B_1 \subseteq B$, there exists a subset $A_1 \subseteq A$, such that $|A_1| = |B_1|$ and $(A - A_1) \cup B_1$ spans the same space as A."

The Replacement Lemma is clearly constructive, in the sense that it tells the reader that a subset of a spanning set exists, but it does not state what it is or even how it could be found. As a result, many students find proofs of the Basis Theorem relying on this lemma to be confusing. Hazzan and Zazkis (2003) designed a series of computer activities to help students understand the Replacement Lemma. These activities allowed students to enter a spanning set A and a set of linearly independent elements B_1 and the computer would then find the elements in the spanning set which could be replaced by the subset B_1. Students who completed these exercises found the subsequent proof of the Basis Theorem to be understandable and meaningful.

Conclusion

In this chapter, we have discussed a number of ways that teachers can use worked examples and employ examples to build undergraduates' understanding of mathematical concepts and proofs. Examples not only illustrate concepts, principles, and proofs, they can help students to explore, expand, generalize, refine, and test their understanding. Students who are exposed to, work with, and generate their own examples are actively engaged in mathematics and learning.

Acknowledgements The authors would like to thank Lara Alcock, the Tall Group for Advanced Mathematical Thinking, and the reviewers and editors for useful comments on earlier drafts of this manuscript.

References

Alcock, L.J. (2004). Uses of example objects in proving. In M. J.Hoines & A. B. Fuglestad (Eds.), *Proceedings of the 28th Conference of the International Group for the Psychology of Mathematics Education*, *2*, 17–24. Bergen, Norway.

Atkinson, R. K., Derry, S. J., Renkl, A., & Wortham, D. (2000). Learning from examples: Instructional principles from the worked examples research. *Review of Educational Research*, *70*(2), 181–214.

Cornu, B. (1991). Limits. In D.O. Tall (Ed.), *Advanced mathematical thinking*. Kluwer: Dordrecht.

Dahlberg, R. P., & Housman, D. L. (1997). Facilitating learning events through example generation. *Educational Studies in Mathematics*, *33*(3), 283–299.

Fukawa-Connelly, T. (2005). Thoughts on learning advanced mathematics. *For the Learning of Mathematics*, *25*(2), 33–35.

Hanna, G. (1990). Some pedagogical aspects of proof. *Interchange*, *21*(1), 6–13.

Halmos, P. (1983). In D. Sarasen & L. Gillman (Eds.) *Selecta: expository writing*. New York: Springer-Verlag.

Harel, G., & Sowder, L. (1998). Students' proof schemes. *CBMS Issues in Mathematics Education: Research in Collegiate Mathematics Education III*, 234–283.

Hazzan, O. & Zazkis, R. (2003). Mimicry of proofs with computers: The case of linear algebra. *International Journal of Mathematics Education in Science and Technology, 34,* 385–402.

Hersh, R. (1993). Proving is convincing and explaining. *Educational Studies in Mathematics*, *24*, 389–399.

Housman, D. L., & Porter, M. K. (2003). Proof schemes and learning strategies of above-average mathematics students. *Educational Studies in Mathematics*, *53*(2), 139–158.

Leron, U. (1985). A direct approach to indirect proofs. *Educational Studies in Mathematics, 16,* 321–325.

Lithner, J. (2000). Mathematical reasoning in task solving. *Educational Studies in Mathematics, 41,* 165–190.

—— (2003). Students' mathematical reasoning in university textbook exercises. *Educational Studies in Mathematics, 52, 29–55.*

—— (2004). Mathematical reasoning in calculus textbook exercises. *Journal of Mathematical Behavior, 23*(4), 405–427.

Mason, J., & Pimm, D. (1984). Generic examples: seeing the general in the particular. *Educational Studies in Mathematics, 15,* 277–289.

Moore, R.C. (1994). Making the transition to formal proof. *Educational Studies in Mathematics, 27,* 249–266.

Peled, I., & Zaslavsky, O. (1997). Counter-examples that (only) prove and counter-examples that (also) explain. *Focus on Learning Problems in Mathematics, 19*(3), 49–61.

Rowland, T. (2002). Generic proofs in number theory. In S. Campell & R. Zazkis (Eds.), *Learning and teaching number theory: Research in cognition and instructions* (pp. 157–184). Westport, CT: Ablex Publishing.

Sowder, L. (1980). Concept and principle learning. In R. J. Shumway (Ed.), *Research in mathematics education* (pp. 244–285). Reston, VA: National Council of Teachers of Mathematics.

Tall, D.O., & Vinner, S. (1981). Concept image and concept definition in mathematics, with special reference to limits and continuity. *Educational Studies in Mathematics, 12,* 151–169.

Tarmizi, R. A., & Sweller, J. (1988). Guidance during mathematical problem solving. *Journal of Educational Psychology, 80*(4), 24–436.

Watson, A., & Mason, J. (2002). Student-generated examples in the learning of mathematics. *Canadian Journal of Science, Mathematics and Technology Education, 2*(2), 237–249.

Weber, K. (2004). Traditional instruction in advanced mathematics courses: A case study of professors' lectures and proofs in an introductory real analysis course. *Journal of Mathematical Behavior, 23*(2), 115–133.

—— (2005a). A procedural route toward understanding aspects of proof: Case studies from real analysis. *Canadian Journal of Science, Mathematics, and Technology Education, 5*(4), 469–483.

—— (2005b). Problem-solving, proving, and learning: The relationship between problem-solving processes and learning opportunities in the activity of proof construction. *Journal of Mathematical Behavior, 24*(3/4), 351–360.

Weber, K., & Alcock, L. (2004). Semantic and syntactic proof productions. *Educational Studies in Mathematics, 54*(3), 209–234.

Zhu, X., & Simon, H.A. (1987). Learning mathematics from examples and by doing. *Cognition and Instruction, 4,* 137–166.

Part II
Cross-Cutting Themes

c. Knowledge, Assumptions, and Problem Solving Behaviors for Teaching

20

From Concept Images to Pedagogic Structure for a Mathematical Topic

John Mason, *Open University*

The principal aim of this chapter is to provide a structure for mathematical topics as an aid to 'psychologizing the subject matter', as Dewey (1933) put it. The secondary aim is to reveal just how complex a matter preparing to teach a topic effectively can be, beyond trying to make the definitions and theorems as clear as possible.

The chapter develops the notion of a *concept image* (Tall & Vinner 1981) into a description of a framework based on a threefold structure of the psyche. Two mathematical topics, quotient groups and L'Hôpital's rule, are used to illustrate how the framework can be used as a reminder to direct attention to structurally different aspects of any topic when preparing to teach it. The framework can be used both at a more abstract level (for example, treating groups or limits as the topic) or at an even more detailed level (for example, quotient groups of cyclic groups or the relation between L'Hôpital's rule and derivatives). When combined with awareness of learners' mathematical powers and with ubiquitous mathematical themes and heuristics, the framework can be used to inform the design of pedagogically effective tasks and interactions with learners.

Concepts and Concept Image

Concepts are not isolated entities floating about in our minds but rather familiar 'lines of thought' triggered by concept labels. For an expert, the mere mention of a technical term in mathematics, such as *coset* or *limit*, gives access to a variety of associations, techniques, ways of speaking, images, symbols and meaningful contexts which David Tall & Shlomo Vinner (1981) summarized as the *concept image* They had found that even when clear definitions of concepts were given, and even where plenty of examples were provided, learners found it difficult to penetrate beneath the surface and really get to grips with what particular concepts were about. Thus the label *concept image* acknowledges psychological experience, in contrast to the mathematically formal concept definition: it is the touchstone, the source of intuition and meaning associated with the concept, not the formal specification; the richness of connections and experience that underlies the honed articulation.

Tall & Vinner (1981) captured this experience in the term *concept image*, which refers to the whole mental structure of interconnected schema associated with a concept. They define a concept-image as

> ... the total cognitive structure that is associated with the concept, which includes all the mental pictures and associated properties and processes. It is built up over the years through experiences of all kinds, changing as the individual meets new stimuli and matures. (p. 152)

An essential feature of the concept-image is that the various aspects are interconnected.

> Many concepts which we use happily are not formally defined at all, we learn to recognise them by experience and usage in appropriate contexts. Later these concepts may be refined in their meaning and interpreted with

increasing subtlety with or without the luxury of a precise definition. Usually in this process the concept is given a symbol or name which enables it to be communicated and aids in its mental manipulation. But the total cognitive structure which colours the meaning of the concept is far greater than the evocation of a single symbol. It is more than any mental picture, be it pictorial, symbolic or otherwise. During the mental processes of recalling and manipulating a concept, many associated processes are brought into play, consciously and unconsciously affecting the meaning and usage. (*op. cit.* p. 152)

The question remains as to how a teacher might make use of their awareness of their own concept image, in planning how to go about teaching that concept to others.

In 1989 a group of us developed a structure or framework to inform and assist teachers in planning how to teach a mathematical topic which articulated our own approach. Originally referred to as *preparing to teach a topic* (Griffin & Gates, 1989; see also Mason, 2002) it has proved useful to refer to it more recently as *the (pedagogic) structure of a topic* (Mason & Johnston-Wilder, 2004). As a framework for planning, it can act as a reminder to make sure when designing tasks that learners encounter various important aspects and elements of each topic. It also acts as a format for accumulating notes and observations so that when a topic is being taught again in the future, the notes provide ready access to issues and concerns and so inform the planning.

Origins and Constituents of Topics

Mathematical concepts are constituents of mathematical topics. But topics are not simply collections of concepts, 'ideas', definitions, theorems and proofs. Rather, they are a complex tapestry of interwoven perceptions, thoughts and images, connections and links, behavioral practices and habits, applications and excitements. Something becomes a topic to be taught when someone recognizes that a class of problems could be solved by using a technique that could be taught. But the technique and hence the topic is based on a subtle shift in ways of perceiving, in ways of stressing some features and ignoring others. Thus, teaching someone a topic is introducing them to a way of perceiving and thinking, as well as ways of acting. This is underlined by the origins of the word *theorem* in the ancient Greek for 'a way of seeing'. When thinking about the pedagogical issues associated with teaching a topic, the ancient notion of a three-stranded psyche provides a practical structure for taking into account the five aspects of mathematical proficiency (Kilpatrick, Swafford, & Findell, 2001) forming the goals of teaching a topic, and comprising conceptual understanding, procedural fluency, strategic competence, adaptive reasoning, and productive disposition.

Three-Stranded Psyche

Ever since the composing of the Bhagavad Gita and the Upanishads twenty-five hundred years ago, people have found it useful to think in terms of the human psyche as made up of an interweaving of intellectual or cognitive functioning, emotional or affective functioning, and behavioral or enactive functioning. Modern psychology has absorbed the three-strand view, using the terms *cognition, affect,* and *enaction.* I find it more helpful to use the terms *awareness, emotion,* and *behavior* since they relate more closely to everyday usage and experience, and because they reflect the insights of ancient psychology more directly. These three interwoven threads can be expanded to inform both preparation for teaching a topic and making choices in the moment while interacting with learners. The next three subsections elaborate on these three strands in the context of mathematical topics, while developing a metaphor for the human psyche that is found in several of the Upanishads: the image of a horse-drawn chariot with driver and owner.

Each of the three strands is associated with a collection of fruitful questions for probing that strand in relation to a topic, and these questions are illustrated for the topics of quotient groups and L'Hôpital's rule.

Awareness The totality of thoughts, images, ideas, associations, related topics, and concepts that come to mind constitute your conscious *awareness* in the moment. Where these are focused on a concept, they form your *concept image*. As thoughts begin to flow, other awarenesses may also come to mind. These are the awarenesses that are dominant for you in association with the term or topic as triggered in that situation. The important pedagogic question is which of these you want learners to have come to their minds as a result of their work on the topic.

For example, you can ask what awarenesses are likely to be activated for learners as a result of working on a particular set of textbook exercises, or constructing examples of mathematical objects subject to certain constraints.

If for some learners there are aspects of a concept or topic that do not readily come to mind, then those learners may be disempowered and prone to error or confusion; perhaps even disheartened and frustrated. Furthermore, some ideas that do come to mind are not always appropriate or even correctly formulated, so associated with *awarenesses* are *absences* (things that learners often forget about), and *obstacles* (misconceptions and mis-construals that you notice learners having to work their way through at various times). Many of these errors are classic in the sense that learners seem prone to making them despite pedagogic attempts to circumvent them. Their existence provides evidence that learners are active construers, not just passive recipients.

There is more to awareness than simply 'coming to mind' since some awarenesses function below the level of consciousness. One of the reasons that experts find teaching novices a challenge is that their own fluent proficiency suppresses conscious awareness of now-automated thinking. Mathematical concepts are often (Johnson, 1987 and Lakoff & Nunez, 2000 might say 'always') based on bodily awareness. It is useful to try to re-encounter these for yourself so as to construct pertinent and meaningful tasks for learners.

Each technical definition in mathematics signals a need felt by someone to articulate a way of thinking that informs the solution of a problem or class of problems, and that now constitutes the topic. It is worthwhile therefore to try to re-enter the shift in thinking indicated by the presence of the term. As Imre Lakatos (1976) suggested, most definitions emerge and are modified as a theorem and its proof is refined so that it makes a theorem or collection of theorems efficient to state and (relatively) easy to prove. So to appreciate a definition, learners will need to experience something of the economy achieved by its formulation. One way for learners to do this is to experience both recognizing and constructing for themselves, examples that fit the definition, as well as examples that do not quite fit it. Watson & Mason (2005) coined the term *example space* to refer to the class of examples which come to mind in association with a concept or technique. Considering the desirable constituents of a learner's example space is another way of considering the core awarenesses forming some of the goals of instruction.

Experiencing, appreciating, and constructing examples and counter examples contributes to becoming aware of what aspects of an example can change while remaining an example. This is what is usually meant by conceptual understanding. Ference Marton (Marton & Booth, 1997; Marton & Trigwell, 2000; Marton & Tsui, 2004) refers to this as becoming aware of *dimensions of variation*, and he suggests that this is the essence of learning. Watson & Mason (2004, 2005) extended this language slightly to speak of *dimensions of possible variation*, since at different times one person is aware of different 'dimensions' that can vary, and very often teacher and learner are not aware of the same possibilities. Furthermore, each aspect that can be varied, can be varied in different ways, so it is useful to refer to the *range of permissible variation* of which someone is aware.

For example, learners may associate the term *group* with symmetries of squares and equilateral triangles but be unaware that this applies to a whole gamut of planar objects. They may be aware that planar objects can include regular polygons (a dimension of possible variation), but be unaware of irregular polygons (a restricted range of permissible change), tessellations, and more abstract settings. They may have some sense of 'planar objects', but be unaware that there are groups that cannot be displayed as the symmetries of such objects. They may even be aware that painting certain features of objects (vertices, edges, etc.) with colors (a dimension of possible variation) reduces the number of (color-preserving) symmetries but still produces a group that is a subgroup of the unpainted object. The range of permissible change due to coloring may not however be as rich as possible; for example, it may not include using multiple colors, and it may not include the fact that *every* subgroup of a group presented as the symmetries of an object can be presented using suitable coloring of that object, and *every* quotient group as a group of symmetries of color classes. By thinking in terms of dimensions of possible variation and associated ranges of permissible change during preparation, choices of examples and of how those examples might be presented can be pedagogically informed.

Useful questions to ask oneself concerning awareness include

(A1) What comes immediately to mind when you hear or read the words ...? What images, what connections to other parts of mathematics? What particular examples come to mind? What dimensions of possible variation (and associated ranges of permissible change) are important in order to appreciate component concepts?

(A2) What sorts of obstacles to or absences of awareness, often manifested as errors, confusions or ignorance, did you once have yourself, and what sorts have you detected in learners in the past?

Initial responses might include, in relation to quotient groups,

(A1) Seeing numbers both as actions and as objects on which actions are performed, and more generally, seeing elements of groups as actions on objects and as objects themselves; integer number-line with all members of each remainder class painted an identifying color; imposing an equivalence relation on actions; link to construction of integers from whole numbers and fractions from integers; normal subgroups; kernel of homomorphism; three isomorphism theorems; the answer to a calculation is independent of which representatives are chosen from the corresponding cosets; Cayley tables and Cayley graphs;

(A2) Learners often struggle with seeing cosets both as subsets and as objects; desire to treat quotient groups as if they were fractions.

Initial responses might include, in relation to L'Hôpital's rule,

(A1) Finding limits of ratios in which the numerator and denominator tend to 0, or both tend to infinity; derivatives; linking limit of $\sin(x)/x$ as derivative of sine at $x = 0$ with evaluating that limit, and seeing L'Hôpital's rule as a generalization.

(A2) Not checking or arranging that both numerator and denominator have 0 as their limits before trying to use the derivatives; not checking for differentiability; not differentiating numerator and denominator separately.

Introduction to the Chariot Metaphor

In describing the structure of the psyche, ancient psychologists found it fruitful to develop the metaphor of the human psyche as a chariot drawn by horses, with the driver being under the direction of an owner. Under the direction of will (the owner), the driver (awareness) is responsible for maintaining the state of the chariot (the body, behavior), and by means of the reins (mental imagery) directing the horses (emotions) that provide the motive energy.

Probing Deeper: richness of awareness

Although the term *cognition* is in popular use, it carries with it a sense of a conscious mind and of rational thinking. The term *awareness* also implies consciousness, but I follow Caleb Gattegno (1987) in using the term to encompass what the body is aware of even if the mind is not. For example, it is often the case that if you watch how you construct a particular example or apply a technique in a particular case, you can discover ways of expressing generality or structure because it is revealed through your actions. Absence of awareness can be debilitating, as, when interpreting graphs, although experts know how to start from a point on the x axis and label other points such as $(x, f(x))$, $(f(x), f(x))$, $(f(x), g \circ f(x))$, $(x, g \circ f(x))$ and so on, using various functions for f and g, because they have an underlying and accessible awareness of how points are coordinated, literally, many learners who have only ever seen graphs as completed curves seem unaware of them as sets of coordinated points and are unable to make sense of cobweb diagrams

Once a functioning is internalized, such as one-to-one matching or interpretation of graphs as both sets of points and curves, very little attention is required in order to make use of it. It is only when something breaks down in normally smooth functioning that we become consciously aware again. For example, the association of 'larger' with 'is a proper subset of' based on experience with finite sets is challenged when the notion of one-to-one matching is used for counting infinite sets. A function that is differentiable at a point and has arbitrarily large slope arbitrarily close to that point challenges previous images of and intuition about continuity and differentiability. One of the problems faced by an expert called upon to teach something very familiar is the need to re-contact awarenesses that are below the surface of consciousness, in order to appreciate what learners face and to construct or choose appropriate tasks to bring relevant awarenesses to the attention of learners.

Gattegno (1987) proposed that awareness is both the origin of a discipline and the product of the emergence and development of that discipline. Thus when someone becomes aware that they can solve not only a single problem, but a class of 'similar' problems, that awareness can be formulated, articulated, made precise, and distilled into a technique that can be recorded and taught to others. He also spoke of awareness of awareness, a notion developed in Mason (1998) to account for the demands placed on an expert who is called upon to teach, and on a teacher called upon to teach others to teach.

Awareness includes more than the concept image, for as well as connections and associations it also includes salient mathematical themes and heuristics that learners could encounter in the topic being considered, influenced by personal mathematical propensities and dispositions that have emotional connections. Thus within the broad category of awareness as an aspect of the psyche, there is the potential for discerning cognitive, behavioral, and emotional components. In other words, the structure of a topic being developed here is potentially recursively complex.

It is convenient to include in the awareness strand of the framework the sorts of difficulties and struggles that learners have shown evidence of in the past, including classic slips, errors and mis-understandings, some of which may constitute what Bachelard (1938) called *epistemological obstacles*. These are fundamental difficulties experienced at the genesis of the topic because of a necessary change in ways of perceiving or thinking. They are usually signaled by the use of specially formulated technical terms and symbols. To make progress learners require more than the use of words and symbols apparently appropriately; they need to extend and alter their ways of seeing and behaving, and hence also their emotional dispositions. Nicolina Malara (2005) has used the term *babbling* to refer to learners' early attempts to articulate relationships and properties using technical terms, because, like young children learning to speak, these articulations are often incoherent at first, as anyone marking homework assignments will recognize.

Awarenesses form the core or the essence of a topic or concept. To *learn* the topic will mean to alter those awarenesses so as to discern details, recognize relationships, and perceive properties that previously went unnoticed and to begin to reason on the basis of those properties.

Probing Deeper: Example Spaces

A significant component of awareness that is of particular importance for teaching is the *example spaces* that could come to learners' minds. Edwina Michener (1978) introduced a taxonomy of examples associated with different mathematical and pedagogic purposes, but her distinctions are sometimes difficult to sustain, so in Watson & Mason (2005) we focused on what the learner does with examples and what features actually make an example exemplary. An *example space* includes not only the particular examples that come immediately to mind, but the whole space of examples that the person is aware they could generate from those by altering different features.

Locally, in the moment, example spaces are person and situation dependent; access to an extended space can be triggered by associations, much as things in the back of a pantry come to view when the pantry is searched for a special ingredient. For example, $|x|$ is traditionally put forward as an example (often even 'the' example) of a function that is differentiable everywhere except at one point. But many learners, even after apparently mastering the calculus, show little sign of being aware that from this one object you can construct infinite classes of similar functions whether by varying the point at which the derivative fails to exist, by scaling, or by compounding with other functions, not to say piecing together non-linear components, or extending the object to be differentiable everywhere except at a finite number of points. Furthermore, learners may meet $x|x|$ as a continuous function through its 'rule' displayed as two functions glued together (x^2 when $x \geq 0$; $-x^2$ when $x < 0$), yet never associate it with $|x|$ and its non-differentiability at 0, nor realize that more generally, for each $\lambda > 0$ there is a whole class of functions f such that whereas $x^\lambda f(x)$ is continuous but not differentiable at $x = 0$, $x^{\lambda+1} f(x)$ is differentiable at $x = 0$. These are further examples of dimensions of possible variation and associated ranges of permissible change.

Pedagogically, it is vital to recognize that what learners attend to and what their teachers attend to is not always the same. In order to pick up on what learners are thinking and doing, and in order to prepare tasks that will provide learners with access to important connections within and beyond a topic, effective teachers find it useful to refresh their awareness of their own awarenesses.

Emotion (Motivation)

Yves Chevallard (1983) introduced the term *didactic transposition* to describe the way in which the intuitions and experiences of an expert are trimmed and edited for teaching purposes, so that what learners encounter is often little more than refined formal definitions, proofs of theorems, and examples of applications of techniques. Expert awareness is transposed or transformed into training of behavior. The result is that no appeal is made to learner's emotions, learners' powers are not called upon, and mathematical themes remain implicit. The pleasure and insight achieved by the expert in organizing the topic and 'making sense' leaks away and is lost to the learner, who experiences merely behavior training.

What for the teacher is a 'motivating example' (Michener 1978) may not actually motivate learners for whom the problem posed is either of little interest or is not perceived as within their grasp. Nor does it necessarily help either those who like a holistic sense of a larger picture, or those who prefer a serialist bottom up development ("just tell us what to do") (Pask 1976). Where access to the original situation that puzzled someone is missing, learners are short-changed and stymied, for this is where much of the motivation lies. In his monumental works, Hans Freudenthal (1973, 1978, 1983, 1991) addressed this by stressing the phenomenological basis of mathematics. Mathematics is a culturally vital means for making sense of a whole range of phenomena, but to appreciate this, learners need to be exposed to and intrigued by puzzling phenomena that can be explained mathematically. Paul Halmos expressed a similar sentiment:

> Let me emphasize one thing ... the way to begin all teaching is with a question. I try to remember that precept every time I begin to teach a course, and I try even to remember it every time I stand up to give a lecture... . [Halmos 1994 p85]

It is useful therefore to try to locate and recreate one or more of the initial problematic situations that spawned the topic in question, or else to create some analogous but contemporary version. It is also useful as a teacher to be aware of the variety of situations and ways in which the topic arises in other contexts. Some of these may be applications outside of mathematics, and some may be within mathematics itself. I believe that this is partly what John Dewey (1933) had in mind when he stressed that the role of the teacher is to *psychologize the subject matter*, that is, to re-formulate and re-organize the concepts, techniques, and awarenesses so that pertinent experiences come to the surface for learners. Boiling down a mathematical exploration through which learners could appreciate a topic, until just the techniques remain, distils out some of the behavior required for competence, but at the expense of desire to obtain that facility and appreciation of its significance.

The emotional strand includes both the effect on learners' dispositions towards mathematics in general and the topic in particular. It is influenced by the ways of working on mathematics developed in the classroom, and it is populated by phenomena (material, imagined or symbolic) that can be used to highlight the concepts and ways of thinking that make the topic what it is (Freudenthal, 1991). The important components for motivational purposes are a combination of initial surprise, of a sense that although initially mysterious an explanation is ultimately within reach, and the use of their own powers rather than having someone else try to do all the work for them. This is where trust is vital, for where learners trust a teacher to be able to make topics accessible, they are much more likely to engage than where that trust is absent.

Useful questions to ask in order to direct attention to the emotional strand include

(E1)   What questions or problems initially led to this topic? Where is the inherent surprise? What sorts of questions do they help resolve?

(E2)   In what contexts might the topic, concept or technique turn out to be useful? What is the range of situations to which they have been applied?

Responses to these questions in relation to quotient groups might include

(E1)   Generalizing the arithmetic of odd and even numbers; connecting clock or modular arithmetic with ordinary arithmetic; characterizing homomorphic images of groups; partitioning Cayley tables; ....

(E2)   Study of structure-preserving maps; constructing groups from components; study of symmetries; physics of particles; crystallography; solutions to differential equations ....

Responses in relation to L'Hôpital's rule might include

(E1)   How to find limits of ratios in which both the numerator and denominator have 0 (or infinity) as limits. Historically, it was probably devised by Johann Bernoulli and included in the textbook that L'Hôpital commissioned from him.

(E2)   Exercises that practice limits and derivatives; studying 'flat' functions such as e^{-1/x^2}.

Notice that there is no attempt here to be definitive and exclusive in the responses. Different people will stress different aspects according to their 'take' on the topic, and this may vary from situation to situation. Over a period of time, and especially in discussion with colleagues, a rich web of responses can accumulate. Regularly updating notes structured according to the six aspects highlighted by these three strands makes it much easier to plan a future session and to develop and improve your teaching on each occasion, instead of repeating habits that have not proved to be maximally effective in the past.

Probing Deeper: relevance and reality

Some people believe strongly that learners are motivated (some would say are only or are best motivated) when the topic is seen to be of direct relevance to their own lives (Mellin-Olsen, 1987; Frankenstein, 1989). Jerome Bruner (1986) pointed out that this worthy aim often translates in school into trivial

> banalities about the home, then the friendly postman and trashman, then the community, and so on. It is a poor way to compete with the child's own dramas and mysteries. (p. 160–161)

At university, these are usually manifested as patently artificial contexts, or inordinately complex applications beyond the reach of learners. Learners' own powers to imagine and to be intrigued are sometimes bypassed by being given apparently 'real' contexts in order to situate a previously distilled topic or technique. The infamous *word problems* are a particular case in point. Some people advocate that the contexts and applications must be authentic uses of mathematics by people outside of education rather than made-up examples, if learners are to be motivated (Brown *et al* 1989). Still others see 'realistic mathematics' as what can become real for learners when they use their powers of imagination and when there is an appeal to their natural curiosity (Gravemeijer, 1994). The expression *zone of proximal relevance* is useful to describe the extent to which learners can become interested and intrigued beyond their immediate concerns and experience (Watson, 2005; Mason & Johnston-Wilder, 2004). But 'application' is not always the chief motivational force. Mathematics can provide a refuge from other concerns, and it can also be a thrilling world of exploration in which 'truth' is validated by logic rather than by reference to experts.

Probing Deeper: harnessing energies

In addition to perceived relevance of a topic and the class of problems it resolves, motivation is also strongly influenced by learners' overall disposition towards engaging with new ideas and new problems. In their description of mathematical proficiency, Kilpatrick *et al* (2001) mention productive disposition as an under-rated but interwoven and vital component. They include under this heading "habitual inclination to see mathematics as sensible, useful, and worthwhile, coupled with a belief in diligence and one's own efficacy" (p. 116; see also p. 131–133, and Goldenberg 1996). Success breeds success by strengthening the disposition to engage with mathematical thinking as a result of making mathematical sense of phenomena, and of successful use of their own developing powers of sense-making to make sense of mathematical topics.

The energies that fuel human behavior arise from the emotions, both positive and negative. For example, sometimes you feel drawn towards or attracted by some goal, while at other times you feel repelled by an anti-goal (Skemp, 1979). Mostly there is tension or conflict between competing goals and anti-goals; for example, you may not want to look foolish but you may nevertheless want to ask a question. Disturbance to the status quo is perhaps the most common trigger for release of energy, whether experienced as cognitive dissonance (Festinger, 1957) or as surprise (Movshovitz-Hadar, 1988) which can be created when expectations of mathematical pattern, form, or structure are suddenly disrupted. They can also be created when social patterns are disrupted, such as by sometimes starting with the general and then specializing, and sometimes with the particular, and then generalizing.

Setting up possibilities for useful dissonance involves provoking learners to anticipate, even to make explicit conjectures, so that they have expectations which can be challenged. To prepare to do this requires teachers to probe beneath their own fluent expertise and to re-enter the topic as novices so as to re-experience for themselves the inherent surprise that contributed to the emergence of the topic in the first place. Sometimes this can be done by starting from first principles yourself; sometimes it can be done by setting yourself a challenge that is analogous in some way to what learners will encounter. For example, reviewing the components of a topic for yourself using an unfamiliar notation for something that is otherwise very familiar (interchange epsilon and delta; write functions on the right rather than on the left of arguments, use right cosets instead of left) can force previously automatic functioning to come to the surface more explicitly. Looking for unusual examples and counter examples can refresh your appreciation of the difficulties as well as the insights that contribute to the significance of the topic. Dissonance can also be provoked through engaging learners in constructing objects that confound their implicit assumptions (Watson & Mason, 2002, 2005). In the context of school mathematics, Alan Bell and colleagues (Bell & Purdy, 1986; Bell, 1986, 1987, 1993) developed what they called *diagnostic teaching* in which learners were confronted with classic learner errors. Learners showed significant gains in performance in both the short and long term as a result.

Generating surprise can initiate activity, but sustaining activity requires more. Treating learners as empty vessels to be filled, as passive recipients of 'clearly presented definitions, theorems, proofs and examples,' does not sustain active learners who have not already developed a proactive stance. Motivation, interest and participation are likely to wane. When learners appear to 'want to be told what to do and how to do it,' it is mainly because their natural powers of sense-making have been ignored in the past, and they have been enculturated into a passive and pragmatic stance towards mathematics. Where learners are prompted to make use of their own natural powers (examples are given in a later section) they experience pleasure and personal growth. They begin to see that not only is understanding possible, but desirable and more efficient for passing tests.

Since the three strands of the psyche function at different speeds, it is not always easy to locate what the trigger has been (Mandler, 1989): all the person is aware of is the complex state they experience. You can be aware of surprise, challenge, and intrigue; you can feel intrigued and energized or you can feel disinterested, daunted, fearful and enervated; you can act as if you are in control at least to some extent, responding freshly to challenge, or you can behave habitually, reacting automatically without taking control or initiating a response. These different emotions can be triggered by the same physiological changes such as increased pulse rate, sweating palms, and adrenalin rushes associated with stimulation and arousal.

One effective way to prompt learners to harness their energies in mathematics, is to take every opportunity to get them to make significant mathematical choices:

- choices about how much practice to do in order to automate a technique ("do as many of these as you need to in order to tell me how to do questions of this type");
- choices about which examples to practice on through engaging in an exploration in which the learners are specializing for themselves;
- explicit tasks to construct mathematical objects that meet certain constraints, perhaps illustrating theorems or the use of techniques, or serving as counter-examples to conjectures and to modifications of theorems (Watson & Mason, 2004, 2005).

Prompting learners to use their own powers of imagining and expressing, specializing and generalizing, conjecturing and convincing, organizing and characterizing, has a strong effect in capturing and maintaining interest and involvement because learners get pleasure from using their powers. It may take some work to develop an atmosphere in which learners begin to respond, especially if they have previously been habituated into being told what to do and given templates for doing it.

More on the Chariot Metaphor

As a strand of the psyche in the image of the chariot, emotion has been likened to the horses, whose energies must be harnessed if the chariot is to reach its intended goal. The horses are often referred to as the senses, because it is sense impressions that activate the emotional energies available to the psyche. Just as, given a chance, the horses will stray into the surrounding fields looking for food, so learner attention will drift if it is not stimulated. In pedagogical terms, emotion is the expression of motivation and interest, the source of drive and energy, expressed through dispositions. When learners' attention is caught, when they experience surprise due to conscious and unconscious anticipation, they are much more likely to direct their energies to resolving and understanding. Ignoring the emotional strand of a topic by assuming that the mathematical beauty will shine through and attract learner attention is rarely effective.

Behavior

Mathematical behavior includes more than the carrying out of techniques (e.g., differentiating and putting to zero as part of finding relative extrema of differentiable functions) and algorithms (e.g., finding the GCD of two integers, or rewriting a rational polynomial in partial fraction form). It includes the practical use of learners' natural powers, elaborated in a later section, such as to specialize and generalize, and the use of mathematical heuristics such as 'working backwards.' Another important aspect of behavior is language: ways of speaking to oneself as well as to others in the mathematical community.

Becoming familiar with concepts means not only having a sense of what they mean, but actually making use of technical terms in order to express and develop your own thinking, both internally and with others. The ways of speaking associated with a topic express relationships and properties and so indicate and support ways of perceiving. The way you speak affects the way you think, disposing you to stress some features and consequently to ignore others. In preparing to teach a topic it is very useful to review the technical language and to look out for potential conflicts between learners' informal use of similar words in ordinary language and their technical use in mathematics. A good case in point arises in group theory where not only is *group* used differently, but *order* is used both for 'the order of a subgroup' and 'the order of an element,' and has little to do with ordinals or placing things in order as in ordinary language.

Every technique or method involves more than a sequence of actions to be performed. It has 'inner incantations' or 'things that you say to yourself' as you carry out the procedures, and it involves choices as to what to do next, which may be situation dependent. The incantations help focus attention and direct activity, alerting the learner to choices that need to be made. This inner world is for some people driven by language and for others by kinesthetic and imagistic triggers. Thus learning the 'language' of a topic is an essential component of the behavior to be enacted. For example, a practice developed at Open University mathematics summer schools was to end each morning with a session in which a lecturer publicly worked through a few typical questions while trying to expose his or her inner incantations and inner thoughts, in order to give learners insight into what lies behind expert behavior.

Useful questions for directing attention to behavioral aspects of a topic include

(B1) What ways of speaking and hence thinking are important in this topic? Where might learners have experienced similar language, notation or other representation, and how does that use correspond or differ with its use in this topic?

(B2) What techniques and methods are associated with it and what inner incantations might be helpful?

Responses in relation to quotient groups might include

(B1) The language of 'equivalence' and of 'equivalence classes'; 'representatives'; 'quotient group;' 'homomorphic image;' 'well-defined,' and so on;

(B2) Writing down Hg and gH in various forms; constructing a homomorphism and checking the kernel; conclusions about the orders of groups and their quotients. Typical utterances include "let g be a representative of the coset gH".

Responses in relation to L'Hôpital's rule might include

(B1) Testing in advance that the limits of both numerator and denominator are 0 or are infinite; calculating the derivatives of numerator and denominator; finding the limit of the ratio of the results (which may involve using the rule again);

(B2) Limit as x goes to a; calculate the derivatives and then apply the limit; …

Probing Deeper: learning from examples

It is common practice, and has been ever since historical records began, for learners to be given worked-out examples and then invited—even urged—to try to do some similar examples (exercises) for themselves. However, where learners are able to use the worked examples as templates for substituting other values, it is unclear how much actual learning can be expected. Using examples as templates is a form of 'going with the grain' (Watson, 2000), that is, following surface patterns, much like splitting logs. By itself this may achieve immediate answers, but it only sets up possibilities for learning. What matters is what learners actually do with the template. To cut across the grain is to expose structure. If they inspect what they have done and link it with the worked example that they followed so as to get a sense of a broader general class of similar examples, then they are 'going across the grain.' They are recognizing structure, extending their potential example space, and enriching their sense of the topic.

Informal discussions with adult learners of mathematics reveals that when looking back to their school experience many had been satisfied to get the answers to assigned homework, while others report having been dissatisfied if they were not confident that they could do a similar question in the future. The former 'pragmatic' stance may lead to short term prowess but endangers long term success. Many learners in university may need to be shown how to 'go across the grain' to make sense of worked examples and exercises so as not to repeat their ineffective learning strategies from the past.

An extensive programme of psychological research has been carried out to try to elucidate the factors that make worked examples useful for learners (Renkl, 1997, Renkl, *et al.* 1998, Sweller *et al.,* 1998). It shows that in mathematics and science, carefully structured worked examples can definitely assist learners, whereas repetitive exercises may not. But what matters most is what learners actually do with the examples they are given. It seems that how well learners explain the examples to themselves significantly influences the depth of their learning. Providing worked examples with some parts omitted or 'smudged', so that learners have to complete missing details, can also be beneficial, especially where the smudged parts are associated with making choices as to the next step.

Probing Deeper: facility & fluency

What is being practiced when learners are set collections of exercises to complete? Are they actually developing facility in the use of particular concepts and techniques (that is, are they learning the ins and outs of particular concepts and the use of particular techniques)? Are they developing fluency by integrating the use of those concepts into the way they function in mathematics so that they no longer need to devote full attention to the details? Are they using technical terms to express their thinking? Too often mathematical exercises resemble their physical counterparts: repetition, justified by the assumption that practices will be confirmed and habituated.

However, mindless repetition does not always result in proficiency, nor even in recall of having carried out that practice: how often do learners deny recall of a topic or technique that the teacher knows they have studied previously (Schoenfeld, 1998)? Whereas physical training of behavior puts little or no value on the extent of the repetition, but rather values immersion in the repetition, learners usually act as though they think that what matters is getting the answers to mathematical exercises. This means that their attention is directed to completion rather than to experiencing a process. Instead of reflecting on how they got an answer, on the dimensions of possible variation and associated ranges of permissible change in an exercise that together create the question space (Sangwin, 2004, 2004a) from which the specific exercise is drawn, learners tend to rush on to the next question. By contrast, when they are asked to "Do as many of these exercises as you need to in order to be able to describe in words how to do a question of this type," and "How would you recognize a question of this type?", learners are prompted towards making sense of the exercises as a whole, as a mathematical object, rather than as isolated hurdles (Watson & Mason, 2005).

Gattegno (1987) described what learners need to do as 'integration through subordination': in order to integrate some behavior into your functioning, you need to subordinate that functioning, to reduce the amount of conscious attention required to carry it out. Therefore learners' attention needs to be directed not towards but away from the technique to be practiced. This is an ancient insight, illustrated in various Eastern martial arts. It follows that an effective way to 'practice' skills and techniques is to engage learners in a task in which they find themselves constructing their own examples on which to use the technique because their attention is directed towards some other more general goal, namely locating, checking and justifying some general conjecture. For example, in relation to cosets, the following exploration calls upon a variety of heuristics and powers:

If G is a finite group and A and B are two subsets, denote by AB the set of all products ab within G where a is in A and b is in B. Characterize those subsets of the power set of G that form a group under this operation.

In constructing examples for themselves in order to see what is going on, learners not only rehearse the use of cosets, but do it because they want to know the answers for a greater reason than just finding cosets. Similarly, the following task calls upon learners to construct functions and to use L'Hopital's rule for their own purposes, not simply to complete a set of exercises.

Construct pairs of functions for which the calculation of the limit of their ratio as x approaches 0 requires the use of L'Hôpital's rule t times, for $t = 1, 2, 3, \ldots$.

In both cases, learners find themselves practicing, but on examples of their own construction, perhaps under the guidance of a lecturer. The motivation level is higher and at the same time attention is constantly being drawn towards the bigger picture and away from carrying out technique.

More on the Chariot Metaphor

In the image of the chariot, the chariot itself is usually taken to represent the body, which must be looked after and maintained. Since behavior is manifested through the body, the chariot can be associated metaphorically with behavior. Maintaining the chariot can be associated with making use of behavior patterns, that is, refreshing the use of old techniques in new contexts. Integration through subordination achieves habits, but habits need to be challenged every so often to make sure that they are maximally efficient. This applies both to learners and to teachers!

The Pedagogic Structure of A Topic

The preceding descriptions can now be summarized in a single framework or pedagogic structure for a mathematical topic in terms of three interwoven strands as shown in Figure 1.

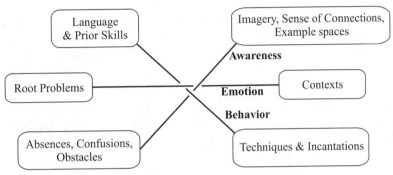

Figure 1. Structure of a Topic framework.

The horizontal strand encompasses emotional components, based around the motivational effects of problematicity, surprise-disturbance and challenge: what was originally problematic or what can be made problematic for learners. It encompasses the virtue of surprising students, whether by challenging a preconception or indicating an unexpected result, and of providing examples of contexts in which the ideas of the topic have proved fruitful that are relevant to learners' current or potential concerns.

The diagonal strand from top left to bottom right represents behavioral components which include terms that students may already know or use in a less formal manner, and language that would offer evidence of competence and understanding. It includes also specific manipulative techniques and any 'inner incantations' that a relative expert may use when carrying out those techniques.

The diagonal strand from top right to bottom left represents awareness components including mental images, associations and connections that the teacher would like the students to develop, and transitions from process to object that are entailed by or employed in the topic or idea, as well as standard absences, confusions that students are likely to display and obstacles they are likely to encounter.

The image of the interweaving threads is intended to emphasize the importance of mutual interaction and support among all three aspects. The framework can be summarized mentally by the expressions 'harnessing emotion,' 'training behavior,' and 'educating awareness,' drawing on the central image of a chariot as a metaphor for the three-fold structure of the psyche.

The Three Only's

Gattegno (1987) coined the somewhat startling slogan that

Only awareness is educable.

He did this in order to stress his observation that learning consists of altering what one is aware of; that is, one's sensitivity to notice and discern, a conclusion also reached by Marton and colleagues (Marton *op cit.*). The Upanishadic image of the chariot then suggested to me two further expressions

Only behavior is trainable, and Only emotion is harnessable,

in order not only to set up contrast with the role of awareness, but to emphasize the need to engage all three aspects of the psyche so that productive and effective learning is likely to take place. Training behavior is effective for establishing and enculturating habits, but trained behavior tends to be inflexible. To be used effectively in novel situations, it needs to be informed by and called upon by awareness as insight, intuition, and sensitivity to notice connections. Development of behavior and of awareness comes about through harnessing the energies released by emotions.

Gattegno enjoyed using challenging assertions that act as *protases,* that is, as the initial statement in a syllogism (Mason, 1998a). When juxtaposed with particular examples from your own experience (the second term of the syllogism) a syllogistic action is set up, and it is through working away at challenging the protasis, accumulating examples, and exploring consequences of the syllogisms that brings about insight, and which, in turn, informs future behavior. Thus the force of the word *only* needs to be juxtaposed with examples that come to mind, whether as illustrations or as potential counterexamples. What you conclude is then due to an action taking place inside you, to your own thinking, not someone else's.

Even when there appears to be considerable agreement amongst colleagues as to the basic components of a topic, beneath the surface there is often considerable variation, as evidenced by the many different textbooks produced, each of which display considerable variety in how topics are presented. The structure being proposed here provides a framework with which to analyze differences in approaches, to locate aspects which are under or over stressed. The framework does not produce a teaching sequence, but it does collect together and highlight in a coherent manner the various components that might be used in any particular approach. In my experience, the teaching sequence itself is best left to the individual teacher who has detailed knowledge of the learners, of the instructional goals, and of course of preferred ways of working with learners on mathematics..

Powers, Themes & Heuristics

Mastering a collection of techniques in a few topics for an examination cannot be the principal aim of learning mathematics, even for students who only need a few mathematical tools for use in their own discipline. While some topics are necessary precursors for further topics (clearly addition and multiplication of integers must precede quotient groups, and differentiation must precede L'Hôpital's rule), mathematical topics are not quite as well ordered as might appear. For example, Denvir & Brown (1986a, 1986b) used mathematical relatedness to construct a partial order for the topics in a high-school mathematics curriculum offered to low attainers. To their surprise, they found that as well as improving in the topic being taught, learners often developed facility in other topics sometimes only distantly related. In other words, learners may be attending to different aspects than the teacher, and some improvement may come about as a result of having attention diverted away from the actual carrying out of techniques, as suggested earlier concerning the use of exercises.

A more holistic view of mathematics education sees topics largely as vehicles for developing learners' natural powers to make sense of phenomena mathematically, and for encountering pervasive themes that provide the warp for a richly interconnected fabric known as mathematics. Beyond the specific and powerful techniques for solving a class of problems, most topics in mathematics, as in any subject, are a matter of choice, certainly in relation to the depth and complexity to which they are pursued with particular learners. In this section it is suggested that individual topics can play an important role in alerting learners to the fact that they already possess natural powers for making sense of mathematics and for mathematical sense making.

Powers

Learners arrive in class having already demonstrated that they have made extensive use of the sorts of powers necessary in order to develop mathematical thinking. Such ideas are not new. For example, John Calkin (1910) summed up an attitude from a Canadian perspective that pervaded European and North American education at the turn of the previous century, echoed by thinkers such as John Dewey (1933 p. 225–6), and before him Herbert Spencer (1878) but that seems recently to have been allowed to slip into the shadows of socio-culturalism: "That the mind is a power to be developed rather than a receptacle to be filled is a sound maxim in education" (Spencer, 1878, p. 18).

Learners have certainly already demonstrated the power to *discern detail* and to distinguish features, as well as to meld or gaze so as to *blur* boundaries. They can *stress,* and thereby they can *ignore,* as Caleb Gattegno (1987) pointed out. They can detect invariance in the midst of change and they can recognize relationships between and amongst features that have been discerned, including refined versions of similarity and difference. They can formulate properties possessed by an object and abstract this to recognize other objects with the same properties. They can also reason on the basis of properties. One of the principal challenges in teaching mathematics is to stimulate learners to reason *solely* on the basis of announced and articulated properties, rather than on the basis of intuitions and accumulated information.

In addition, and more evidently mathematically, learners have already shown that they can, in the words of George Pólya (1945, 1962), *specialize* and *generalize*, or more fully, see the general through the particular and the particular in the general. This echoes Alfred Whitehead who suggested, "To see what is general in what is particular and what is permanent in what is transitory is the aim of scientific thought" (Whitehead, 1911, p. 4). As Paul Halmos (1980) put it:

> An intrinsic aspect of [teaching] at all levels, elementary or advanced, is to concentrate attention on the definite, the concrete, the specific. … We all have an innate ability to generalize; the teacher's function is to call attention to a concrete special case that hides (and, we hope, ultimately reveals) the germ of conceptual difficulty. (p. 852)

The issue is whether learners are offered tasks that call upon those powers, or whether, from a mistaken desire for efficiency or concern that learners should 'know what they are expected to do', those powers are bypassed by teacher and author who provide particular examples and articulated generalizations without directing learner attention to make effective use of them. As a learner once said to me "I see now that my job is to generalize when you give me a particular and to particularize when you give me a generality."

Learners can also *imagine* and anticipate what is not present, and they need to do this in order to specialize and to generalize. They can also *express* what they imagine in various media including words, diagrams and symbols. However, how often are they invited to imagine something, to articulate what they are imagining, to locate and express relationships and to articulate these as general properties, instead of having the author or teacher try do it all for them? Learners can *conjecture* and *convince* (themselves, a friend, a skeptic[1]: see Mason *et al*, 1982). Finally, they can *classify* and *characterize* objects by the properties they perceive. All of these are vital for doing mathematics; all need developing and refining if mathematics is to be learned effectively. Indeed, honing these powers equips learners to deal flexibly and effectively with new topics or new problematic situations. When their attention is directed to the ways that their powers are developing, they are learning how to learn.

Using their own powers in a supportive atmosphere not only makes significant learning possible, but also reveals to the individual the fact of their powers, something learners may not even be aware of. Gattegno echoed the same theme:

> What is it, then, that will allow us to teach mathematics to anyone with a functioning mind and an inclination to learn? Simply, finding a way to make the learner aware of the powers of his mind–the powers he uses every day, those which allowed him to learn his native language and to use imagery and symbolism. This means that the job of teaching is one of bringing about self-awareness in learners through whatever means are available in the environment: words, actions, perceptions of transformations, one's fingers, one's language, one's memory, one's games, one's symbolisms, one's inner and outer wealth of perceived relationships, and so on. (Gattegno, 1974, p. 111 postscript)

There is nothing so strongly motivating as realizing you can do something that is valued and valuable. The exercise of your own powers, independently, is a major source of pleasure for human beings, whereas dependency on others breeds discontent. Learners can become frustrated when the teacher or text usurps their role by doing things for them: roles they are on the edge of assuming or actions they are already able to do for themselves. All too often learners decide that their powers are not wanted in the mathematics classroom, and so they stop using them even where there is an opportunity. It is highly de-motivating and disempowering to find that your own powers are not called upon, not encouraged, not used; it is a source of pleasure and empowerment to find that you can use your own powers to make sense of phenomena, situations and ideas. Furthermore, the more you are called upon to use your powers, the more developed and sophisticated they are likely to become; the less they are called upon, the more likely they are to atrophy, or at least to be parked at the classroom door.

---

[1] David Tall (personal communication) replaces the original 'enemy' with 'sceptic,' drawing attention to the pedagogic observation that an important contribution to learning to convince others is learning to be skeptical yourself.

It is not sufficient simply to tell learners about their powers: they have to experience their productive use, because it is only through continued use that these powers will develop. As Whitehead put it:

> the apprehension of general ideas, intellectual habits of mind, and pleasurable interest in mental achievement can be evoked by no form of words, however accurately adjusted. ... There is no royal road to learning through an airy path of brilliant generalizations. (Whitehead, 1932, p. 9–10)

It all comes down to choices of tasks and choices about ways of working with learners on mathematics, rather than simply getting them to work their way through sets of repetitive exercises.

Themes & heuristics

Mathematical topics are held together not only by the use and development of learners' powers but by the presence of mathematical themes and the use of common heuristics in making sense of problematic situations. In this section I mention some of these themes in order to round out the use of the structure of a topic framework.

Invariance in the Midst of Change

One of the pervasive themes in mathematics is the notion of *invariance in the midst of change*. Many if not most mathematical theorems can usefully be thought of as articulating what it is that can change while some relationship remains invariant. For example, the fundamental theorem of algebra states that no matter what polynomial of degree d with integer coefficients you construct (what can change) the number of complex roots is d (invariance); the intermediate value theorem states that no matter what continuous function you choose on whatever closed interval $[a, b]$ you like and no matter what values v such that v lies between $f(a)$ and $f(b)$ (what can change), there is a value c with $a < c < b$ such that $f(c) = v$. Although this seems pretty obvious, it is often the case that learners are unaware of the full scope of the permissible change. This observation led Mason & Watson (see chapter in this book) to develop and exploit the notion of *dimensions of possible variation* (discerning what can vary) and associated *ranges of permissible change* (in what way they can vary).

Similarly, to appreciate a concept is to be aware of what aspects of an example of that concept are necessary to make it an example, and what features can change. If learners are unaware of what makes an example exemplary, that is, of what can change but allow it to remain an example, then the example fails to be exemplary for them. The result is that learners either ignore or memorize the example, leaving it unconnected with their concept image.

Doing & Undoing

Another theme that pervades pure mathematics is the notion of *inverse*, or of *doing & undoing* (and its more sophisticated version, conjugation: see Melzack, 1983 and www.inverse-problems.com). It is often very fruitful mathematically to take a technique that yields an answer, and to ask yourself what other input data would give the same answer, and then to try to characterize all the possible answers. In other words, interchanging input and output opens up mathematical exploration, leading to characterizing and to experiencing mathematical creativity. This idea applies to mathematical operations and also to ordinary exercises. For example: what numbers can arise as one more than the product of four consecutive integers? what linear functions integrated over the interval $[0, 1]$ (or more generally $[a, b]$) have an answer of zero?

Note the strong connection with invariance in the midst of change: changing a 'doing' into an 'undoing' is effectively asking what can be changed about the input while leaving the output invariant. Furthermore, this perspective corresponds closely to the use of people's natural power and desire to characterize and classify: solving an 'undoing' problem is tantamount to characterizing the class of inputs that yield a given output. Seeing the technique as a function, all you are doing is asking for the inverse image of an element in the codomain of the function. A related characterization problem is to characterize the range of the technique: what sorts of objects can arise as the result of using the technique?

Extending and Restricting Meaning

One of the powerful features of mathematics is the multiplicity of interrelated meanings of objects. For example, an expression such as $2x + 3$ can be seen as a rule for performing a calculation, as the answer to performing that calculation on an as-yet-unknown value, as the answer in general to such a calculation, and as an expression of generality. Similarly, 2/3 can mean an operator, the result of applying that operator to 1, a fraction, a division to be performed, the answer to that division, the value of a ratio, a position on a number line, and the value of all fractions equivalent to itself. In school, learners encounter the word *number* used to mean whole number, then integer, then fraction, then decimal (real), and perhaps even complex number. They also meet essential ambiguities such as that of −3 (is it 0 − 3 or is it the number −3, and how are these related?). In moving from school to university, learners encounter trigonometric functions as ratios, circular functions, solutions to differential equations, power series, and perhaps as solutions to integral equations, as well as the building blocks for Fourier series.

The meaning of mathematical terms is extended metaphorically with rarely any comment or mention. These extensions of meaning are often reduced to behaviors to be picked up by learners, without harnessing their emotions through the use of their own powers. Furthermore, sometimes it is necessary to restrict meaning, such as when discussing continuity of a function on a restricted domain, or when attention is directed towards a particular subset of a familiar set (e.g., the numbers with remainder 1 on dividing by 3 when discussing the significance of the unique factorization theorem for integers). Extending and restricting meaning pervade the way mathematics develops, and provides some of the power of mathematical notation. However, unthinking and implicit extension of meaning can act as an obstacle for learners who are not expecting to have to modify what they thought they already knew.

Freedom and Constraint

One further theme of considerable importance in mathematics is *freedom and constraint*. In order to appreciate a concept it is necessary to be aware of what features of an object can be varied freely, and which features are constrained in some way. Most mathematical problems can usefully be thought of in terms of starting with an object with considerable freedom, and then imposing a collection of constraints. Mathematics can then be seen as a constructive activity, trying to find out what freedom remains for construction when a collection of constraints are imposed. Looking at topics through this lens presents mathematics as support for constructing objects that meet constraints, rather than as a collection of tools for producing answers on tests.

Making Pedagogical Choices

Teaching involves making choices, both in planning, and in the moment by moment conduct of sessions. Because everyone depends on habits and routines, on patterns of behavior and on preferred ways of working, these choices depend on choices made in the past. In order to respond freshly every so often rather than always reacting habitually, in order to be present to learners so that they are in the presence of mathematical thinking rather than simply acting as dictator to clerks recording what is displayed in front of them, it is necessary to be making fresh choices in the moment. This requires being aware of the structure of your own attention. The main thrust of the Structure of a Topic Framework is to provide an informative focus for attention when preparing to teach a topic.

Having clarified the structure of a topic, it becomes much easier to consider the pedagogic choices available. Only when you are clear about the awarenesses that you want learners to develop does it make sense to consider what phenomena to expose them to. Only then does it make sense to consider whether such phenomena will serve those particular learners most effectively as initiating phenomena to raise questions (perhaps informed by your sense of the source questions for the topic), as ongoing phenomena by means of which to encounter and experience details of the topic, or as summarizing phenomena by means of which to review and reflect upon the topic as a whole and appreciate some of the contexts in which it has proved useful.

Only by considering what powers, themes and heuristics are likely to be stimulated, and which ones could usefully be employed, can sets of exercises be analyzed and either modified or constructed so as to prompt effective learning. By choosing to prompt learners to make mathematically significant choices, you can help them harness their emotions to take initiative and be mathematically proactive. By provoking learners to construct mathematical objects, you can stimulate them to enrich their example spaces, educate their awareness, and train their behavior.

Only when you are clear about the sorts of 'absences' and confusions that learners sometimes develop and the struggles they sometimes experience, are you in an informed position to modify or augment any textbook you are using in order to get learners to confront possible errors directly, or to try to avoid learners having similar struggles.

Specific mathematical objects can be chosen to serve as examples by highlighting salient dimensions of possible variation and drawing attention to ranges of permissible change. Particular worked examples can be chosen with the same view in mind.

Only when you have an enlivened sense of the topic yourself are you in a position to provide what Jerome Bruner (1986) called *consciousness for two*, that is, to be able to attend to prompting learners to make use of their own powers in order to make sense of the topic, rather than trying to do all the work for them by dotting every *i* and crossing every *t*. If the textbook and lecturer do all the *specializing* and *generalizing*, the *conjecturing* and *convincing*, the *imagining* and *expressing*, the *organizing* and *characterizing*, then learners are placed in a passive role, at best accepting what they are told and shown. Pedagogically effective teaching provokes learners to take an active and assertive stance, using and developing their own powers.

For example, in carrying out a 'worked example', only if you are aware of the general as you make use of the particular, only if you are aware of the connections which enrich your appreciation of what you are doing, only if you are aware of the example space(s) which this particular case exemplifies, only if you have access to how this particular relates to the problems which spawned the topic and the contexts in which the topic can be applied, and only if you are aware of how it is you know what to do next in carrying out the technique, are you in a position to direct learner attention fruitfully. Without these, your learners are left immersed in trying to make sense of your overt behavior.

Kilpatrick, Swafford, & Findell (2001) proposed a useful five-fold structure for mathematical proficiency, comprising conceptual understanding, procedural fluency, strategic competence, adaptive reasoning, and productive disposition which provides a structure for the goals of teaching a topic. The Structure of a Topic framework informs choices which pursue these goals.

Underpinning the interwoven strands of the psyche that provide structure for each mathematical topic is the endemic question of whether, through engaging in the tasks provided, learners actually encounter and make sense of the important ideas. By being aware of what awarenesses learners already have and through contemplating how these might be brought to the surface, recognized, and articulated in words and symbols, teachers can hope to construct pedagogically effective tasks. But behind the doing of tasks lies the pervasive and endemic question of what learners are attending to. If the teacher is attending to relationships amongst details or features while learners are trying to discern the details that the teacher is referring to, if the teacher is presenting properties while learners are still trying to recognize relationships, and if the teacher is reasoning on the basis of defining properties alone while learners are still trying to perceive properties as general properties of a class of objects and not simply of particular objects being presented, then miscommunication is all too likely. Learners are likely to get the feeling of being lost, of being left behind.

It makes sense, when preparing to teach a topic, to make notes of awarenesses, behaviors (language and techniques) and problematic sources, contexts of use, and embedded surprises, so as to make preparation easier in the future. By noting down unusual things that learners say and do, you can build up a rich repository of details of a topic that would be difficult to reconstruct from memory every year when needed. The notes then provide a quick reminder so that you can re-enter two worlds: the world of the learner and the world of the sensitized teacher.

Summary

The aim of this chapter has been to offer a rich and informative structure for preparing to teach any mathematical topic, while at the same time revealing that there is more to preparing to teach a topic than organizing the definitions, theorems, proofs and examples so as to be maximally clear. It is possible to enrich learners' experience by taking into account the structure of the psyche, and to accustom learners to take responsibility for their own learning.

Claims made in the chapter include the following:

- training behavior without educating awareness is no more useful than educating awareness without training behavior: neither alone constitutes learning, so concentrating on just one strand of the three strand view can cripple learning;

- effective teaching prompts learners to harness their emotional energy in order both to train their own behavior and to educate their own awareness by making use of and developing their natural sense-making powers, and calling upon pervasive mathematical themes and heuristics;

- in preparing to teach a mathematical topic it can be pedagogically helpful to use the three-strands (awareness, behavior and emotion) as a reminder of key aspects of the topic;
- a useful way to learn from and build upon experience in order to improve one's teaching is to review and update notes structured along the lines of the three strands.

Trying to train other people's behavior, to educate their awareness, or to motivate them, without their consent and involvement is, at best, unfruitful, if not ultimately impossible. Teaching cannot force, necessitate or guarantee learning, but teaching can make learning more likely and more effective if it makes use of learners' powers and dispositions, and exposes them to significant and fruitful ways of thinking and perceiving.

What the Structure of a Topic framework offers is not a formal mechanism or checklist like a packing list to be used before going on a journey. Rather it offers a background structure to remind you to ask yourself certain pedagogically important questions. Since it was first introduced in 1989, it has been found to be sufficiently informative to continue to be used by teachers and teacher educators. When it informs how you think, then its details only need to come to mind when you find yourself running out of ideas or becoming stale. Then it can serve as a reminder to ask questions. It provides a structure to fall back on when preparing to teach a topic, and a basis for learning from experience through collecting notes under the various headings that comprise it.

Recursive Complexity

It was mentioned briefly that each of awareness, behavior and emotion, considered as strands of the psyche, can be thought of as having an internal structure also comprised of awareness, behavior and emotion. Thus awareness in the full sense in which Gattegno used it has a conscious or 'awareness' component in what comes to mind, but it also includes emotional dispositions that are triggered, and automated behaviors that are accessed at the same time along with the concept image. For example, the mere mention of L'Hôpital can trigger someone into a state of alert sensitivity to finding limits of certain kinds of ratios and some degree of pleasure or foreboding depending on past success. So too, emotions harnessed through appreciation of root problems from which a topic originated, and uses to which the topic can be put, can be thought of as having a similar three-strand structure. There are the conscious awarenesses (cognition) of specific problems associated with the topic, there are emotion-driven habitual behaviors associated with eagerness or reluctance to engage with the topic, and the emotions themselves generate further emotions associated with awareness and behavior. For example, Triandis (1971) and Ajzen (1988) both refer to cognitive, affective, and conative aspects of attitudes. These consist of

expressions of beliefs about an attitude object (cognitive aspect of emotion)

expression of feelings towards an attitude object (affective aspect of emotion)

expressions of behavioral intention (conative or enactive aspect of emotion).

Attribution theory (Heider, 1958; Weiner, 1986) also suggests that reinforcing learners' attribution of qualities to themselves (affective aspect of emotion) can be much more effective in building their self-image than trying to persuade through argument (cognitive) or even role-modeling behavior (Miller *et al*, 1975).

Finally, behavior too can be thought of as three-stranded. For example, there is conscious awareness of specific behavior such as techniques, while they are being carried out, however subordinated and automated the performance. There are dispositions and attitudes which are triggered by patterns of behavior and which influence behavioral choices as they are made. There are behaviors that come to the surface triggered through resonance with consciously enacted behaviors, and these may not always be positive and beneficial.

References

Ajzen, I., & Fishbein, M. (1980). *Understanding attitudes and predicting social behaviour*. New York: Prentice Hall

Bachelard, G. (1938, reprinted 1980). *La Formation de l'Esprit Scientifique*, Paris, France: J. Vrin.

Bell, A., & Purdy, D. (1986). Diagnostic Teaching. *Mathematics Teaching, 115* 39–41.

Bell, A. (1993). Principles for the design of teaching. *Educational Studies in Mathematics, 24,* 5–34.

—— (1986). Diagnostic teaching 2: Developing conflict: discussion lesson, *Mathematics Teaching, 116,* 26–29.

—— (1987). Diagnostic teaching 3: Provoking discussion, *Mathematics Teaching, 118,* 21–23.

Brown S., Collins A., & Duguid P. (1989). Situated cognition and the culture of learning. *Educational Researcher, 18* (1), 32-41.

Bruner, J. (1986). *Actual minds, possible worlds.* Cambridge, MA: Harvard University Press.

Calkin, J. (1910). *Notes on Education: A practical work on method and school management.* Halifax, Nova Scotia, Canada: Mackinlay.

Chevallard, Y. (1985). *La Transposition Didactique.* Grenoble, France: La Pensée Sauvage.

Denvir, B., & Brown, M. (1986a). Understanding number concepts in low attaining 7–9 year-olds. *Educational Studies in Mathematics, 17*(1), p. 15–36.

—— (1986b). Understanding number concepts in low attaining 7–9 year-olds: Part II. *Educational Studies in Mathematics, 17*(2), 143–64.

Dewey, J. (1933). *How we Think.* London: D.C. Heath & Co.

Festinger, L. (1957). *A theory of cognitive dissonance.* Stanford, California: Stanford University Press.

Frankenstein, M. (1989). *Relearning mathematics: A different R—radical math(s)* London: Free Association.

Freudenthal, H. (1973). *Mathematics as an educational task.* Dordrecht, The Netherlands: Reidel.

—— (1978). *Weeding and sowing: preface to a science of mathematical education.* Dordrecht, The Netherlands: Reidel.

—— (1983). *Didactical phenomenology of mathematical structures.* Dordrecht, The Netherlands: Reidel.

—— (1991). *Revisiting mathematics education: China lectures.* Dordrecht, The Netherlands: Kluwer.

Gattegno, C. (1974). *The common sense of teaching mathematics.* New York: Educational Solutions.

—— (1987). *The science of education Part I: Theoretical considerations.* New York: Educational Solutions.

Goldenberg, P. (1996). 'Habits of mind' as an organizer for the curriculum. *Journal of Education, 178*(1) 13–34.

Gravemeijer, K. (1994). *Developing realistic mathematics education.* Utrecht, The Netherlands: Freudenthal Institute.

Griffin, P., & Gates, P. (1989). *Project mathematics update: pm753a,b,c,d, preparing to teach angle, equations, ratio and probability.* Milton Keynes, United Kingdom: The Open University.

Halmos, P. (1980). The heart of mathematics. *American Mathematical Monthly, 87* (7), 519-524.

—— (1994). What is teaching? *American Mathematical Monthly, 101*(9) 848–854.

Heider, F. (1958). *The psychology of interpersonal relations.* New York: Wiley.

Johnson, M. (1987). *The body in the mind: the bodily basis of meaning, imagination, and reason.* Chicago: University of Chicago Press.

Kilpatrick, J., Swafford, J., Findell, B. (Eds.). (2001). *Adding it up: Helping children learn mathematics.* Washington, DC: National Academy Press.

Lakatos, I. (1976). *Proofs and refutations: The logic of mathematical discovery.* Cambridge, UK: Cambridge University Press.

Lakoff, G., & Nunez, R. (2000). *Where mathematics comes from: How the embodied mind brings mathematics into being.* New York: Basic Books.

Malara, N. (2005). Dialectics between theory and practice: Theoretical issues and aspects of practice from an early algebra project. In N. Pateman, B. Doughterty & J. Zilliox (Eds.), *Proceedings of PME27, 1,* 33–48.

Mandler, G. (1989). Affect and learning: Causes and consequences of emotional interactions, in D. McLeod & V. Adams (Eds.), *Affect and mathematical problem solving: A new perspective.* London, UK: Springer-Verlag, pp. 3–19.

Marton, F., & Booth, S. (1997). *Learning and awareness.* Mahwah, NJ: Lawrence Erlbaum Associates.

Marton, F., & Trigwell, K. (2000). Varatio est mater studiorum, *Higher Education Research & Development, 19* (3), 381–395.

Marton, F., & Tsui, A. (Eds.). (2004). *Classroom discourse and the space for learning.* Mahwah, NJ: Lawrence Erlbaum Associates.

Mason, J., Burton, L., & Stacey, K. (1982). *Thinking mathematically.* Boston, MA: Addison-Wesley.

Mason J. (1998). Enabling teachers to be real teachers: Necessary levels of awareness and structure of attention. *Journal of Mathematics Teacher Education, 1* (3), 243–267.

—— (1998a). Protasis: A technique for promoting professional development, in C. Kanes, M. Goos, & E. Warren (Eds.). *Teaching mathematics in new times: Proceedings of MERGA 21, 1.* Mathematics Education Research Group of Australasia, 334–341.

—— (2002). *Mathematics teaching practice: A guide for university and college lecturers.* Chichester, UK: Horwood Publishing.

Mason, J., & Johnston-Wilder, S. (2004). *Fundamental constructs in mathematics education,* London: Routledge Falmer.

Mellin-Olsen, S. (1987). *The politics of mathematics education.* Dordrecht, The Netherlands: Reidel.

Melzak, Z. (1983). *Bypasses: a simple approach to complexity.* New York: Wiley.

Michener, E. (1978). Understanding understanding mathematics. *Cognitive Science, 2* 361–383.

Miller, R., Brickman, P., & Bolen, D. (1975). Attribution versus persuasion as a means of modifying behavior. *Journal of Personality and Social Psychology, 3,* 430–441.

Movshovitz-Hadar, N. (1988). School mathematics theorems—An endless source of surprise. *For the Learning Of Mathematics, 8* (3), 34–40.

Pask, G. (1976). Styles and strategies of learning. *British Journal of Educational Psychology,* 46, 128–148.

Pólya, G. (1945). *How to solve it: A new aspect of mathematical method,* Cambridge, MA: Princeton University Press.

Pólya, G. (1962). *Mathematical discovery: On understanding, learning, and teaching problem solving.* Combined edition. New York: Wiley.

Renkl, A. (1997). Learning from worked-out examples: A study on individual differences. *Cognitive Science, 21,* 1–29

Renkl, A., Stark, R., Gruber, H., & Mandl, H. (1998). Learning from worked-out examples: The effects of example variability and elicited self-explanations. *Contemporary Educational Psychology, 23,* 90–108.

Sangwin, C. (2004). Encouraging higher level mathematical learning using computer aided assessment. In J. Wang and B. Xu, (Eds.). *Trends and challenges in mathematics education.* Shanghai: East China Normal University Press.

—— (2004a). Assessing mathematics automatically using computer algebra and the internet. *Teaching Mathematics and its Applications, 23* (1),1–14.

Schoenfeld, A. (1998). Making mathematics and making pasta: From cookbook procedures to really cooking. In J. Greeno & S. Goldman (Eds.). *Thinking practices in mathematics and science learning.* Mahwah, NJ: Lawrence Erlbaum.

Skemp, R. (1979). *Intelligence, learning and action.* Chichester, Sussex, UK: Wiley.

Spencer, H. (1878). *Education: intellectual, moral, and physical.* London: Williams & Norgate.

Sweller, J. van Merrianboer, J. & Paas, F. (1998). Cognitive architecture and instructional design. *Educational Psychology Review, 1,* 251–296.

Tall, D. & Vinner, S. (1981). Concept image and concept definition in mathematics with particular reference to limits and continuity. *Educational Studies in Mathematics, 12* (2) 151–169.

——. Concept image and concept definition. Retrieved March 6, 2005, from www.warwick.ac.uk/staff/David.Tall/themes/concept-image.html

Triandis, H. (1971). *Attitude and attitude change.* New York: Wiley and Sons.

Watson, A. (2000). Going across the grain: Mathematical generalisation in a group of low attainers. *Nordisk Matematikk Didaktikk (Nordic Studies in Mathematics Education)*, 8 (1) 7–22.

—— (2005). Red herrings: Post-14 'best' mathematics teaching and curricula. *British Journal of Educational Studies*, 52 (4) 359–376.

Watson A. & Mason, J. (2002). Student-generated examples in the learning of mathematics, *Canadian Journal of Science, Mathematics and Technology Education*, 2 (2) 237–249.

—— (2004, June). The exercise as mathematical object: Dimensions of variation in practice. Presentation at British Society for Research in the Learning of Mathematics, Leeds, Yorkshire, UK.

—— (2005). *Mathematics as a constructive activity: The role of learner-generated examples*. Mahwah, NJ: Lawrence Erlbaum Associates.

Weiner, B. (1986). *An attributional theory of motivation and emotion*. New York, NY: Springer-Verlag.

Whitehead, A. (1911, reset 1948). *An introduction to mathematics*. London: Oxford University Press.

—— (1932). *The aims of education and other essays*. London: Williams & Norgate.

21

Promoting Effective Mathematical Practices in Students: Insights from Problem Solving Research

Marilyn Carlson, Irene Bloom, Peggy Glick
Arizona State University

Mathematicians and mathematics educators have been curious about the processes and attributes of problem solving for over 50 years. As mathematics teachers at any level of education, we want to know what teaching practices we can employ to help our students develop effective problem solving abilities. This curiosity has led to numerous investigations of the attributes and processes of problem solving. In this chapter, we describe insights from a study we conducted of the mathematical practices of 12 research mathematicians. We believe these insights are useful to teachers striving to promote mathematical practices in students at all levels—from first-grade mathematics to beginning algebra, calculus, and abstract algebra.

Our chapter begins by inviting you to work a problem that our research study posed to 12 mathematicians and to reflect, as they did, on your own problem solving behavior as you attempt to solve this problem. In inviting you to work this problem, our intent is to raise your awareness of the processes, emotions, knowledge, heuristics, and reasoning patterns that you use when working a novel problem. Our research suggests that by reflecting on our own mathematical practices, instructors can become more attentive to the development of problem solving attributes in students (Bloom, 2004).

This exercise should make the remaining sections of our chapter more meaningful. In particular, it is our hope that our description of the *Multidimensional Problem Solving Framework* is more accessible. After describing how one of our subjects attempted the same problem, our chapter provides an overview of the research literature on problem solving in mathematics. We then describe our own study in more detail and conclude with suggestions for developing students' problem solving practices. We believe that this chapter illustrates the importance of exploring the mathematical practices and behaviors that lead to mathematical proficiency. We also believe that the chapter explicates a way of thinking about problem solving that posits a reflexive relationship between the development of students' content knowledge and their mathematical practices.

Reflect on Your Attempt to Solve a Novel Problem

We invite you now to work the "Paper-Folding Problem" (Figure 1), provided it is a novel problem for you. This is one of several problems we gave the mathematicians in our study (Carlson and Bloom, 2003). If you already know how to work this problem, select another one that you expect will require multiple attempts and some persistence for you to complete. As you attempt the problem, observe and take notes on what you are thinking and feeling. What knowledge and techniques do you access? What is the basis for the decisions you make as you attempt a solution? What emotional responses do you have when you first confront the problem, and how does your emotional state change as you move toward a solution? Can you describe the reasoning patterns that emerge as you work?

When you have finished recording your observations, read our description of the approach taken by Paul, one of our subject mathematicians, and compare your experience to his.

> A square piece of paper $ABCD$ is white on the front side and black on the back side and has an area of 3 square inches. Corner A is folded over to point A' which lies on the diagonal AC such that the total visible area is ½ white and ½ black. How far is A' from the fold line?

Figure 1. The Paper-Folding Problem

Consider how Paul Solved the Paper-Folding Problem

Paul, an active research mathematician, was one of 12 mathematicians whose problem solving behavior we investigated in the 1998–2004 research study that we describe in detail later in this paper.

In the observation of Paul, one interviewer observed and questioned him while he attempted four novel problems in his university office. The session was audio taped. When the researcher presented Paul with the paper-folding problem, he initially read the text without comment, while repeatedly clicking his pen. He then lifted his head and gazed out the window with his eyes fixed ahead and his face expressionless. After about three minutes of silence, he redirected his eyes toward the interviewer. Grinning slightly, he said, "The distance of A' from the fold line is 1."

When prompted to verbalize his thinking, Paul explained, "I quickly deduced that after unfolding the paper, the new fold line would divide the paper into three equal areas. Thus, each of the pieces has an area of 1." When asked to provide more detail, he said, "Transitivity." The interviewer waited for an explanation. Paul continued, "...let the three areas be x, y and z; then since $x = y$, and $y = z$, then $x = z$, and since the total area is 3, each area would be 1." When asked to explain how he determined the value of 1 for the distance of A' from the fold line, he responded that he had used his knowledge of the relationships among the sides of a 45-45-90 degree triangle to arrive at the answer.

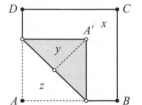

Figure 2. Paul's Sketch

This prompting provided some insights into the reasoning patterns that Paul employed to arrive at the answer; however, it does not explain how Paul was able to quickly observe that all three areas were equal; nor does it explain why Paul was so efficient in computing the exact value of the distance of A' from the fold line. What sequence of mental actions led Paul to his simple and elegant solution? Were all of his initial ideas fruitful, or did he reject some on the way toward a correct response? Why did Paul click his pen repeatedly when reading the problem? Was it a nervous habit, an expression of frustration, or something else? Why did Paul grin when he looked up to tell the interviewer his answer? Was this grin an expression of pride, delight in the elegance of his solution, or enjoyment of the simple pleasure of solving it?

We cannot say exactly what Paul was thinking, nor can we see exactly how he arrived so efficiently at a correct answer. However, through formal investigations of Paul and other mathematicians as they engage in solving mathematical problems, we have gained many insights into the thinking, actions, and behaviors that these accomplished problem solvers typically display.

Overview of the Research Literature on Problem Solving

A brief overview of research into problem solving reveals many insights about problem solving behaviors and attributes. While early work in problem solving focused on describing the problem solving process (Pólya, 1957), more recent investigations (including our own) have identified specific abilities of the problem solver that seem to contribute to problem solving success. Reviewing the problem solving literature from 1970 to 1994, Lester (1994) noted that problem solving performance appears to be a function of interdependent factors such as knowledge, control, beliefs, and sociocultural contexts. These factors overlap and interact in various ways. More recent studies point to planning and monitoring as key discriminators in problem solving success (Carlson, 1999a; DeFranco, 1996; Geiger

& Galbraith, 1998; Schoenfeld, 1992) and to the influence of such affects as beliefs, attitudes, and emotions (DeBellis & Goldin, 1997).

Phases of Problem Solving

Investigations of problem solving suggest that the solver moves through distinct phases as he completes a novel mathematics task. Pólya's early work (1957) stated that the solver must i) understand the problem; ii) develop a plan; iii) carry out the plan; and iv) look back. Pólya described the problem solving process as a linear progression from one phase to the next and advocated that when solving a problem,

> [First,] we have to see clearly what is required. Second, we have to see how the various items are connected, how the unknown is linked to the data, in order to obtain the idea of the solution, to make a plan. Third, we carry out our plan. Fourth, we look back at the completed solution, we review it and discuss it. (Polya, 1957, pp. 5–6)

Twenty-five years later, Garofalo and Lester (1985) again described problem solving behavior as consisting of four phases of distinctly different metacognitive activities: orientation, organization, execution, and verification.

The Influence of Affect on the Problem Solving Process

Affective variables such as beliefs, attitudes, and emotions have also been observed to have a powerful influence on the behavior of the problem solver (DeBellis & Goldin, 1997; Lester, Garofalo, & Kroll, 1989; McLeod, 1992; Schoenfeld, 1989). Although emotions are more evident than beliefs during problem solving, beliefs (deep-seated convictions such as "learning mathematics is mostly memorization") also play an important role (Carlson, 1999a, 1999b; Schoenfeld, 1989, 1992). Belief systems help to explain why solvers persist in mathematics course taking (Carlson, 1999a) and succeed or fail in their attempts to solve mathematics problems (Schoenfeld, 1992). An individual's beliefs determine the perspective with which she approaches mathematics and mathematical tasks, and therefore contribute to her success and failure in solving problems. Effective problem solvers consistently express beliefs that: the solution process may require many incorrect attempts; problems that involve mathematical reasoning are enjoyable; mathematical ideas should be understood instead of just memorized; learning mathematics requires sorting out information on one's own; and verification is a natural part of the problem solving process (Carlson, 1999).

Emotions play an important role as well. Positive feelings such as satisfaction and pride and negative ones such as anxiety and frustration appear throughout the problem solving process. Cycles of struggle, success, and elation spark new motivation, while cycles of struggle, frustration, and failure create anxiety. The failure/anxiety cycle has been shown to result in students' choosing to abandon a solution attempt (DeBellis & Goldin, 1997; 1999; Hannula, 1999). *Intimate* mathematical experiences have been characterized as a bonding between the individual and his mathematics (DeBellis & Goldin, 1997). When problem solvers feel this strong bonding, they often devote large amounts of time to thinking about a problem and trying to solve it. We (Carlson & Bloom, 2003) have observed instances when a solver's intimacy with a problem led to him thinking about it incessantly, to the point of obsession. We have also observed that mathematicians are very honest about their understanding of a situation or problem; they do not offer up solutions that do not have a logical foundation. We have labeled these traits as mathematical integrity; expert problem solvers are honest about their understandings and possess standards for their reasoning.

The Importance of Decision Making and Monitoring During Problem Solving

In the context of problem solving, *global decisions* include actions such as selecting a particular approach or choosing to abandon another. Such decisions have a profound influence on a problem solver's efficiency and effectiveness. *Local decisions*, such as selecting particular resources and strategies, also influence problem solving effectiveness. How well the solver monitors his thinking and products during the solution attempt is also significant, because a solver's skill in monitoring determines the efficiency with which facts, techniques, and strategies are exploited. In fact, a poorly managed solution path frequently results in poor decisions and a failed solution attempt (Schoenfeld, 1992).

Not surprisingly, more effective problem solvers make wiser decisions during the problem solving process (De Franco, 1996). In a study involving professional mathematicians, De Franco also found that mathematicians who enjoyed more professional success and recognition exhibited more effective monitoring than their peers. In contrast,

Goos, Galbraith, and Renshaw (2000) found that inexperienced students fail to act on "red flags" such as lack of progress, error detection, and anomalous results. Despite these findings, prior research has not fully characterized the nature of the reasoning patterns, regulatory behaviors, knowledge, and decision-making that contributes most to problem solving success.

Describing Our Study

We were interested in knowing why some individuals emerge as highly effective problem solvers, while others do not. We chose to work with professional mathematicians because we hypothesized that observing individuals with a broad, deep knowledge base and extensive problem solving experience would reveal unique insights into the problem solving process, the interactions of various problem solving attributes (e.g., thought processes, monitoring and reflecting behaviors, emotional responses), and the reasoning patterns and attributes that contribute to success. We further believed that this knowledge could yield valuable practical information for informing course design, curriculum development, and classroom instruction.

In our study we observed 12 mathematicians as they completed four different problems. The subjects verbalized their thought processes to the interviewer as they worked. The problems required only knowledge of concepts that are initially taught at the secondary level (e.g., Pythagorean theorem, rate of change), although they were complex enough to elicit multiple solution paths and strong affective responses, even from mathematicians. For more details about the methods for collecting and analyzing our data refer to sections 4.1 and 4.2 of Carlson and Bloom's 2003 article.

Gerald Solves the Paper-Folding Problem

To begin the description of our study, we invite you to look in on one more mathematician as he solves the paper-folding problem. As you read the transcript and analysis of Gerald's verbalizations in Table II, see if you can follow his solution path and identify specific reasoning patterns and attributes that he employed. How does his solution compare to your own? How does it compare to Paul's approach? In particular, we suggest that you attend to Gerald's sense-making, monitoring behaviors, reasoning patterns, use of heuristics and conceptual knowledge, procedural fluency, and expressions of emotion. In the table, the first column contains the words and gestures Gerald used as he worked the problem. The second column represents our characterization of Gerald's behaviors (e.g., sense making, accessing a heuristic, monitoring) as manifested in his language, constructions and gestures. The third column indicates his transition from one type of general cognitive activity or behavior to another (e.g., orienting, planning, executing, checking).

Table 1 illustrates the close analysis we performed on the interview transcripts of our 12 mathematicians. Using agreed-upon definitions and terms, we carefully identified and labeled the statements, actions, and behaviors recorded in the transcripts. Analysis of Gerald's approach revealed that he engaged in behaviors of sense making and organizing as he constructed both a mental image and physical sketch of the situation. Gerald initially conjectured that the black and the white areas were each equal to 3/2 in². He continued by drawing on heuristics and specific knowledge of the relationships in a 45-45-90 degree triangle and realized that this knowledge was in conflict with his initial conjecture. This led him to review the problem statement, after which he quickly verified his new conjecture that each of the three equal pieces of his sketch has an area of 1 in². Throughout his solution attempt he appeared to monitor his thinking and approach, while also accessing a large reservoir of conceptual and procedural knowledge that provided a reliable basis for his decisions and actions. We also observed that as negative emotions emerged (frustration and impatience, Line 23), he did not allow these to shake his confidence. Rather, his pride and ego appeared to motivate him to reengage with the problem.

Analysis of the collection of interviews from this problem solving study (Carlson & Bloom, 2003) revealed that these mathematicians were able to overcome what appeared to be strong emotional responses. Even as their emotions intensified, the subjects kept their focus and employed various coping mechanisms to continue working toward a solution to the problem. This response has not been commonly observed in students.

Our analysis also revealed that the subjects engaged in cyclic reasoning patterns that were highly effective in helping them to determine a solution approach. Our observations have shown that the phases of *orienting, planning, executing,* and *checking* are linked in a cycle, a cycle our subjects executed repeatedly until they arrived at a solution or abandoned the problem (Figure 3). While these phases are similar to those first observed by Pólya, we did not begin

Table 1. Coded Transcript

	Excerpt	Behavior	Phase
(1)	Okay, a square piece of paper is white on the front side and black on the back side has an area of 3 square inches [*He continues to read the problem very slowly*]	• Initial Engagement • Sense Making	
(2)	Corner *A* is folded...so the total visible area is half white and half black. [*He sketches the square and labels the corners A B C D. Then he sketches in the fold and labels the corner A'.*]. How far is *A'* from the fold line? *He then constructs the line that represents the distance he wants to find.*]	• Heuristics—makes a sketch	Orienting
(3) (4)	*Long pause* Okay, so each side is square root 3. [*He labels the two sides of the square as* $\sqrt{3}$.*]	• Sense Making—organizing information • Mathematical Knowledge —side of a square given the area	
(5)	And then fold it over, so that each of these guys are the same [*He shades in the triangle that represents the folded region.*]	• Heuristic—makes a sketch	
(6)	So the total area here is 3/2 [*He incorrectly labels the two visible areas as 3/2, ignoring the area of the back side of the triangle.*]	• Conjecture	
(7)	So, this is *x* and this is *x*, then this is *x* squared over 2. [*He labels the sides of his shaded triangle as x.*]	• Imagine/Verify • Strategy—find area of shaded region • Mathematical Knowledge—area of a triangle	
(8)	What's that supposed to mean?	• Monitoring—does this make sense?	
(9)	Total area is 3. I fold it over so this is half. And so...the whole square is 3.	• Sense Making	
(10)	I fold it in, so that is 3/2.	• Conjecture	Planning
(11)	I don't ....oh....[*He then retraces the fold on the paper and labels the sides of the smaller square x and labels the point A' on the diagonal.*]	• Testing Conjecture • Self Monitoring—quality of thinking	
(12)	I'm sorry, that's not correct. It's not 3/2.	• Rejecting Conjecture	
(13)	This area is a, this area is a, and that area out there is supposed to be a [*Gestures towards figures on his diagram.*]	• Re-engagement • New Conjecture	
(14)	So, we're supposed to have half white and half black.	• Sense Making • Conjecture/Imagine/Verify	
(15)	So, this area x squared is 2a. [*Pointing to sketch*]	• Strategy	

(16)	2a plus a is 3a. So, a is going to be 1. So, x squared is 1/2…	• Executing Strategy	*Executing*
(17)	How far is A′ from the fold line, so I want from A′ to the fold line.	• Monitoring Progress	
(18)	So, x is 1 over the square root of 2. And then, that is that divided by that, so you have…..1 over the square root of 2 divided by the square root of 2 is 1/2…	• Executing Strategy • Mathematical Knowledge—side of a square given the area • Mathematical knowledge—altitude of an isosceles right triangle	
(19)	Let me check this and make sure…so if x is 1 over the square root of 2, this area is 1/4. *[Pointing to the area of the triangle formed from folding corner A over to A′]*	• Verifying Work • Mathematical Knowledge—area of triangle	*Checking Cycles back*
(20)	No,	• Rejecting Solution • Mathematical Knowledge	
(21)	that is, is 1/2…*[pointing to sketch again]*….	• New Conjecture	*Planning*
(22)	1/2… That's 1/2, that's 1/2, that's 1/2, so that's 1/2 so this area would be 1/2.	• Conjecture • Tests Conjecture	
(23)	No!	• Rejects Conjecture • Affect—frustration, impatience	
(24)	What am I doing wrong?	• Reflects on Thinking • Affect—pride, ego, frustration	
(25)	Okay, a square piece of paper is white on the front side and black on the back side has an area of 3 square inches. Corner A is folded…so the total visible area is half white and half black. How far is A from the fold line?	• Re-engages with problem text	
(26)	What am I,	• Conjectures	
(27)	You can't…	• Tests Conjecture	*Executing*
(28)	*[Pushes paper aside]*	• Rejects Conjecture • Affect–frustration • Affect— intimacy	*Checking Cycles back*
(29)	There's nothing wrong with my brain, it's my calculations.	• Affect—aha!	*Planning*
(30)	The total area is 3. That's the total…Yes.	• Sense Making	
(31)	Now I fold it *[makes a folding movement with his hand]* and then this area, which is black, is the same as this area. So, this is some area A.	• Heuristic—modeling the problem	
(32)	And that's A. That area is the same as this area. This is the lost area.	• Sense Making	
(33)	So, 3a equals 3. The area is 1.	• Conjecture	
(34)	So, I want the area of this animal here to be 1	• New Strategy	
(35)	So, if that's x and that's x, the area of the whole square is 2…x is the square root…	• Executing Strategy • Mathematical Knowledge—diagonal of a square	*Executing*
(36)	That makes much more sense…	• Self Monitoring	
(37)	So this one I can do in my head…	• Affect—ego/pride	
(38)	….let's see …if that's 1 and that's 1, that's 1/2, that's 1/2 and that's 1 that's 1 that's 1.	• Executing • Mathematical Knowledge—relationships in a square	

(39)	So, it's, and the answer was how far is A' from the fold line, so I take it by that you mean this line and that distance is 1.	• Verifying Solution	*Checking Completion*
(40)	Sloppy calculations…	• Affect—embarrassment	

our analysis with a priori categories; rather, they emerged from coding and analyzing the data. Moreover, we also found another distinct cycle embedded in the *planning* phase of the larger cycle. We have labeled this the *conjecture—imagine—evaluate* subcycle. This subcycle usually was signified subtly; the mathematicians might pause and contemplate briefly, for example, before actively pursuing a solution approach. Other mental actions may well occur during these cyclical phases, but we believe that the labels we have chosen characterize well the primary and most general form of cognitive activity that our subjects exhibited as they cycled through these phases enroute to a solution.

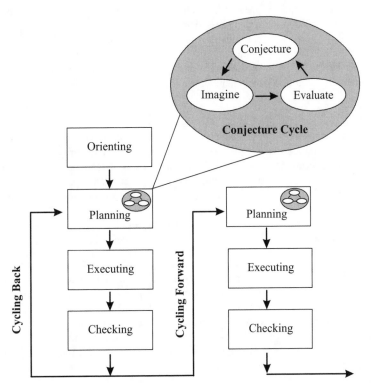

Figure 3. The Cycles of Problem Solving

Having identified and labeled the cyclical phases of our subjects' problem solving process, we next attempted to identify when and how these problem solvers called on and applied their mathematical knowledge, heuristics, and monitoring, as well as how they managed their emotional responses. We reanalyzed the data (Table 1) in search of answers to these questions and revisited the audio recordings to characterize expressions of anger, anxiety, frustration, pleasure, and other emotions that surfaced during their problem solving attempts.

The subjects' consistent *reflections on* and *regulation of* their thought processes and products were particularly striking. The effectiveness of their monitoring was highly dependent upon their fluency in accessing both conceptual and procedural knowledge. Strong emotional responses (both positive and negative) emerged. Even their small successes were followed by exhibitions of joy or pride, while exhibitions of frustration were followed by coping mechanisms that included their diverting attention from the problem. They sometimes interjected idle chat such as talking about the view out the window or their favorite sport. What emerged from our analysis was a more structured, coherent, and descriptive characterization of the interplay between the problem solving phases and various problem solving attributes (resources, heuristics, affect, and monitoring). We illustrate this characterization in the form of a multidimensional problem solving framework (Table 2).

Table 2. A Multidimensional Problem Solving Framework

Phase (Behavior)	Resources	Heuristics	Affect	Monitoring
Orienting • Sense making • Organizing • Constructing	Mathematical concepts, facts and algorithms are accessed when attempting to make sense of the problem. The solver also scans her knowledge base to categorize the problem.	The solver often draws pictures, labels unknowns and classifies the problem. (Solvers were sometimes observed saying, "This is an X kind of problem.")	The curiosity and interest level of the solver affects the solver's motivation to make sense of the problem. If the solver is not interested, he may lack motivation and stall before starting.	Self-talk and reflective behaviors serve to keep the mind engaged. The solvers were observed asking: "What does this mean?"; "How should I represent this?"; "What does that look like?"
Planning • Conjecturing • Testing • Strategizing	Conceptual knowledge and facts are needed to construct conjectures and make informed decisions about strategies.	Specific heuristics were accessed and considered while evaluating and choosing a solution approach.	Beliefs about the methods of mathematics and one's own abilities influence conjectures and decisions. Signs of intimacy, anxiety, and frustration are also displayed.	Solvers monitor their strategies and plans. They ask themselves: "Will this take me where I want to go?"; "How efficient will approach x be?"
Executing • Computing • Constructing	Conceptual knowledge, facts, and algorithms are essential for executing, computing, and constructing. Without conceptual knowledge, monitoring of constructions is misguided.	Fluency with a wide repertoire of heuristics, algorithms, and computational approaches are needed for the efficient execution of a solution.	Intimacy with the problem, integrity in constructions, frustration, joy, defense mechanisms, and concern for aesthetic solutions emerge in the context of constructing and computing.	Conceptual understandings and numerical intuitions are employed to monitor both the solution progress and products while constructing statements.
Checking • Verifying	Resources, including well-connected conceptual knowledge, informs the solver as to the reasonableness or correctness of the solution attained.	Computational and algorithmic shortcuts are used to verify the correctness of answers and to ascertain the reasonableness of computations.	As with the other phases, there were a number of affective behaviors displayed. It is often at this phase that frustration overwhelmed the solver, causing him to abandon the task.	Reflections on the efficiency, correctness, and aesthetic quality of the solution provide useful feedback to the solver.

Primary Findings from Our Study

Each problem solving phase and the predominant behaviors exhibited during that phase are listed in the cells on the far left column of Table 2. For example, in the orienting phase, the subjects predominantly engaged in behaviors of sense making, organizing, and constructing. The primary role of each problem solving attribute (e.g., resources, heuristics) during that phase is illustrated in the cells to the right. For example, when orienting themselves to the problem, the experts typically exhibited behaviors of sense making, organizing, and constructing, while accessing resources that included concepts, facts, and procedures. During this phase, the mathematicians in our study also applied a variety of heuristics such as drawing pictures, labeling unknowns, and classifying the problem in a specific category of problems. The primary affective behaviors demonstrated in this phase were curiosity and interest—the intensity of the subjects' efforts to make sense of the problem was typically influenced by their interest in the problem type and curiosity about its solution. Various monitoring behaviors such as self-talk and reflections about the productiveness of their orientation behaviors were also observed to be influential in keeping their thinking moving in productive directions. Reviewing the decomposition of the collection of behaviors revealed a diverse classification of behaviors during each phase of the problem solving process.

In addition to their global decisions about their solution approach—deciding to cycle forward or cycle back, or to act on a conjectured approach—the mathematicians were making decisions about their moment-to-moment constructions and actions. They were continually asking themselves: Is this approach getting me anywhere? What does this tell me? Does that calculation make sense? The effectiveness of this monitoring, like their sense making, was highly dependent on their knowledge base, including their ability to recognize what facts, algorithms, and concepts would be useful for moving their thinking and solutions in productive directions. Effective monitoring played a powerful role in the efficiency and effectiveness of their solution attempts.

We found that these attributes of problem solving (e.g., resources, heuristics, affect, and monitoring) were evident during each problem solving phase. Consistent with the findings of DeBellis (1998) and Hannula (1999), we observed that local affective pathways play a powerful role in the problem solving process. We found that frustration occurred frequently during the solution attempts of these mathematicians; however, unlike what has been reported when observing students, these mathematicians effectively employed a variety of coping mechanisms to manage frustration and anxiety and to persist in pursuing a solution. Our experts were also observed displaying joy and satisfaction when they were successful. Other common expressions of affect included mathematical integrity and intimacy. When we reviewed the interviews, it became apparent that all the mathematicians in our study based their decisions on a logical foundation; they also did not pretend to know something when they didn't. We refer to these behaviors as *mathematical integrity*. During the interviews the mathematicians also frequently reported becoming so obsessed with a problem that it occupied much of their waking hours. Others told of being unable to "let go" of a problem for years, even though rationally they had concluded that the problem was beyond their abilities. These behaviors were characterized as exhibitions of *mathematical intimacy*.

Reflections on the Complexities of Acquiring Expert Mathematical Practices

Our investigations have persuaded us that practicing mathematicians rely on what appear to be acquired dispositions and reasoning patterns when solving difficult problems. In turn, this leads us to suggest that all of us who teach mathematics at any K–20 level need to make more explicit efforts to support the development of students' problem solving abilities. We must also consider that school mathematics as it is now taught may have a *negative* effect on young students' ability to solve problems. For example, in a 1984 study that followed children from their entry into first grade through their exit in third grade, Carpenter and Moser (1984) found that children were better problem solvers on the first day of first grade than they were on the last day of third grade. This is not to say that they solved more sophisticated problems as first graders. Rather, it says that they exhibited greater creativity, persistence, and flexibility as first graders than they did as third graders. By third grade, the children had learned that one mustn't *think* when facing a math problem. The approach they had learned was to try to remember the steps that the teacher had demonstrated. Of course, if their memory failed, their solution did not advance. This pattern of attacking problems by trying to recall and apply a rote algorithm has also been observed in undergraduate students. In other studies (e.g., Carlson, 1998; Schoenfeld, 1989), undergraduate students in mathematics have been observed memorizing content without really understanding what it means. When confronting a problem they do not already know how to solve, they often act in ways that reveal low confidence, limited persistence, and little evidence of sense making. Rather than utilizing the imagine-conjecture-verify cycle described in the MPS framework, they tended to fall back on memorized facts and algorithms. Such studies call attention to what appears to be an alarming lack of focus on the development of expert mathematical practices at the classroom level. It is possible, we think, to do better. By promoting in students the practices we and other researchers have observed in expert problem solvers, mathematics instructors would likely contribute to more students continuing their study of mathematics. To illustrate what this promotion looks like in practice, we conclude this chapter with two brief scenarios drawn from one of the authors' undergraduate classrooms. As you read the scenarios, observe how the instructor attempts to encourage students to use the practices we observed in our 12 mathematicians and described in our MPS framework. [For a detailed description of an entire course aimed at promoting effective problem solving behaviors in undergraduate students, see the sequence of articles that describe Schoenfeld's problem solving course for undergraduates (Arcavi, Kessel, Meira & Smith,1998; Santos-Trigo, 1998; Schoenfeld, 1998).]

How Do Undergraduates Compare to the Experts?

Researcher Bloom set out to design and implement instructional strategies that would foster effective behaviors in undergraduate students. As a first step, she used the MSP framework to characterize the problem solving behaviors of undergraduate students (Bloom, 2002; 2004). That investigation revealed a sharp contrast between undergraduate and expert behaviors. When initially confronting a problem, students often *appeared* to be attempting to orient themselves. However, a closer examination of their thinking revealed that they were actually scanning the problem statement for *key words* that might provide hints for a solution approach. They rarely engaged in sense making; nor did they attempt to access their conceptual knowledge and other mathematical resources they possessed. The undergraduates in this investigation also lacked the vocabulary and logical fluency in communication that was observed in mathematicians. It was also observed that students rarely took time to contemplate their solution approach. Their planning frequently involved random employment of different heuristics (e.g., plugging in numbers, writing down a formula) with little thought given to what approach might be more effective or efficient.

Consistent with findings of Schoenfeld (1989), Bloom (2002) reported that undergraduate students are easily frustrated and quick to stop trying when encountering a dead end. They rarely engage in on-line monitoring, resulting in their making computational and reasoning errors that go unnoticed. They also exhibit low confidence in their abilities, as revealed in their regularly looking to an external source, such as the interviewer, teacher, or answer key for verification of their thinking or answer.

As a next step, Bloom designed instructional strategies to support undergraduates in acquiring the practices of expert problem solvers (such as accessing appropriate concepts and mathematical resources at the right moment when solving a problem). Following are two scenarios from her classrooms that illustrate the strategies she developed and continues to refine.

Incorporating Knowledge of the Problem Solving Process into Classroom Instruction: Scenario 1

On the first day of a College Algebra class the instructor directed her students to attempt a problem that required access and use of fundamental concepts from their past courses in algebra.

The Rhombus Problem

In rhombus *EFGH* the coordinates of *E* and *G* are (–6,–3) and (2, 5) respectively. Find the perimeter of the rhombus if the slope of segment *EH* is 2.

Since this problem was used to begin the first day of class for the semester, the instructor posed questions that she believed would prompt her students to enact effective problem solving behaviors. Her promotion of orienting behaviors included her asking them to read the problem and make sense of the situation. When some appeared stumped she suggested that they create a drawing of the situation. The ensuing discussions led to their generating a list of relevant mathematics concepts, facts, and algorithms, including the definition and properties of a rhombus (for example, that sides are equal length, opposite sides are equal, and diagonals are perpendicular and bisect each other). She also prompted them to articulate the definition of slope, distance formula, midpoint formula, equation for determining a line, and methods for finding the point where two lines intersect. From this list, the class was encouraged to select the definitions and formulas they believed may apply. Continuing the orienting phase, the instructor encouraged the students to employ the heuristic of labeling their sketch. As they moved to the planning phase, she prompted them to devise a strategy for determining the coordinates of F and H. As she visited each group, she prompted them to monitor the effectiveness of their solution plans by asking, "Why will that help?" or "What will you do with that?" When a group encountered a dead end, she suggested that they return to their original set of useful concepts and recycle through the *conjecture-imagine-evaluate* cycle again. Their work continued—as the instructor roamed from group to group she reminded them to prepare a logical description of their solution, and the reasoning that led up to their solution, so they could present it to the rest of the class. After several groups had completed their solutions, she asked for a volunteer to explain each group's solution approach. As the student presented their group's solution and thinking, the instructor prompted the student to articulate her reasoning and the rationale for her approach. After all groups had shared their solutions, the instructor asked the students to reflect on the efficiency and aesthetic quality of their solutions—she also prompted them to consider if it was necessary to find H and F, as well as the length of all four sides. After several minutes of discussion, she asked if they were pleased with their final solution.

In this scenario the instructor acted in ways to convey that she values reasoning and sense-making and that she is committed to helping them develop effective problem solving abilities. She also made explicit efforts to assure that her students eventually succeeded in their attempts to solve a novel problem. It was helpful that the problem was aligned with their current ability level. She also gave encouragement and careful thought to scaffolding her classroom instruction so that her students were expected to access familiar mathematics concepts, procedures, and facts when attempting a problem that initially appeared too difficult for them to solve. When the students displayed negative emotions, she intervened as needed to promote their use of various coping mechanisms that were so effectively displayed by mathematicians.

Below, we offer one more scenario from Bloom's classroom that demonstrates the pedagogical approach our research supports.

Scenario 2

To launch a class of preservice mathematics teachers into a geometric construction problem (Figure 4), the instructor reviewed some fundamentals of geometric construction by prompting them to respond to questions, such as "how do you bisect an angle?" and "what constructions are needed to copy an angle?"

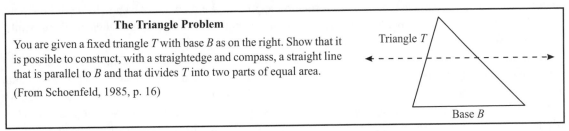

The Triangle Problem

You are given a fixed triangle T with base B as on the right. Show that it is possible to construct, with a straightedge and compass, a straight line that is parallel to B and that divides T into two parts of equal area.

(From Schoenfeld, 1985, p. 16)

Figure 4. The Geometric Construction Problem

She then instructed them to work in groups of three. Initially, they were asked to complete a simple construction; this was followed by progressively more complex tasks. As she observed negative emotions surface, she acknowledged the progression in difficulty and encouraged them to stick with it. She was observed saying, "It requires a proof," she says, "Proof is iffy, but we are math geeks. We can handle it." The students began to sort out and record what they knew as they engaged in acts of sense making. The instructor roamed from group to group. One group conjectured that they needed to find the centroid, and she encouraged monitoring by asking, "How will that help?" She continued to watch and listen as the groups worked. Another student asked how the areas of similar triangles are related. She suggested that he, "Calculate a couple and see." As another student offered an alternative conjecture she asked, "How could you test your conjecture?" More time passed and the noise and frustration levels rose. The instructor acknowledged the emotion and asked, "Are you ready for a hint? She paused and when no one responded she said, "Look for similar triangles. Will this help you?" Those moving in that direction appeared to be encouraged, and others moved back to the planning phase. One group was observed measuring an angle. She urged these students to monitor their reasoning. "What are you going to use the measure for?" she asked, and "What have you found out and how could you verify it?" She then suggested another technique: "Could you try it out on a simpler triangle, like a right triangle?" As the frustration level increased the instructor related anecdotes of students in another class who spent four weeks on this problem. She also recounted her own frustration with a problem that nearly ended her mathematics career. "I want you to really take ownership of it," she said. She also encouraged them to become intimate with the problem so they could think about it while running or driving. Only by working through the struggle, she told them, will you gain the ability, confidence, and persistence to continue getting better at solving challenging problems. As she continued to roam the classroom one group asked if they could verbalize their thinking to her. One member of the group indicated that he didn't agree with the solution that was presented. Others in the group began questioning him and challenged his thinking, while one member elected to take another tack at explaining the logic behind the solution. When the last hold-out was persuaded of the logic of the group's solution, the instructor pointed out to the group what had happened: "He got it. Did you see the light go on?" The students in this group smiled and appeared to become more relaxed as they pushed their chairs back from the table. They also appeared to be experiencing satisfaction from having persisted in arriving at a logical and correct solution. The instructor encouraged the students to experience the pride and satisfaction from their work to complete a difficult problem—she told them to try to feel this as intensely as they felt the frustration that was part of their journey in arriving at a correct solution.

Discussion

The body of research on problem solving suggests that effective problem solving ability does not spontaneously emerge in most students just by taking courses in mathematics (e.g., Carlson, 1998; 1999a; Schoenfeld, 1989; 1992). Rather, we believe that it requires instruction that is designed to promote the reasoning patterns and attributes described in the multidimensional problem solving framework. Regardless of the course you are teaching, we recommend that you integrate challenging problems that are appropriately linked to the content focus of your course. We also recommend that you initiate class discussions of appropriate problem solving strategies and that you regularly model practices such as the imagine-conjecture-verify reasoning cycle. It is also helpful to prompt students to consider the viability of an approach before moving forward with their computations. Encourage use of heuristics (drop an altitude, solve a simpler problem) as appropriate. We also recommend that you assist students in learning to manage the emotions that emerge during problem solving—students should adopt the belief that frustration, disappointment, and elation are all natural responses to the problem solving process. We also share the techniques we have observed mathematicians use to manage frustration and continue persisting in a solution approach. We draw on our knowledge of how powerful the *struggle-success-elation* cycle is for students, and we make explicit efforts to stretch their thinking, while also engineering success after periods of struggle and frustration.

Finally, we advocate that both instructors and curriculum developers reflect explicitly on the nature of the knowledge and reasoning abilities that they want their students to learn. According to Thompson (1985) anything we might call knowledge is a structure of thinking, and this structure is a structure of processes. Acquisition of knowledge that is meaningful requires that students engage in processes that build structured understanding. If one takes this perspective in teaching content, then, the ideas of calculus, for example, would best be learned by engaging in carefully scaffolded problem solving activities that draw from literature on cognitive aspects of learning concepts of calculus, and that engage students in using and acquiring mathematical practices described in the MPS framework. This is what we mean when we say earlier in this paper that there is a reflexive relationship between developing students' content knowledge and developing their *mathematical practices*. This volume contains several chapters that provide insights into the cognitive processes that promote both coherent understandings (e.g., Oehrtman, Carlson, and Thompson; Thompson and Silverman, this volume) and mathematical practices. We believe it is the responsibility of curriculum developers and instructors to gain knowledge of what coherent and meaningful understandings look like—ones that provide a solid foundation for building future mathematical knowledge and ones that translate to students' ability to access these ideas when solving problems. This should result in movement away from instructional practices that have a primary focus on imparting knowledge to students, and movement toward instruction and curriculum that attend to the mathematical practices that students acquire en route to building coherent understandings of central ideas. "The task of the curriculum developer is to select problematic situations that provide occasions for students to think in ways that have generative power in regard to the objectives of the lesson" (Thompson, 1985).

Acknowledgement   Research reported in this paper was supported by National Science Foundation Grants No. EHR-0412537 and EHR-0353470. Any conclusions or recommendations stated here are those of the authors and do not necessarily reflect official positions of NSF.

References

Arcavi, A., Kessel, C., Meira, L., & Smith, J. P. (1998). Teaching mathematical problem solving: An analysis of an emergent classroom community. In J. Kaput, A. H. Schoenfeld & E. Dubinsky (Eds.), *Research in collegiate mathematics education III* (Vol. 7, pp. 1–70). Providence, RI: American Mathematical Association.

Bloom, I. (2002). *The problem solving behavior of mathematics majors.* Paper presented at the Twenty-Fourth Annual Meeting of the North American Chapter of the International Group for the Psychology of Mathematics Education, Athens, GA.

Bloom, I. (2004). *Mathematics for teaching: Facilitating knowledge construction in prospective high school mathematics teachers.* Paper presented at the Twenty Sixth Annual Meeting of the North American Chapter of the International Group for the Psychology of Mathematics Education, Toronto.

Carlson, M., & Bloom, I. (2003). The cyclic nature of problem solving: An emergent framework. *Educational Studies in Mathematics, 58,* 45–76

Carlson, M. P. (1998). A cross-sectional investigation of the development of the function concept. In A. H. Schoenfeld & E. Dubinsky (Eds.), *Research in collegiate mathematics education III* (Vol. 7, pp. 114–162). Providence, RI: American Mathematical Association.

Carlson, M. P. (1999a). The mathematical behavior of six successful mathematics graduate students: Influences leading to mathematical success. *Educational Studies in Mathematics, 40,* 237–258.

Carlson, M. P. (1999b). *A study of the problem solving behaviors of mathematicians: Metacognition and mathematical intimacy in expert problem solvers.* Paper presented at the 24th Conference of the International Group for the Psychology of Mathematics Education, Haifa, Israel.

Carpenter, T. P., & Moser, J. M. (1984). The acquisition of addition and subtraction concepts in grades one through three. *Journal for Research in Mathematics Education, 15*(3), 179–202.

DeBellis, V. A. (1998). *Mathematical intimacy: Local affect in powerful problem solvers.* Paper presented at the Twentieth Annual Conference of the North American Group for the Psychology of Mathematics Education.

DeBellis, V. A., & Goldin, G. A. (1997). *The affective domain in mathematical problem solving.* Paper presented at the Twenty-first Annual Meeting of the International Group for the Psychology of Mathematics Education, Lahti, Finland.

DeBellis, V. A., & Goldin, G. A. (1999). *Aspects of affect: Mathematical intimacy, mathematical integrity.* Paper presented at the Twenty-third Annual Meeting of the International Group of for the Psychology of Mathematics Education, Haifa, Israel.

DeFranco, T. C. (1996). A perspective on mathematical problem-solving expertise based on the performances of male PhD. mathematicians. In *Research in collegiate mathematics II* (Vol. 6, pp. 195–213). Providence, RI: American Mathematical Association.

Garofalo, J., & Lester, F. K. (1985). Metacognition, cognitive monitoring and mathematical performance. *Journal for Research in Mathematics Education, 16,* 163–176.

Geiger, V., & Galbraith, P. (1998). Developing a diagnostic framework for evaluating student approaches to applied mathematics problems. *International Journal of Mathematics, Education, Science and Technology, 29*(4), 533–559.

Goos, M., Galbraith, P., & Renshaw, P. (2000). A money problem: A source of insight into problem solving action. *International Journal for Mathematics Teaching and Learning.*

Hannula, M. (1999). *Cognitive emotions in learning and doing mathematics.* Paper presented at the Eighth European Workshop on Research on Mathematical Beliefs, Nicosia, Cyprus.

Lester, F. K. (1994). Musings about mathematical problem solving research: 1970–1994. *Journal for Research in Mathematics Education, 25*(6), 660–675.

Lester, F. K., Garofalo, J., & Kroll, D. L. (1989). Self-confidence, interest, beliefs, and metacognition: Key influences on problem-solving behavior. In D. B. McLeod & V. M. Adams (Eds.), *Affect and mathematical problem solving: A new perspective* (pp. 75–88). New York: Springer-Verlag.

McLeod, D. B. (1992). Research on affect in mathematics education: A reconceptualization. In D. A. Grouws (Ed.), *Handbook of research on mathematics teaching and learning* (pp. 575–595). New York, NY: Macmillan Publishing Company.

Pólya, G. (1957). *How to solve it; a new aspect of mathematical method.* (2nd ed.). Garden City: Doubleday.

Santos-Trigo, M. (1998). On the implementation of mathematical problem solving instruction: Qualities of some learning activities. In A. H. Schoenfeld & E. Dubinsky (Eds.), *Research in collegiate mathematics education III* (Vol. 7, pp. 71–80). Providence, RI: American Mathematical Association.

Schoenfeld, A. (1985). *Mathematical problem solving*. Orlando Florida: Academic Press.

Schoenfeld, A. H. (1989). Explorations of students' mathematical beliefs and behavior. *Journal for Research in Mathematics Education, 20*, 338–355.

Schoenfeld, A. H. (1992). Learning to think mathematically: Problem solving, metacognition and sense-making in mathematics. In D. A. Grouws (Ed.), *Handbook for research on mathematics teaching and learning* (pp. 334–370). New York: Macmillan Publishing Company.

Schoenfeld, A. H. (1998). Reflections on a course in problem solving. In A. H. Schoenfeld & E. Dubinsky (Eds.), *Research in collegiate mathematics education III* (Vol. 7, pp. 81–113). Providence, RI: American Mathematical Society.

Thompson, P. W. (1985). Experience, problem solving, and learning mathematics: Considerations in developing mathematics curricula. In E. A. Silver (Ed.), *Teaching and learning mathematical problem solving: Multiple research perspectives* (pp. 189–243). Hillsdale, NJ: Erlbaum.

22

When Students Don't Apply the Knowledge You Think They Have, Rethink Your Assumptions about Transfer

Joanne Lobato, *San Diego State University*

Teaching so that knowledge generalizes beyond initial learning experiences is a central goal of education. Yet teachers frequently bemoan the inability of students to use their mathematical knowledge to solve real world applications or to successfully tackle novel extension problems. Furthermore, researchers have been more successful in showing how people fail to *transfer* learning (i.e., apply knowledge learned in one setting to a new situation) than they have been in producing it (McKeough, Lupart, & Marini, 1995). Because we are most frequently prompted to reflect upon transfer when it doesn't occur, this chapter begins with an undergraduate teaching vignette in which the students did not appear to apply the knowledge that the teacher thought they had developed.

If we presented a vignette of mathematics instruction dominated by the presentation of decontextualized formulas, it would come as little surprise if students struggled to solve real world applications. Instead, the vignette is drawn from a specially designed two-semester course in calculus for biology majors, with several features considered to promote the transfer of learning. First, major concepts were developed using biological contexts, followed by homework problems and on-line worked examples drawn from multiple contexts. Second, explicit connections were drawn between real world situations and abstract representations such as formulas and graphs. Finally, the course materials emphasized conceptual development, not just procedural competency. Specifically, many applets were created to help students develop underlying concepts and to explore dynamic mathematical models. The teacher reported that the students had become accustomed to working on application problems and often performed quite well. However, he also reported a number of surprising incidents in which he expected students to be able to transfer their understanding to a given homework problem, yet they appeared unable to do so. One such instance from the section on linear regression is presented below.

Teaching Vignette from a Calculus Class for Biology Majors

To introduce the need for regression, students explored the notion of a line of best fit for a set of biological data (involving the C period for *E. coli* bacterium) by manipulating the slope and y-intercept values of an equation in an interactive Java applet. The concept of absolute error was developed visually by relating points in the scatter plot to the line of best fit. The idea was then introduced that the line of best fit can be found by finding the minimum value of the sum of the squares of the errors function $J(a,b) = e_1^2 + e_2^2 + e_3^2 + \cdots + e_n^2$. Using scatter plot data of the average height of a child depending on age, students explored the relationship between changes in a and b in $y = ax + b$ and $J(a,b)$ via a Java applet. Finally, the simpler case of $y = ax$ was investigated in greater depth in order to better demonstrate its relationship to the quadratic function $J(a)$, the sum of the squares of the errors. Data showing the rate of mRNA synthesis/cell (denoted r_m) were utilized (see Figure 1), where the rate r_m depends upon the length of time it takes for a cell to double (the doublings/hr are denoted by m).

m	0.6	1.0	1.5	2.0	2.5
r_m	4.3	9.1	13	19	23

Figure 1. The rate of mRNA synthesis/cell (r_m) related to the number of cell doublings/hr (m).

The graphed data suggest a linear mathematical model of the form $r_m = am$. To determine the value of a, the sum of the squares of the errors was computed using the formula presented previously. Specifically, the error terms were calculated $[e_1^2 = (4.3 - 0.6a)^2, e_2^2 = (9.1 - a)^2, e_3^2 = (13 - 1.5a)^2, e_4^2 = (19 - 2a)^2,$ and $e_5^2 = (23 - 2.5a)^2]$, expanded, and summed, resulting in the function $J(a) = 13.86a^2 - 253.36a + 1160.3$. Students explored an applet in which the slope, a, of the line could be manipulated while observing what happens to the value of the quadratic function $J(a)$. Figure 2 shows one of the screens that the students could see in their exploration with the applet. By manipulating the value of a, students could observe that the line of best fit is determined by finding the minimum value of $J(a)$. Several additional examples from other contexts were worked in class in order to help students focus on the similar structure across examples.

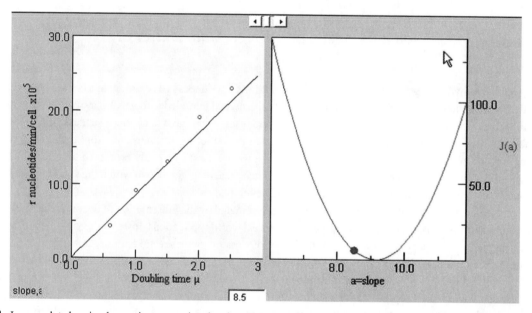

Figure 2. Java applet showing how a in $r_m = am$ is related to $J(a)$, the function determined by the sum of the squares of the errors.

The application problem shown in Figure 3 was then assigned as homework. The teacher thought it would be straightforward. After all, students were likely to have experience with scaling in secondary school, and this was basically a scaling situation with some error in the measurements. He thought students could solve the problem easily by making a table of data as they had seen in previous examples (e.g., Figure 1), finding the error values $[e_1^2 = (20 - 3.3p)^2; e_2^2 = (2 - 0.5p)^2; e_3^2 = (12 - 2p)^2]$, expanding and summing the terms to generate the function $J(k) = 15.14k^2 - 182k + 548$, and then using algebra techniques to determine the vertex, slope, and least sum of square error values $(-b/2a, J(-b/2a) = (6.01, 1.03)$.

Much to the teacher's surprise, the students did not know where to begin. It appeared that the students had not transferred what they had learned from the lessons on linear regression.

Because this episode did not occur in the context of a research study, I was unable to interview the students or investigate their work on a range of tasks. While I do not know why these students found the photography problem so difficult, I have some conjectures based on a relevant high school study and on a reconceived view of transfer.

Related Research Results from a High School Study

In this study, five classes of high school Algebra I students participated in a 6-week replacement unit on slope and linear functions. The curricular materials were developed as part of the activities of a research group which was

In looking through some old photos, a woman finds a picture of her great-grandfather standing near the family home, where she now lives. In the photograph, she measures the height of the roofline, which she knows to be 20 ft, as 3.3 cm. The 2 ft wide window measures 0.5 cm on the photo, and the distance from the front door to the oak tree at the driveway is 12 feet, which is 2 cm in the photograph.

    a. The conversion of the measurements in the photo p to measurements in actual distance d is given by the formula $d = kp$. Write a formula for the quadratic function $J(k)$ that measures the sum of squares error of the line fitting the measurements in the photo. Find the vertex of this quadratic function. This gives the value of the best slope k, while the $J(k)$ value of the vertex gives the least sum of squares error.

    b. In the photograph, her great-grandfather is 1 cm tall. Her mother remembers her great-grandfather as a tall man of about 6 ft, whereas her father thinks he was shorter, about 5.5 ft. Use the model (with the best value of k) to predict the height of the great-grandfather and determine whether the mother or father better remembers the height of her great-grandfather.

Figure 3. Application problem from a homework assignment.

investigating the connections fundamental for students to understand the domain of linear functions (see for example, Schoenfeld, Arcavi, & Smith, 1996).

Several features of the curricular approach are similar to those of the biology calculus materials. First, the mathematical topic (slope) was explored in multiple contexts (e.g., real staircases, piles of beans, and velocity) in order to help students see past superficial differences and extract a common mathematical structure. This practice is supported by current research in mainstream cognitive science (see for example, Fuchs et al, 2003). Second, explicit connections were drawn between explorations of realistic contexts and abstract representations. Specifically, students' informal knowledge of real stairs was linked to a more abstract representation of stairs using dynamic computer staircases and was later connected to conventional symbol systems including graphical displays of lines, the slope formula, and linear equations. This is consistent with a large body of literature indicating that overly contextualized knowledge can reduce transfer and abstract representations can help promote transfer (National Research Council, 2000). Third, conceptual understanding was emphasized along with procedural competency, in part because of the notion that students will transfer that which they deeply understand (Schroeder & Lester, 1989).

Results from paper-and-pencil exams given to the entire population of 139 students indicated that the students were able to find the slope of objects encountered in the experimental curriculum, such as staircases (87% correct) and lines (80% correct). However, transfer was poor — 40% to a playground slide task and 33% to a roof task (Lobato, 1996). In the slide task, a drawing of a playground slide was presented and students were asked to describe how they would determine the slope of the slide and what measurements they would take. In the roof task, a diagram of a house was presented in which various measurements were labeled, (e.g., the height of the house, the height of the roof, the length of the roof top, and so on). Students were asked to calculate the slope of the roof and to circle the measurements that they used to determine the slope.

Interviews were also conducted, using fifteen students selected from across the five classes. Two typical responses to the slide task are shown in Figure 4. Both students correctly recalled the slope formula as "rise divided by run" and treated the slope formula as relevant in the novel slide situation. However, the students made incorrect rise and run choices. Jarek's response is particularly striking because the rise and run seem disconnected from the part of the apparatus that is steep.

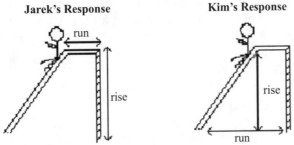

Jarek and Kim: You divide the rise by the run to find the slope.

Figure 4. Responses by two students to a transfer task.

The disappointing results from both the high school slope study and the opening calculus vignette invoke a desire to respond and fix the problem. In the case of the slope study, we could attribute the disappointing results to poor prerequisite understanding or a lack of student motivation. Yet these are unsatisfactory explanations because the students' scores on the initial learning tasks were high. The researchers and teachers could respond by revising the instructional conditions the next year in order to promote better performance on transfer tasks. However, on what principles would these changes be based, without further understanding of the apparent lack of transfer? For example, the curriculum designers could include the slide and roof situations in the next revision of the curriculum, but then would better performance simply be a result of "teaching to the test"? More real world situations could be included in the revised curriculum, but on what basis would these examples be chosen?

An alternative response—one that is explored in this chapter—is to rethink one's assumptions about transfer. I present an alternative approach to transfer called the *actor-oriented transfer perspective* and then use this perspective to reanalyze the slide data. Because the actor-oriented transfer approach emerged in response to critiques of the classical approach to transfer, a brief history of the transfer of learning in educational research is presented first.

Brief History of the Transfer of Learning

The classical transfer approach refers to the family of *common elements* theories that have dominated the 20th century, starting with Thorndike's (1906) theory of identical elements. According to Thorndike, transfer occurs to the extent to which original learning and transfer situations share identical elements — typically interpreted as shared features of physical environments. In the cognitive approaches to transfer of the latter quarter of the 20th century, the notion of identical elements was reformulated as mental representations. That is, people construct symbolic representations of initial learning and transfer situations. Transfer occurs if these two representations are identical, if they overlap, or if a mapping can be constructed which relates features of the two representations (Anderson, Corbett, Koedinger, & Pelletier, 1995). As summarized in a report by the National Research Council (2000), "Transfer between tasks is related to the degree to which they share common elements, although the concept of elements must be defined cognitively" (p. 78).

In a typical study of transfer, researchers generate transfer tasks that share some structural features with the initial learning tasks (e.g., a common solution approach) but have different surface forms (e.g., different word problem contexts). When performance improves between the learning and transfer tasks, researchers infer that students have applied knowledge that they gained during the learning experience to the transfer tasks.

During much of the 20th century, there were only a few isolated intellectual challenges to the mainstream common elements approach to transfer. However, dissatisfaction gained momentum in the 1980s and 1990s, when researchers began questioning the dominant conceptualization of transfer by bringing to bear the assumptions about knowing and learning from the theoretical perspective of situated cognition. Three critiques of transfer are summarized here (for more details, see Lobato, 2006).

First, transfer experiments rely on models of expert performance, often becoming an "unnatural, laboratory game in which the task becomes to get the subject to match the experimenter's expectations," rather than an investigation of the "processes employed as people naturally bring their knowledge to bear on novel problems" (Lave, 1988, p. 20). Second, transfer researchers often treat context as the task presented to learners and analyze the structure of tasks independently of the learners' purposes and construction of meaning in these situations (Cobb & Bowers, 1999; Greeno, 1997). Finally, the classical transfer perspective does not adequately account for the contribution of the environment, artifacts, and other people to the generalization of learning (Beach, 1999). The numerous critiques of transfer have contributed to a growing acknowledgment that, "…there is little agreement in the scholarly community about the nature of transfer, the extent to which it occurs, and the nature of its underlying mechanisms" (Barnett & Ceci, 2002, p. 612).

Rethinking Assumptions of Transfer

The Actor-Oriented Transfer Perspective

The *actor-oriented transfer perspective* emerged in response to the critiques presented above (Lobato, 2003). From this perspective, the conceptualization of transfer shifts from what MacKay (1969) calls an *observer's* (expert's) viewpoint to an *actor's* (learner's) viewpoint. By adopting an actor's perspective, we seek to understand the ways in which people generalize their learning experiences rather than predetermining what counts as transfer using models of expert performance. That is, *psychological similarity*—how a new situation is connected with a person's experience of a previous situation—rather than similar elements of task features—serves as the basis of transfer. From the actor-oriented perspective, transfer is defined as the generalization of learning. In what follows I will not use the term "generalization" as it is often used in mathematics to refer to the development of a rule or relationship that holds for all *n* within a specified range of *n* values. Instead, when I speak of actor-oriented transfer as generalization, I am referring more broadly to any influence of prior experiences on learners' activity in novel situations. For example, I make a case in the next section of this chapter that the students from the high school study were in fact generalizing their learning experiences when they were working on the playground slide task. I will present evidence that particular features of the curricular unit influenced the ways in which the students perceived the slide situation.

The actor-oriented transfer perspective differs from the classical transfer perspective in many ways. The actor-oriented perspective examines the generalization of learning, even if that learning results in incorrect or non-standard performance. In contrast, the classical transfer perspective, because of the particular methods employed and the reliance on expert models, emphasizes the formation of particular, highly-valued generalizations (Lobato, in press). The actor-oriented perspective seeks to understand the processes by which people connect learning experiences with new situations. This connection-making between situations most predominately involves the process of similarity-making, but it can also involve the processes of discerning differences and modifying situations (Lobato, Clarke, & Ellis, 2005; Lobato & Siebert, 2002; Marton & Tsui, 2003).

Table 1 summarizes the differences between classical and actor-oriented transfer perspectives. The assumptions that have bearing on the argument presented in this chapter are briefly described below; see Lobato (2003) for a more detailed description of an earlier version of this table.

Table 1. Theoretical assumptions of actor-oriented transfer compared to the classical approach.

Dimension	Classical transfer approach	Actor-oriented transfer
1. Definition	The application of knowledge learned in one situation to a new situation	Generalization of learning; "similarity-making" is a primary process but generalizing can also involve constructing differences and modifying situations
2. Perspective	Observer's (expert's)	Actor's (learner's)
3. Research method	Experimental methods to identify improved performance between learning and transfer tasks	Ethnographic methods to look for the influence of prior activity on current activity and how learners see situations as similar
4. Research questions	Was transfer obtained? What conditions facilitate transfer?	What relations of similarity are created? How are they supported by the environment?
5. Surface v. structure	Paired learning and transfer tasks share structural features but differ by surface features	Researchers examine learners' construal of "transfer" settings, acknowledging that a surface feature for an expert may be structurally substantive for a learner
6. Location of transfer	Transfer measures a psychological phenomenon	Transfer is distributed across mental, physical, social, and cultural planes
7. Transfer processes	Overlapping symbolic abstract mental representations (schemes)	Multiple processes including "focusing phenomena"
8. Metaphor	Static application of knowledge	Dynamic production of relations of similarity
9. Content domain	Mathematics is often treated as a set of procedures, and transfer is as a decontextualized ability	The conceptual sense that people make of mathematics is central to the nature of transfer
10. Abstraction	Inductive and individualistic—a common property is extracted	Constructive and involves social and individual processes

Shifting perspectives from the classical to the actor-oriented approach has methodological implications. The quantitative experimental methods for measuring transfer utilized in the classical approach tend to underestimate what counts as transfer from an actor-oriented perspective. If transfer is taken as the generalization of learning experiences, then qualitative methods are needed to scrutinize a given activity for any indication of influence from previous activities and to examine the particular ways in which people construe situations as similar. Furthermore, researchers operating from the classical approach typically predetermine "what" will transfer rather than making the "what" an object of investigation. Even when researchers do identify the particular knowledge that is transferred, they typically rely on models of expert performance (see for example, Gentner, 1989). In contrast, researchers from an actor-oriented perspective of transfer endorse knowing as fundamentally interpretative in nature, and as a result, accept students' idiosyncratic and even mathematically incorrect ways of connecting situations as evidence of transfer.

From the actor-oriented perspective, transfer is a distributed phenomenon. The generalization of learning involves individuals who create personal connections across activities, material resources that enable certain connections while constraining others, people who are oriented toward helping individuals see particular similarities, and the practices in which activities take place. For example, Lobato, Ellis, and Muñoz (2003) have advanced the notion of *focusing phenomena* to demonstrate a link between the ways in which features of instructional environments focus students' attention on particular mathematical properties and the ways in which individuals generalize their learning experiences (more details will be provided later in this chapter).

Finally, the metaphor for transfer in the classical approach is that of application or transportation. Knowledge is acquired in one setting and transported to another situation where it is applied. This suggests that the knowledge being applied remains unchanged, as do the tasks across which transfer occurs. However, the static nature of the application metaphor does not account for the ways in which people change transfer situations until they become similar to something they know or how people reconstruct their understanding of initial learning situations in order to make connections to the transfer situation ((Bransford & Schwartz, 1999; Carraher & Schliemann, 2002). From an actor-oriented perspective the metaphor of *production* replaces that of *application*, suggesting a more dynamic process of connecting situations in which relationships are produced or constructed. Indeed, transfer situations can be dynamic sites for invention and reorganization (Lobato & Siebert, 2002).

Reanalysis of the Data from an Actor-Oriented Perspective

According to classical measures of transfer, the students in the high school study showed little evidence of transfer. However, all of the interview participants correctly recalled the slope formula. Their rise and run choices, although often incorrect, did not appear to be random (e.g., no one designated a primarily horizontal quantity as the rise). This suggested that the students may have generalized their experiences from the instructional unit in ways that were not captured by the transfer measures. Therefore, we began to look carefully at the students' experiences in the curricular unit. The instructional activities related to slope had been dominated by staircases. Students measured the "treads" and "risers" of real staircases, explored dynamic staircases on the computer, and used mathematical "stairs" to determine the slope of a line and other objects. With this in mind, the data were re-analyzed from an actor-oriented perspective by looking for ways in which these experiences with staircases may have influenced the students' comprehension of the transfer situations.

The re-analysis revealed significant evidence of the generalization of learning experiences for each interview participant (Lobato, 1996). The two most prevalent ways in which students connected initial learning and transfer situations were through what we called the "stair step" and the "height/width" connections. For example, Jarek's choices for rise and run suggest that he was looking for a *stair step* in the slide setting (i.e., something with visually-connected "up" and "over" components that suggests climbing in an imagined state of affairs). He appeared to find a stair step on the right side of the slide apparatus (see Figure 5). The platform may have held particular appeal as the run, because it is the only tread-like feature in the diagram. A correct run, on the other hand, needs to be constructed with the use of an auxiliary line. Kim's work suggests that she related the activities in the instructional unit with the transfer situation through a "height/width" connection (see Figure 5). The rise and run can be conceived as either the height and width of the entire staircase or the height and width of a single step. Thus, slope may be correctly conceived as height divided by width in the staircase situation but generalized to the slide situation as the height and width of the entire apparatus.

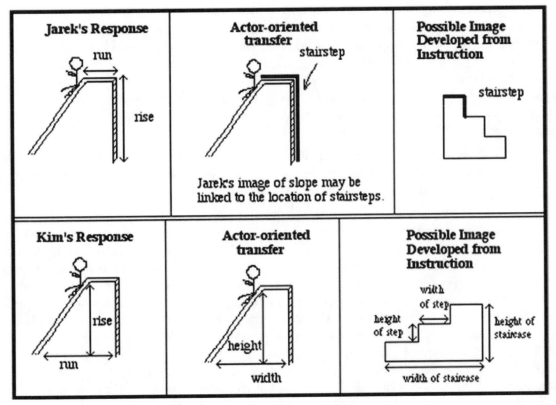

Figure 5. Actor-oriented transfer inferred from the work of two students.

In sum, use of the classical transfer approach in the high school study resulted in a concealment of the ways in which students generalized their learning experiences. While only 40% transfer was observed using classical transfer measures, each of the 15 interview participants demonstrated evidence of the generalization of their learning experiences from an actor-oriented transfer perspective. The analysis illuminated how the new situation might be connected with the thinkers' conceptions of previous situations even though the particular connections were unexpected and non-standard.

Significance of the Actor-Oriented Perspective

A critical reader may be concerned that the actor-oriented perspective seems to validate incorrect mathematical ideas that students generate. While mathematical correctness should not be ignored or de-emphasized, there are three reasons why examining transfer from an actor-oriented perspective is important even when it illuminates incorrect performance.

First, the actor-oriented transfer perspective helps guard against false conclusions regarding the degree to which humans generalize. Classical transfer studies often fail to demonstrate transfer in the laboratory (Kirshner & Whitson, 1997). This has led some to conclude that transfer is rare (Detterman, 1993). Others have interpreted the poor transfer results as evidence that reasoning is hopelessly context-bound, or "that our cognitive apparatus simply does not incline very much to transfer" (Perkins & Salomon, 1989, p. 22). However, the classical transfer approach underestimates the generalization of learning by accepting as evidence of transfer only specific correspondences defined a priori as the "right" mappings. It is not surprising then that we find overwhelming evidence for the lack of transfer from the classical perspective, since we know that novices do not make the same set of connections as experts (see National Research Council, 2000, for a summary of expert-novice studies).

Second, information gained through an actor-oriented transfer investigation can usefully inform revisions of curriculum materials and pedagogical approaches. We use the notion of *focusing phenomena* to demonstrate how seemingly random generalizations (such as those shown in Figure 5) are supported by features of instruction (Lobato & Ellis, 2002; Lobato, Ellis, & Muñoz, 2003). Specifically, focusing phenomena are aspects of instructional environments that regularly direct students' attention toward certain mathematical properties when a variety of information competes for students' attention. Focusing phenomena emerge not only through the instructor's actions but also through mathematical language, features of the curriculum, and the use of physical materials such as graphing calculators. By understanding and altering the nature of the focusing phenomena, we have been able to positively affect the nature of students' generalizations (Ellis & Lobato, 2004; Lobato, 2005). Consequently, we can profit from an investigation even when students produce mathematically incorrect generalizations.

Third, the actor-oriented transfer perspective provides a way to examine how novices generalize their learning experiences. According to Bransford and Schwarz (1999):

> Prevailing theories and methods of measuring transfer work well for studying full-blown expertise, but they represent too blunt an instrument for studying the smaller changes in learning that lead to the development of expertise. New theories and measures of transfer are required. (p. 24)

By better understanding how novices generalize their learning experiences, we may be able to identify increasing levels of sophistication in generalizing activity and consequently support students' development of expertise.

An important aspect of understanding how novices generalize their learning experiences is being aware of how novices make sense of typical transfer situations. For example, it is important to understand what made the playground slide situation so difficult for the students in the high school study. When I created the slide task, I had assumed that it differed from the situations explored in class (such as staircases, lines, and piles of beans) by only a surface feature (namely the playground slide setting). In the next section, I report the results of a series of studies in which we set aside our expert assumptions regarding surface and structural features and made students' comprehension of transfer situations the object of inspection. The results indicate that what is typically considered to be a surface feature can, in fact, represent structural complexities for novices. I also outline how we used the knowledge gained from these investigations to inform a conceptual analysis of the mathematical ideas involved in more sophisticated ways of generalizing. This, in turn, informed the revision of the slope curriculum. The findings are also relevant to the calculus vignette.

Relevant Findings from Research

Students' Comprehension of Transfer Situations

In two follow-up studies to the high school study, we investigated students' comprehension of typical transfer situations for slope such as wheelchair ramps, speed, dripping water rates, and the protein concentration in nutrition bars. The results indicate that making meaning of these contexts involves the following elements of substantive mathematical reasoning: (a) isolating measurable attributes, (b) determining the effect of changing various quantities, and (c) constructing two quantities as independent and of equal status. Each component will be described briefly and illustrated for the ramp situation (for details, see Lobato, Clarke, & Ellis, 2005; Lobato & Siebert, 2002; Lobato & Thanheiser, 1999, 2002).

In the first follow-up study, we interviewed 17 high school students enrolled in a traditional introductory algebra course. Students were asked how they would create a way to measure the steepness of a wheelchair ramp (see Figure 6). Although we designed the task to investigate slope as a measure of the steepness of a ramp, we recognize that it can be solved by using the angle of inclination or by creating other ratios such as that of the slant height to the length of the base. Consequently, we also probed students' reactions to measures created by hypothetical students, such as the height divided by the length of the base. These follow-up questions helped us explore students' association of slope with the measure of steepness.

Over half of the students struggled to isolate the attribute of steepness from other attributes, such as "work required to climb the ramp" or "materials needed to construct a ramp" (Lobato & Thanheiser, 1999; 2002). For example, many students talked about the importance of including the distance of the slanted part of the ramp in their measure because a longer ramp is more difficult to climb (i.e., a person is slowed down as he or she moves up the ramp). This makes sense if one considers that steepness is only one attribute involved in the ramp situation and may not be the most salient characteristic for students who have had everyday experiences with skateboards and hills. A student may not see a wheelchair ramp as having the same steepness throughout if that student is focused on the attribute of work.

Indeed as one climbs a ramp, more work is required as one proceeds. Furthermore, students may conflate work and steepness, thus concluding that the steepness is not constant throughout the ramp and that the attributes affecting work should also be included in a measure of steepness. [For a paper that conveys the intellectual struggle of students who eventually sorted out these attributes in a teaching experiment, see Lobato, Clarke & Ellis, 2005.]

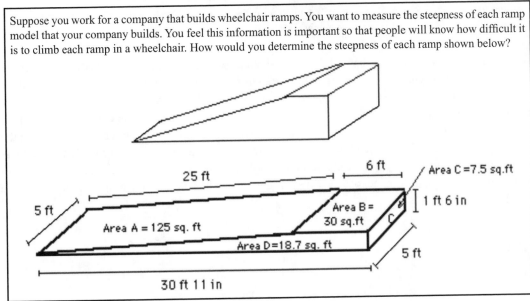

Figure 6. Wheelchair ramp task.

Additionally, a majority of the interview participants had difficulty determining the effect of changing various quantities (such as the length of the base) on the steepness of the ramp. Students were able to reason correctly that increasing the height increased the steepness and similarly, that decreasing the height made the ramp less steep. Surprisingly, over half of the students were unable to correctly determine the effect of increasing or decreasing the length of the base or the platform. For example, one student argued that if you make the base shorter, then the ramp will become less steep, and if you make the base longer than the ramp will become steeper (Lobato & Thanheiser, 1999; 2002).

Finally, in a second follow-up study, we designed an instructional environment to help students develop the types of reasoning that we thought would support successful performance in situations such as the wheelchair ramp (Lobato & Siebert, 2002). By examining a case study of a student who changed the way he comprehended the wheelchair ramp situation, we gained additional insight into the conceptual complexity of the aspects of transfer situations that are often considered surface features by researchers. Specifically, this study highlighted the difficulty of conceiving of the height and length of the wheelchair ramp as independent, equally important quantities. This conception is important because it appears to be prerequisite to forming a ratio between the height and the length.

Prior to instruction, Terry (an 8th grader who had earned a B in his Algebra I class) appeared to view height as more important than length and length as dependent upon height. When asked about the effect of lengthening the base on the steepness of a wheelchair ramp, Terry explained that "it [referring to the steepness of the ramp] basically matters on the height." When asked how decreasing the length of the base would affect the steepness, Terry moved the rightmost part of the ramp to the left (see Figure 7). Then he used his understanding of the effect of changing the height to reason that he had made the ramp taller and steeper. Because the length of the base of the new ramp was shorter than the base of the original ramp, decreasing the length of the base had resulted in a steeper ramp. Rather than reasoning directly with the length, Terry reasoned about the effect of height changes on steepness and treated length as dependent upon changes in height.

After instruction Terry appeared to have constructed length and height as independent quantities of equal status. Terry was asked to solve a proportion problem that had eluded him prior to instruction, namely to create a new ramp with a height of 3 ft so that the ramp has the same steepness as an original ramp with a height of 2 ft and base length of

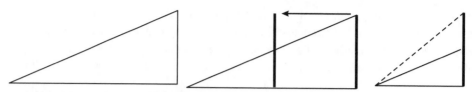

Figure 7. Author's representation of Terry's method for indirectly reasoning about the effect of decreasing the length of the base by first increasing the height.

15 ft. First, he drew the incline and the base of the original ramp (see steps 1 and 2 in Figure 8). Then Terry extended the incline (see step 3 in Figure 8), explaining that the height would get bigger until it eventually got up to 3 feet. He then drew the base and height of the new ramp (steps 4 and 5 in Figure 8). In contrast to his pre-instructional approach, Terry attended to the "same steepness" constraint prior to attempting to change the height. Furthermore, he acknowledged length as one of the important quantities involved in changing the ramp, and he generated the length of the new ramp prior to creating the height. As a result, Terry was able to reason proportionally to achieve a correct answer to the ramp problem. Terry was also able to successfully solve a harder problem, namely to determine the height of a ramp with the same steepness as the original ramp when the length of the original ramp is extended from 15 ft to 16 ft.

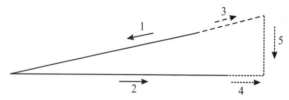

Figure 8. Author's reconstruction of Terry's effort to increase the height and length of a ramp yet retain the same steepness. Steps 1 and 2: Terry draws the slant height and length, respectively. Step 3: Terry extends the slant height to make the ramp taller without changing the steepness. Steps 4 and 5: Terry extends the length and draws in the height, respectively.

These results demonstrate how surface features for experts may in fact involve structural complexities for students. Furthermore, the investigation of the students' comprehension of transfer situations can help researchers make explicit the mathematical ideas that are often implicit in their own expertise but are critical for the design of effective learning experiences for students.

Using the Results to Inform a Conceptual Analysis

We used the knowledge gained from the investigations presented in the previous section to generate a conceptual analysis of the mathematical ideas involved in forming more sophisticated connections between learning and transfer situations than those formed by the students in the high school study. Specifically, we argued that productively generalizing experiences with the slope formula to situations like that of the wheelchair ramp involves the following: (a) isolating steepness from other attributes; (b) understanding how increasing and decreasing various quantities in the situation affect the steepness of the ramp; (c) conceiving of length and height as independent quantities of equal status; (d) constructing a multiplicative relationship between height and length; (e) understanding that if a ramp has a given height relative to a given length, then any segment of the height requires a proportional segment of the length in order to maintain the same steepness; (f) connecting the arithmetic operation of division to the partitioning of the "height-length" unit; (g) conceiving of slope as an appropriate measure of the steepness of a ramp; (h) understanding that corresponding changes in height and length correspond to "rise" and "run" in the slope formula; and (i) connecting the operation of division in the slope formula to the identification of a ramp with a height of $1/L \cdot H$ and a length of 1 unit, while understanding that the smaller ramp has the same steepness as the larger one (Lobato & Siebert, 2002). This understanding of the mathematical content differs significantly from the conceptualization of the content that grounded the development of the curricular materials for the initial high school study.

Implications

Revisions to the Slope Curriculum

We utilized the conceptual analysis presented above, along with the results of the studies we conducted from an actor-oriented transfer perspective, to make substantial revisions to the original slope curriculum. We have used the revised materials in two teaching experiments with better results. In several papers, we report the ability of students to generalize their learning experiences in more productive ways than those reported in the original study (Lobato, 2005; Lobato, Clarke, & Ellis, 2005; Lobato & Siebert, 2002). These papers indicate that the revised curricular materials tend to focus attention on the coordination of co-varying quantities rather than on the location of something vertical and something horizontal in each new problem. We briefly sketch three features of the revised curriculum. This is not intended to be an exhaustive list but rather to serve as a useful comparison with the features of the original slope curriculum.

First, in the original slope curriculum (and in the biology calculus materials), each new concept was exemplified using multiple contexts. By engaging with multiple contexts, it was thought that learners would extract a common structure by deleting superficial details across examples. In the revised slope curriculum, we made a radical departure with this approach and explored a single contextual setting (namely speed). We no longer conceive of the process as "extracting a structure" but rather as constructing a ratio as the measure of a given attribute (what Simon and Blume, 1994, call a "ratio-as-measure"), which involves building a complex set of related understandings. We believe this can be accomplished by in-depth exploration in a limited number of contextual settings.

Second, the original slope curriculum (and apparently the biology calculus materials) assumed that transfer is most likely when students experience particular mathematical notions in a range of real world settings, while at the same time learning to connect features of the contexts to abstract representations such as formulas and graphs. Consequently the conventional representations of equations and graphs were presented early and attention to them was interspersed with connections to realistic situations. However, in the high school study, it became apparent that many students conceived of a linear graph as a physical object (e.g., thinking of a line on a graph as a piece of wire) rather than as a mathematical object. In our revised approach, we postpone the introduction of graphical representations until students have formed an understanding of a rate (such as speed) as an equivalence class of ratios. Once rate has been formed mentally, then the graph of $y = mx$ can be understood as a representation of the equivalence class of ratios.

Finally, the revised slope curriculum is organized by a different conceptualization of the mathematical ideas. The new unit begins by showing students a character walking across a computer screen (using a modified version of the *Mathworlds* software developed for the SimCalc Project, Roschelle & Kaput, 1996). Students are asked to determine a way to measure how fast the character is walking. The activity and subsequent discussion allows students to isolate the attribute of how fast the character moves through space (speed) from other attributes such as how fast the character moves his feet (leg locomotion). This is important because each attribute is measured by a different ratio (e.g., distance over time versus number of steps taken over time). Once students isolate the attribute of motion through space, they explore how changing time and distance affect speed by entering distance and time values for various characters in the simulation software.

Once students understand the effect of increasing and decreasing distance and then time, they are in a position to construct speed as a ratio of distance to time. We conceive of the mental construction of ratio either as the formation of a multiplicative comparison or as the formation of a composed unit (Olive & Lobato, in press). In the new curriculum, ratio construction occurs in response to a discussion of "same attribute" tasks. Specifically, students use the speed simulation software to generate a family of distance and time pairs that make characters walk the same speed. We document how the development of speed as a ratio is surprisingly challenging, even for beginning high school students (Lobato & Siebert, 2002; Lobato & Thanheiser, 2002). The use of the "same attribute" task represents a major shift from the original slope curriculum. In the original curriculum, we focused on vertical and horizontal components on a graph without adequately linking them to what was being measured. For example, Jarek's work indicates a disconnection of the "run" from the part of the ramp that is steep. In contrast, the "same attribute" task focuses attention on the attribute that is measured by slope. An equivalence class of ratios is then constructed, typically by iterating and partitioning ratios. At this point students are better prepared to make sense of a line as the expression of the equivalence class of ratios and to link the equation $y = mx$ to its meaning in speed situations. Once the $y = mx$ case has been explored, situations represented by $y = mx + b$ are considered.

Revisiting the Calculus Vignette

There are three major implications of the research on actor-oriented transfer for the calculus vignette. The first implication involves the assumptions made regarding surface and structural features of transfer tasks. The teacher assumed that the classroom tasks and the photography task shared the same structure. All were situations in which nearly linear data were produced. The teacher apparently conceived of the photography situation as representing a surface feature, given that students likely had experiences with scaling in secondary school and given that photography is a part of everyday experience for most students. However, it is very likely that the photography situation presented structural complexities for students just as the wheelchair ramp and playground slide situations presented substantive mathematical challenges for the students in the high school studies. For example, the line of best fit directs students to a two-dimensional plane, but there is a three-dimensionality lurking in the perspective of the photograph with which students may be struggling. Students have likely had many everyday experiences with photographs like that shown in Figure 9, in which the photographic measurement of the flowers exceeds that of the mountains. Thus, depending upon the situation, a photo may not be approximately linear in terms of the relationship between d and p. In other words, the photo obscures the projective geometry operative in the problem. Not much information was given in the word problem that addressed the relative placement of the objects in the photograph or that addressed other, presumably, "surface" features.

Figure 9. Photograph illustrating the lack of proportionality between photographic and actual measurements of objects.

A second implication from the actor-oriented perspective is that one should not take poor performance on transfer tasks as an indication of the failure of students to generalize their learning experiences. Indeed transfer is in the eye of the beholder. When we adopt a classical transfer perspective, we are unlikely to find overwhelming evidence for transfer, since we already know that novices do not make the same set of connections as experts. However, if transfer is examined from an actor-oriented perspective, then results will likely demonstrate conceptually substantive ways in which students make sense of new experiences in terms of past experiences. Rather than assuming that the calculus students weren't generalizing, a teacher could ask the following questions to explore the nature of students' generalizations:

1) What conditions would need to be met in order for you to use the least squares approach and the line of best fit?

2) Could the question in the photograph situation be modified in order for you to think that the technique we learned today would be useful?

3) What's different about the photography situation than the situations you saw in class?

In fact, the teacher reported that he gained some important information about students' generalizations when he began to present the solution to the photography problem. As he graphed the data points (shown in Figure 1), many students appeared to know how to make use of the least squares method. In other words, the students' generalizing activity appeared to be modulated by the appearance of an approximately linear scatter plot of data. Learning about this generalization is important and suggests some revisions that can be made to the instructional approach, as I will articulate below.

The third implication is that instructional environments can unwittingly focus students' attention on features that lead to undesirable generalizations. In the high school study, the instruction appeared to unintentionally direct students' attention toward the vertical and horizontal components of stair steps without paying adequate attention to the relationship between steepness and the corresponding changes in vertical and horizontal components of the given object (slide, roof, ramp, etc). In the calculus example, it appears that the instruction drew attention toward the concept of linearity as the existence of a straight line on a graph, which is a limited notion of linearity. For example, the worked classroom example with the doublings (see Figures 1 and 2) enabled students to decide that the relationship was linear by determining that a line approximately covered the data without considering whether it makes sense to form a multiplicative (as opposed to an additive) comparison between the two quantities in the situation.

The research studies of students' comprehension of typical transfer situations for slope indicated that meaning-making in these situations involves reasoning with measurable attributes (e.g., sorting out the relationships between a large number of quantities in the wheelchair ramp situation such as the length of the platform, the height of the ramp, work required to climb, the height of the ramp, steepness, and so on). Similarly there are important relationships among measurable attributes in the photography situation that need to be addressed. We want students to consider whether it makes sense in the context of photography to form a multiplicative comparison between p and d. In fact, this only makes sense if the objects are flat or nearly flat. Students need to discuss the notion of visual perspective and sources of error. Students would also benefit from a discussion of linearity as the formation of a multiplicative relationship between two quantities (in the $y = mx$ case) and as a multiplicative relationship between the corresponding changes between two quantities (in the $y = mx + b$ case). Decisions of the appropriateness of the formation of multiplicative relationships need to be made without necessarily relying on the visual appearance of graphs.

Concluding Remarks

Very often the educational research considered to be most practical is that which demonstrates how well a particular curriculum, tool, or teaching method works. Indeed, "what works" research can be valuable in terms of informing decisions such as whether to adopt particular curricula, technology, or pedagogical methods. This chapter demonstrated how research that prompts a rethinking of one's (often implicit) assumptions about transfer can also lead to improvements in teaching and learning. Specifically, this chapter illustrated how adopting the actor-oriented transfer perspective can affect teaching practices by suggesting alternative ways for teachers to interpret students' work, conceive of the mathematical content differently, and create new instructional activities.

The actor-oriented transfer perspective is useful at any level and for any mathematical topic. This point was made, in part, by including a range of topics and instructional levels in this chapter—from slope at the secondary school level to linear regression at the undergraduate level. Additionally, my colleagues and I are finding the actor-oriented transfer perspective useful as we move from an investigation of linear functions to quadratic functions at the high school level. For example, Ellis & Grinstead (2007) found that two-thirds of the students they interviewed from a high-school intermediate algebra class identified the parameter a in $y = ax^2 + bx + c$ as the "slope" of a parabola in a series of novel "transfer" tasks. By analyzing the classroom data from an actor-oriented transfer perspective, they identified three focusing phenomena in the classroom environment that inadvertently supported a focus on "slope-like" properties of quadratic functions: a) the use of linear analogies, b) repeated use of the rise over run method, and c) viewing a as dynamic rather than static. Their results transcended a previously reported tendency of students to inappropriately draw on their linear-function understanding to try to solve tasks for nonlinear functions (Hershkowitz & Schwarz, 1997; Zaslavsky, 1997), by demonstrating particular ways in which students, textbook materials, and teacher actions interacted to create inappropriate connections between the meaning of a and the notion of slope for linear functions. As was the case in our earlier work with linear functions, their examination of transfer from an actor-oriented perspective informs the interpretation of students' work, a different conceptualization of the mathematical content, and new instructional activities for subsequent teaching experiments.

Acknowledgments This research was supported in part by the National Science Foundation under grant REC-9733942. The teaching experiment discussed in the chapter was conducted in a computer laboratory funded by the

National Science Foundation under grant DUE-9751212. The views expressed do not necessarily reflect official positions of the Foundation. I thank Anthony E. Kelly for his thoughtful comments on a draft of this chapter. I am also grateful to the editors of this volume, Chris Rasmussen and Marilyn Carlson, for their excellent feedback and guidance in revising the manuscript.

References

Anderson, J. R., Corbett, A. T., Koedinger, K. R., & Pelletier, R. (1995). Cognitive tutors: Lessons learned. *The Journal of the Learning Sciences, 4(*2), 167–207.

Barnett, S., & Ceci, S. J. (2002). When and where do we apply what we learn? A taxonomy for far transfer. *Psychological Bulletin, 128*(4) 612–637.

Beach, K. D. (1999). Consequential transitions: A sociocultural expedition beyond transfer in education. In A. Iran-Nejad & P. D. Pearson (Eds.), *Review of Research in Education,* (Vol. 26, pp. 101–140). Washington, DC: American Educational Research Association.

Bransford, J. D., & Schwartz, D.L. (1999). Rethinking transfer: A simple proposal with multiple implications. *Review of Research in Education, 24,* 61–100.

Carraher, D., & Schliemann, A. D. (2002). The transfer dilemma. *The Journal of the Learning Sciences, 11*(1), 1–24).

Cobb, P., & Bowers, J. (1999). Cognitive and situated learning perspectives in theory and practice. *Educational Researcher, 28*(2), 4–15.

Detterman, D. K. (1993). The case for the prosecution: Transfer as an epiphenomenon. In D.K. Detterman & R.J. Sternberg (Eds.), *Transfer on trial: Intelligence, cognition, and instruction* (pp. 1–24). Norwood, NJ: Ablex.

Ellis, A. B., & Grinstead, P. (2007). Hidden lessons: How a focus on slope-like properties of quadratic functions encouraged unexpected generalizations. Manuscript under review.

Ellis, A. B., & Lobato, J. (2004, April). Using the construct of "focusing phenomena" to explore links between attentional processes and "transfer" in mathematics classrooms. In J. Lobato (Chair), *Attentional processes, salience, and "transfer" of learning: Perspectives from neuroscience, cognitive science, and mathematics education.* Symposium conducted at the annual meeting of the American Educational Research Association, San Diego, CA.

Fuchs, L. S., Fuchs, D., Prentice, K., Burch, M., Hamlett, C. L., Owen, R., Hosp, M., & Jancek, D. (2003). Explicitly teaching for transfer: Effects on third-grade students' mathematical problem solving. *Journal of Educational Psychology, 95(2),* 293–305.

Gentner, D. (1989). The mechanisms of analogical learning. In S. Vosniadou & A. Ortony (Eds.), *Similarity and analogical reasoning* (pp. 199–241). Cambridge, England: Cambridge University Press.

Greeno, J. G. (1997). Response: On claims that answer the wrong questions. *Educational Researcher, 26*(1), 5–7.

Hershkowitz, R., & Schwarz, B.B. (1997). Unifying cognitive and sociocultural aspects in research on learning the function concept. In E. Pehkkonen (Ed.), *Proceedings of the 21st International Conference for the Psychology of Mathematics Education* (Vol. 1, pp. 148–164). Lahti, Finland: University of Helsinki.

Kirshner, D., & Whitson, J. A. (1997). Editors' introduction. In D. Kirshner, & J. A. Whitson (Eds.), *Situated cognition: Social, semiotic, and psychological perspectives* (pp. 1–16). Mahwah, NJ: Erlbaum.

Lave, J. (1988). Cognition in practice: Mind, mathematics, and culture in every day life. New York: Cambridge University Press.

Lobato, J. (1996). Transfer reconceived: How "sameness" is produced in mathematical activity. (Doctoral dissertation, University of California, Berkeley, 1996). *Dissertation Abstracts International*, AAT 9723086.

Lobato, J. (2003). How design experiments can inform a rethinking of transfer and vice versa. *Educational Researcher, 32*(1), 17–20.

Lobato, J. (2005, August). Attention-focusing and the "transfer" of learning. In J. Emanuelsson & M. F. Pang (Chairs), *Contrasting different perspectives on the object of learners' attention.* An invited SIG symposium conducted at the 11th Biennial Conference of the European Association for Research on Learning and Instruction (EARLI), Nicosia, Cyprus.

Lobato, J. (2006). Alternative perspectives on the transfer of learning: History, issues, and challenges for future research. *Journal of the Learning Sciences, 15*(4), 431–449.

Lobato, J. (in press). Research methods for alternative approaches to transfer: Implications for design experiments. In A. Kelly, R. Lesh, & J. Baek (Eds.), *Design research in education.* Mahwah, NJ: Erlbaum.

Lobato, J., Clarke, D., & Ellis, A. (2005). Initiating and eliciting in teaching: A reformulation of telling. *Journal for Research in Mathematics Education, 36*(2), 101–136.

Lobato, J., & Ellis, A.B. (2002). The focusing effect of technology: Implications for teacher education. *Journal of Technology and Teacher Education, 10*(2), 297–314.

Lobato, J., Ellis, A. B., & Munoz, R. (2003). How "focusing phenomena" in the instructional environment afford students' generalizations. *Mathematical Thinking and Learning, 5*(1), 1–36.

Lobato, J., & Siebert, D. (2002). Quantitative reasoning in a reconceived view of transfer. *The Journal of Mathematical Behavior, 21*(1), 87–116.

Lobato, J., & Thanheiser, E. (1999). Re-thinking slope from quantitative and phenomenological perspectives. In F. Hitt, & M. Santos (Eds.), *Proceedings of the Twenty-first Annual Meeting of the North American Chapter of the International Group for the Psychology of Mathematics Education* (Vol. 1, pp. 291–297). Columbus, OH: ERIC.

Lobato, J., & Thanheiser, E. (2002). Developing understanding of ratio as measure as a foundation for slope. In B. Litwiller (Ed.) *Making sense of fractions, ratios, and proportions: 2002 Yearbook* (pp. 162–175). Reston, VA: National Council of Teachers of Mathematics.

MacKay, D. (1969). *Information, mechanism, and meaning.* Cambridge, MA: MIT Press.

Marton, F. & Tsui, A.B.M. (2003). *Classroom discourse and the space of Learning.* Mahwah, NJ: Lawrence Erlbaum Associates.

McKeough, A., Lupart, L., & Marini, A. (Eds.). (1995). *Teaching for transfer: Fostering generalization in learning.* Mahwah, NJ: Erlbaum.

National Research Council, Committee on Developments in the Science of Learning. (2000). Learning and transfer. In J. D. Bransford, A. L. Brown, & R. R. Cocking (Eds.), *How people learn: Brain, mind, experience, and school* (Exp. ed., pp. 51–78). Washington, DC: National Academy Press.

Olive, J., & Lobato, J. (in press). The learning of rational number concepts using technology. In M. K. Heid & G. Blume (Eds.), *Research on technology and the teaching and learning of mathematics.* Greenwich, CT: Information Age Publishing, Inc.

Perkins, D. N., & Salomon, G. (1989). Are cognitive skills context-bound? *Educational Leadership, 18*(1), 16–25.

Roschelle, J., & Kaput, J. (1996). SimCalc *MathWorlds* for the mathematics of change. *Communications of the ACM, 39*(8), 97–99.

Schoenfeld, A. H., Smith, J. P., & Arcavi, A.A. (1996). Learning: The microgenetic analysis of one student's evolving understanding of a complex subject matter domain. In R. Glaser (Ed.), *Advances in instructional psychology, Volume 4* (pp. 55–175). Hillsdale, NJ: Erlbaum.

Schroeder, T. L., & Lester, F.K. (1989). Developing understanding in mathematics via problem solving. In P.R. Trafton (Ed.), *New directions for elementary school mathematics* (pp. 31–42). Reston, VA: National Council of Teachers of Mathematics.

Simon, M., & Blume, G. (1994). Mathematical modeling as a component of understanding ratio-as-measure: A study of prospective elementary teachers. *The Journal of Mathematical Behavior, 13*, 183–197.

Thorndike, E. L. (1906). *Principles of teaching.* New York: A. G. Seiler.

Zaslavsky, O. (1997). Conceptual obstacles in the learning of quadratic functions. *Focus on Learning Problems in Mathematics*, *19*(1), 20–44.

23

How do Mathematicians Learn to Teach?
Implications from Research on Teachers and Teaching for
Graduate Student Professional Development

Natasha Speer, *Michigan State University*
Ole Hald, *University of California at Berkeley*

Scenario 1[1] At a pre-semester orientation session for mathematics graduate students, the following question and student work were presented:

$$\text{If } y = (3 - x^2)^3, \text{ what is } \frac{dy}{dx}?$$

$$\text{Student work: } \frac{dy}{dx} = 3(3 - x^2)^2 \cdot (-2x) \cdot (-2).$$

The graduate students were asked to describe what a student might have been thinking when producing such an answer. After a few moments, the question was repeated, but none of the graduate students offered a potential explanation. Then, a professor who was sitting in the room said, "Well, it's not a bad answer." He then explained how the student's answer showed a pretty solid understanding of the chain rule, but that the student had applied the rule repeatedly instead of just the one time required.

Why could the professor explain what the student had done but the graduate students were unable to do so? Does the professor know more about the chain rule than the graduate students? Will the graduate students know more about the chain rule when they finish their degrees and then be able to make sense of students' answers in the way the professor was able to? Or does the professor know other things from years of experience?

Scenario 2 A graduate student was grading calculus exams. In order to calculate the area between two curves, one problem required that students solve $3y + 4 = 6x - 8$ for y. Once the equation was transformed into y as a function of x, the problem could be solved by taking the appropriate difference and computing the integral. The graduate student showed the course professor one student's work:

$$3y + 4 = 6x - 8 - 4 = 6x - 12 = \frac{6x - 12}{3} = 2x - 4$$

The graduate student asked how many points to deduct since the student, "had not taken the time to write out each step on a separate line and the resulting solution was sloppy."

The professor examined the work and said, "The student wasn't sloppy, she's just confused about the equals sign. She probably learned one thing in elementary school and didn't notice that things changed when she got to algebra."

---

[1] These scenarios are taken from the experiences of the first author.

Is the professor's ability to make sense of the student's work a function of the amount of mathematics known? Is there a course that the graduate student could take that would teach the things about the equals sign that the professor knows?

The goal of this chapter is to present findings from research on teachers and teacher professional development to help people think about how to prepare graduate students to teach. In particular, we examine the role that teachers' knowledge plays in shaping instructional practices and student outcomes. The claim that "what students learn is a function of what their teachers know" turns out to be difficult to establish empirically. From research that has been accumulated in this area, it appears that a subset of "knowledge" has more explanatory power. We review findings from research on teachers' knowledge as well as from successful teacher professional development programs and we discuss potential implications of these findings for professional development of mathematics graduate students.

As the other chapters of this volume demonstrate, there is a substantial body of literature about student learning of college mathematics. Less research exists on the teaching of college mathematics. In the history of educational research, studies of teachers and teaching have typically lagged behind in number compared with those of student learning. Research at the K–12 levels has a long history and has resulted in significant findings about teachers and teaching. At this stage of development in research on undergraduate mathematics teaching (where research is relatively scarce), it may be productive to look to areas of K–12 research on teaching to inform peoples' thinking about how best to assist graduate students as they learn to teach.

There is another reason why research on K–12 teachers and teaching is relevant at the undergraduate level: a significant part of the mathematics taught in colleges (e.g., college algebra, pre-calculus, calculus) is also taught in high school. While there are certainly factors unique to teaching this content at the college level, it is reasonable to assume that research on teaching this content has relevance to undergraduate teaching. We seek to draw from research at K–12 levels to inform work at the undergraduate level—both in terms of practices (preparation for future faculty) and directions for future research. Of course, for graduate students to benefit from what is known about factors that shape teachers' practices (and their students' learning) there need to be opportunities for them to participate in professional development programs and activities. The hope is that the remainder of this chapter provides both an argument for the importance of graduate student professional development as well as suggestions for those who have the responsibility for providing that professional development.

What Makes Someone A "Good" Teacher?

For a long time, educational researchers have asked, "What makes someone a good teacher?" While much is known about what teachers do and why, connecting teachers' traits to students' learning is a complex task and no simple answers have emerged (Darling-Hammond, 1999; Mewborn, 2003). Traits examined in the history of educational research include teachers' general academic ability, subject matter knowledge, and teaching skills. Research on professors and graduate students is sparse in comparison, but similar issues are starting to be examined (Speer, Gutmann, & Murphy, 2005).

In this chapter, we chose to focus on knowledge for three reasons. First, it seems obvious that what teachers know should determine, at least in part, what their students learn. Second, teachers' knowledge has been the subject of much investigation and there have recently been significant findings in this area. Lastly, some of the findings from K–12 professional development programs that focus on aspects of knowledge seem especially well-suited to adaptation for the professional development of mathematics graduate students.

Does What You Know Determine What Your Students Learn?

While it is natural to assume that what students learn is influenced by what their mathematics teachers know, finding empirical support for this claim has been extremely difficult (Ball, Lubienski, & Mewborn, 2001). Such research has examined the *amount* of mathematics that teachers know (as measured by number of courses taken, number of credits earned, etc.) as a measure of teachers' mathematical knowledge. If mathematical knowledge is measured in these ways, there is no definitive relationship between teachers' knowledge and their students' learning of mathematics.

One of the most-cited studies in this area is from Begle (1979). He reported findings from a meta-analysis of studies conducted from 1960 through the mid-1970s that examined effects of teacher traits on student performance.

His meta-analysis found no significant positive relationship between schoolteachers' highest academic degree, post-bachelor's course credits, or majoring in mathematics, and student achievement. If only courses beyond calculus are considered, only 10% of the time did teachers' taking such courses produce greater student performance. Even more stunning was his finding that about 8% of the time, having more courses post-calculus led to lower levels of student achievement.

More recently, however, Monk (1994) found "positive relationships between the number of undergraduate mathematics courses in a teacher's background and improvement in students' performance in mathematics" (p. 130). The courses under consideration in this study were those at the sophomore and junior levels. This encouraging picture is tempered a bit by specifics of the findings: taking an additional mathematics course translates into, at most, a 1.2% increase in student performance (0.2% for sophomores). It is also important to note that increases of this size are only apparent for the first five undergraduate mathematics courses teachers take—taking additional courses beyond five is associated only with a 0.2% gain in student performance.

These counter-intuitive findings prompted the educational research community to seek other measures of teacher knowledge that correlate with student learning. After all, finding such correlations could inform the design of teacher preparation and professional development programs with some confidence that such programs would increase student achievement. This has resulted in two lines of research, both of which are refinements to more traditional definitions of mathematics content knowledge. One line of research expanded the scope of what is taken as knowledge, while the other proposes a particular kind of content knowledge that is closely connected to teaching. Both of these areas of research are discussed below.

Different kinds of knowledge

Several alternative ways of characterizing knowledge appear to explain more about student learning opportunities than the measures used in the research summarized in the previous section. Two of these refinements to examinations of knowledge are: pedagogical content knowledge (PCK) and mathematical knowledge tied specifically to teaching. For more extensive reviews of research on teachers' knowledge see, (Ball, Lubienski, & Mewborn, 2001; Calderhead, 1996; Mewborn, 2003). In this section, these two kinds of knowledge are described and research about the roles of this knowledge in teaching and student learning are discussed.

Pedagogical content knowledge overview

For much of the history of mathematics education research, two areas of knowledge were examined: knowledge of subject matter and knowledge of pedagogy. As noted above, however, standard measures of knowledge do not have much predictive power for how effective teachers will be. Neither do the myriad measures of teachers' general pedagogical knowledge and skills (classroom management, organization, etc.).

In the mid-1980s, researchers identified another kind of knowledge possessed by teachers. This includes much of what teachers draw on while teaching, planning for teaching, and making sense of student thinking. In particular, it includes information about typical student difficulties, typical ways in which students approach particular tasks (both unsuccessfully and successfully), examples that are especially illuminating of the ideas, etc.

This kind of knowledge is referred to as *pedagogical content knowledge* (Grossman, Wilson, & Shulman, 1989; Shulman, 1986) and combines subject-matter knowledge and knowledge about teaching that subject matter. In mathematics, for example, there are things that teachers know that are specific to particular topics (e.g., typical errors students make when they first learn the quotient rule in calculus; common misunderstandings of the definition of limit) that are not part of what they were taught in mathematics courses. Pedagogical content knowledge is in large part what enables teachers to understand why what students write (or say) *makes sense to them*, even if it is not correct.

Pedagogical content knowledge in undergraduate teaching and learning

In the first vignette of this chapter, the professor was able to make sense of the student's work on the derivative problem because he knew that students sometimes "over generalize" and carry out the chain rule process on expressions more times than is appropriate. To the student, applying the chain rule in this repeated fashion may make sense—it may just be that the student has not yet developed the ability to distinguish between situations where it applies and where it does not. In such situations, students are making an effort to apply what they know and the work they produce provides clues to what they do (and do not) understand. You will not find this "over generalizing" difficulty described in any calculus textbook. Knowing this about student thinking is not knowledge of the chain rule (although having knowledge of the chain rule may play a role in how the professor responds to the student's error).

In the second vignette we saw a graduate student and a professor confronting a student's algebraic reasoning while solving part of a calculus problem. The student produced the following string of computations:

$$3y + 4 = 6x - 8 - 4 = 6x - 12 = \frac{6x - 12}{3} = 2x - 4$$

when the task required solving for y in $3y + 4 = 6x - 8$. Instead of beginning with the original equation and writing the result of the transformation of each line separately (thereby maintaining the equality), the student had created a trace of only the transformations of the right-hand side of the equation. This gives the correct expression for y in the final term, but creates a string of expressions that are not equal to one another.

Why do students produce work like this? Are they just failing to be careful and systematic with their work? Do such errors arise just because students are trying to take a short cut by not writing down each step? This kind of error is actually fairly common and the result of how students interpret the use of the equals sign from their study of arithmetic and/or their weak understanding of how the equals sign is used in algebra. For most of their elementary schooling, students encountered problems such as $5 + 7 = \_\_$. In this example, and most others involving basic arithmetic, students are asked to perform a calculation and put the answer to the right of the equals sign. Repeated exposure to these kinds of tasks instills in students the belief that the equals sign is a "do something" operator (Behr, Erlwanger, & Nichols, 1980; Kieran, 1981; Saenz-Ludlow & Walgamuth, 1998).

When students encounter expressions such as $3 + x$ in algebra they often have difficulty leaving them as is and feel compelled to "do something," resulting in expressions such as $3 + x = 3x$. If students do not develop a different view of the equal sign, when they are faced with solving equations for a particular variable, they may manipulate expressions in ways that violate the equality. In the example from the vignette, the student understands the processes for solving the equation for y, but is carrying out those processes only on the right-hand side of the equation. At each step, the student is "doing something" to the expression, without regard to how what they are doing alters the equality of the expressions.

Knowing that the equals sign is treated in a particular way in elementary arithmetic and that some students carry that understanding into their study of algebra and beyond is an example of pedagogical content knowledge. This is knowledge that is both about mathematics (in this case, about equality and algebraic transformations) and about student understanding of that mathematics (including errors that are symptoms of students' under-developed knowledge of the equals sign).

There are other examples of pedagogical content knowledge (PCK) possessed by professors. Similar to the chain rule example, "over generalizations" happen when students apply L'Hopital's Rule multiple times as they evaluate limit problems. Although the rule states that it is only applicable in situations that meet certain criteria (e.g., the expression being evaluated must be a quotient of some indeterminant form), students sometimes keep applying the rule even after the expression they have obtained is no longer an indeterminant form. This kind of thinking results in chains of expressions such as:

$$\lim_{x \to 0} \frac{(1 - \cos x)}{x^3} = \lim_{x \to 0} \frac{\sin x}{3x^2} = \lim_{x \to 0} \frac{\cos x}{6x} = \lim_{x \to 0} \frac{-\sin x}{6} = 0.$$

The source of these kinds of errors is different than the source of other errors (e.g., "simplifying" $\frac{\infty}{\infty}$ and getting 1, evaluating $\frac{0}{0}$ as 0, etc.) and being able to distinguish among these types of errors draws on a professor's pedagogical content knowledge.

Other examples of pedagogical content knowledge include: knowing that when students are first learning to construct delta-epsilon proofs for limits, they are likely to have difficulty understanding and producing expressions that involve absolute values (e.g., $|x - 3| < \varepsilon$); knowing that students will have difficulty understanding the relationship between a function being continuous and being differentiable (mixing up "continuous functions are differentiable" with "differentiable functions are continuous"); knowing that errors that students make when using partial fractions to evaluate integrals may come more from their skills with fractions than from their understanding of integration; etc.

Research on pedagogical content knowledge

In the time since the specialized knowledge described above was named "pedagogical content knowledge," researchers have sought to document the extent to which teachers possess such knowledge and how that knowledge relates to their teaching practices and their students' learning. A particular generative line of investigations has come from researchers associated with a project called "Cognitively Guided Instruction." Research has included investigations of elementary school teachers' knowledge of how students think about particular content and examinations of the role such knowledge plays in teachers' instructional practices (Carpenter, Fennema, Peterson, & Carey, 1988; Carpenter, Fennema, Peterson, Chiang, & Loef, 1989). Researchers have developed instruments to evaluate the extent of teachers' knowledge of student difficulties and student strategies. It has also been possible to assess the extent to which teachers use their knowledge of student thinking as they teach.

In conjunction with such work, professional development that focuses on developing teachers' knowledge of student thinking (described in more detail in a later section) appears to be a promising approach to improving teaching (Carpenter, Fennema, Peterson, Chiang, & Loef, 1989; Fennema et al., 1996; Fennema & Scott Nelson, 1997). In short, research findings demonstrate that it is possible to help teachers increase the depth and breadth of their knowledge of student thinking and that such changes in knowledge can be linked to positive changes in teaching practices. These changes create more opportunities for students to think about and understand mathematical ideas.

Researchers have also taken this line of work one step farther and examined the consequences of such changes in teachers' practice on student achievement. It appears that the more use that teachers make of their knowledge of student thinking while teaching, the more mathematics their students will learn (Fennema et al., 1996).

Similar lines of work are just beginning to appear at the college level. While this research is relatively sparse, researchers have begun to examine mathematics graduate students' knowledge of student thinking for some key concepts from calculus (e.g., limit, derivative). Findings indicate that graduate students do not necessarily have extensive knowledge of student strategies and difficulties for these topics (Speer, Strickland, & Johnson, 2005) but that it is possible for graduate students to develop rich and detailed understanding of student thinking (Kung, in press).

In the next section, a second kind of knowledge is described and research is discussed that examines this kind of knowledge in teachers and its role in students' learning.

Mathematical knowledge for teaching overview

In addition to pedagogical content knowledge, researchers have proposed other refinements to the basic categorizations of knowledge. In particular, researchers have found evidence that there are *certain kinds of knowledge of mathematics* that play important roles in how teachers teach.

This type of knowledge is distinct from pedagogical content knowledge discussed above: it includes knowledge of mathematics content and is not inherently related to student learning or understanding. This knowledge may come, in part, from the special kind of mathematical work that teachers engage in—a kind of work in which those who use mathematics outside of teaching are unlikely to engage. This kind of knowledge, as well as how knowledge is used in teaching, has received considerable attention in both the K–12 and undergraduate mathematics education community in recent years (Ball & Bass, 2000; Ferrini-Mundy, Burrill, Floden, & Sandow, 2003; Hill, Rowan, & Ball, 2004, 2005; Hill, Schilling, & Ball, 2004; Ma, 1999).

Ma (1999) provided a window into the rich and connected knowledge of mathematics possessed by some school teachers by investigating the extent to which teachers had knowledge of the complex relationships among mathematical ideas that arise in elementary mathematics teaching. By comparing how Chinese and U.S. teachers responded to mathematics tasks and teaching-related questions associated with those tasks, she highlighted the depth and breadth of mathematical knowledge that teachers can bring to their work.

Ma investigated teachers' knowledge of mathematics in certain domains (subtraction, multiplication, fractions, area and perimeter) and teachers' knowledge that is linked to the teaching of topics in those domains. For example, teachers in her research completed division of fractions tasks and then constructed word problem examples to reflect particular division of fractions computations. While most teachers were able to carry out the computations accurately, some struggled to generate a word problem that correctly represented the division of fractions process. Teachers who were successful in creating a word problem that modeled the mathematics displayed a type of knowledge of fractions and division that is distinct from what is typically gained through regular schooling (at least in the U.S.). Such teachers also knew alternative strategies for solving division of fractions tasks that could be used to explain the ideas to students and could be used as a basis for the design of word problems.

These findings demonstrate that computational fluency is not necessarily an indicator of deep understanding of mathematical processes. While this finding is not unique, what Ma's research also showed was that teachers make use of a *particular kind of knowledge of mathematics* when they engage in the work of teaching. Moreover, this kind of knowledge for teaching is not necessarily a natural and automatic by-product of knowledge of mathematical content. Ma concludes that having such knowledge enables teachers to make sense of student thinking and contributes to the learning opportunities that teachers can create for their students.

Mathematical knowledge for teaching in undergraduate teaching and learning

To date, analogous studies have not been conducted with people who teach undergraduate mathematics. It remains to be seen what kinds of specialized knowledge of mathematics graduate students and professors develop as a result of interpreting students' ideas and engaging in other teaching-related mathematical activities. The next section describes research on these issues for K–12 teachers.

Research on mathematical knowledge for teaching

Complementary to the line of work (described above) that identified particular examples of knowledge connected to teaching, others have extended this research by identifying other examples of knowledge for teaching, creating assessments of that knowledge, and examining connections between having such knowledge and student achievement (Hill, Rowan, & Ball, 2005; Hill, Schilling, & Ball, 2004). It appears that it is possible to create assessment items that tap into knowledge that is particular to teaching—knowledge that people with extensive non-teaching backgrounds in mathematics are unlikely to possess. In this work, a distinction is made between content knowledge and knowledge particular to teaching. This distinction has been described by Hill, Schilling, & Ball (2004) as follows:

> One way to illustrate this distinction is by theorizing about how someone who has not taught children but who is otherwise knowledgeable in mathematics might interpret and respond to these items. This test population would not find the items which tap ordinary subject matter knowledge difficult. By contrast, however, these mathematics experts might be surprised, slowed, or even halted by the mathematics-as-used-in-teaching-items; they would not have had access to or experience with opportunities to see, learn about, unpack and understand mathematics as it is used at the elementary level. (p. 16)

The following sample (from Hill, Rowan, & Ball, 2005) illustrates the kind of assessment item generated in this research program:

Imagine that you are working with your class on multiplying large numbers. Among your students' papers, you notice that some have displayed their work in the following ways:

Student A	Student B	Student C
35	35	35
× 25	× 25	× 25
125	175	25
+ 750	+ 700	150
875	875	100
		+600
		875

Which of these students would you judge to be using a method that could be used to multiply any two whole numbers?

	Method would work for all whole numbers	Method would NOT work for all whole numbers	I'm not sure
a) Method A	_____	_____	_____
b) Method B	_____	_____	_____
c) Method C	_____	_____	_____

The researchers contend that, "To respond in such situations, teachers must draw on mathematical knowledge: inspecting the steps shown in each example to determine what was done, then gauging whether or not this constitutes "a method," and, if so, determining whether it makes sense and whether it works in general" (p. 388). They go on to say that doing so is not common work for adults who do not teach, however, this work "is entirely mathematical, not pedagogical; to make sound pedagogical decisions teachers must be able to size up and evaluate the mathematics of these alternatives" (p. 388).

Similar work is also being conducted specifically in the domain of algebra. Researchers are developing assessment items and frameworks for analyzing the nature of the knowledge that is utilized in the teaching of algebra (Ferrini-Mundy, Burrill, Floden, & Sandow, 2003).

In conjunction with some of the projects described above, researchers have looked for connections between teachers possessing such knowledge and the learning of their students. Findings from this work indicate that, "teachers' content knowledge in mathematics, as measured by items designed to be close to the content and its uses that teachers deploy, is positively related to student achievement" (Hill, Rowan, & Ball, 2004, p. 35). Findings from the same study also, "offer evidence that such effects are due to more than general intelligence, or mere course-taking" (p. 35).

How do Teachers Acquire Knowledge for Teaching?

The research programs involving experienced K–12 teachers are just beginning. Among the long-term goals are: figuring out how teachers develop this kind of knowledge, and figuring out how to best support teachers in the acquisition of this knowledge. Very little is known about precisely how this knowledge is acquired, but we can speculate about what is not the source: formal course work in mathematics. This type of content is not part of the curriculum and so it is likely that it is acquired through on-the-job learning.

This section contains descriptions of research on professional development for teachers and explores the question of how teachers develop knowledge for teaching. With that as background, the remainder of the chapter includes a discussion of the issue of what these research findings might mean for the professional development of people who teach college-level mathematics.

For elementary and high school teachers, preparation for teaching consists of a mixture of mathematics content courses, teaching methods courses, and supervised experiences teaching. Each of these elements of teacher preparation contributes to the learning opportunities that teachers are able to create for their students. Content knowledge is obtained from mathematics courses, teaching methods courses expose teachers to some pedagogical content knowledge, and additional knowledge develops during experiences teachers have in conjunction with their preparation programs. When teachers emerge from preparation programs, however, there is still a great deal to be learned.

Much of that learning occurs as teachers plan and carry out lessons, interact with students, and examine students' homework and exams. Teachers develop extensive mental catalogs of the difficulties that students have, of errors that occur during particular chapters, and of examples that are especially illuminating for certain topics. This learning in the context of teaching is a major way in which teachers develop pedagogical content knowledge and mathematical knowledge for teaching.

For mathematics professors, the process of becoming a teacher is different. Professors do not participate in preparation programs and do not typically take courses in education as part of their graduate schooling. Some professors, however, have supervised teaching experiences while in graduate school, often as a teaching assistant. It is during these experiences (teaching discussion sections, possibly teaching full courses, talking with students during office hours, grading exams, etc.) that graduate students begin to develop the pedagogical content knowledge and pedagogically useful mathematical knowledge we saw in the examples above.

Some people have written about the preparation of university faculty to teach, but reports of *research* in this area are scarce. In most articles or volumes about teaching and learning at the university level, there is much information about how particular kinds of teaching take place and what the outcomes are for teachers and students (e.g., Holton, 2001, "The Teaching and Learning of Mathematics at University Level: An ICMI Study.") Such pieces about teaching, however, speak mostly about what teachers do or might do and how these practices relate to particular learning goals for students. Implicit in such work is the assertion that professors need to acquire the abilities and knowledge to carry out the particular kind of instruction described in the reports. It is possible to read articles and chapters about university teaching with an eye toward the kind of knowledge that might underlie the teaching practices being described or proposed, but such information is not typically an explicit part of such reports. For a particular interesting venue in which to attempt this exercise of inferring the knowledge needed to teach, see Mason (2001). In this chapter ("Mathematical teaching practices at tertiary level: Working group report,") working group members describe approaches to teaching that they value. Nearly all place particular emphasis on anticipating and finding out what students already know about a topic and organizing class activities in ways that provide students opportunities to learn mathematics and also provide the professor with a window into how students are thinking about the mathematical ideas in question.

Given the substantial demands on teachers' knowledge that teaching well entails, many programs exist that provide K–12 teachers with opportunities to acquire more knowledge after they begin their teaching careers. Similar opportunities for mathematics professors are scarce (NSF, 1992). In the next section, a particularly effective form of professional development for elementary schoolteachers is described and in a later section, we discuss how elements of this program might be incorporated into graduate student professional development.

Providing teachers with opportunities to develop knowledge

If factors other than traditional content knowledge and basic pedagogical skills influence what students learn, what kinds of efforts have been made to provide teachers with opportunities to learn these things? Educational research on pedagogical content knowledge has a longer history than analogous work on mathematical knowledge for teaching. As a result, programs for teachers and associated research are more extensively developed for pedagogical content knowledge. Among various aspects of PCK, knowledge of student understanding (including knowledge of how students typically understand particular concepts, how ideas from prior courses interact with their learning of new content, typical student strategies for solving problems, and common student difficulties/errors) has been the focus of some programs with effective results.

One of the best-documented and most extensively researched programs is Cognitively Guided Instruction (CGI). Teachers participating in this program acquire knowledge about student thinking and learning of particular mathematics topics. Teachers participate in workshops where mathematics educators present research on how students think about specific topics (e.g., addition, subtraction, etc.). Teachers also read reports of research about student strategies and common difficulties for particular kinds of problems. During the workshop, teachers also watch videos of students working on problems and being interviewed about how they are thinking about the problems. In addition, teachers investigate how students think about particular topics by conducting interviews and investigate how findings from educational research relate to what students did and said during the interviews. These activities give teachers opportunities to see and understand the (sometimes surprising) variety of ideas that students have about particular problems.

Researchers who study participants in these workshops have found that teachers develop richer knowledge of student thinking and they also change their teaching practices to include more requests for students to explain or justify their answers (Carpenter, Fennema, & Franke, 1996; Carpenter, Fennema, Peterson, & Carey, 1988; Carpenter, Fennema, Peterson, Chiang, & Loef, 1989). In addition, research on CGI has documented significant gains in student achievement as a result of this kind of professional development (Carpenter, Fennema, & Franke, 1996; Carpenter, Fennema, Peterson, & Carey, 1988; Carpenter, Fennema, Peterson, Chiang, & Loef, 1989; Fennema et al., 1996). Student problem solving and concept knowledge increased, and in many cases, overall student achievement improved by as much as one standard deviation. Unlike the research on teachers' content knowledge, these findings indicate a strong relationship between teachers' pedagogical content knowledge (in this case, knowledge of student thinking and learning) and their students' learning.

Themes in Research and Implications for Graduate Student Professional Development

In this section we discuss three themes evident in the research reviewed earlier in this chapter and describe how professional development for graduate students might be designed in light of these themes. We describe each theme, discuss what makes the theme suited to implementation with mathematics graduate students, and provide some specific suggestions for implementing professional development consistent with the theme.

This section is inherently speculative — research into the professional development of mathematics graduate students is scarce (Speer, Gutmann, & Murphy, 2005). What we do know, however, is that most graduate students receive preparation for their first job as a teacher. This may take the form of university-wide orientation sessions for graduate students or it may be a program designed specifically for mathematics. While these programs help graduate students adapt to their new role as a teacher, the emphasis is typically on administrative responsibilities and mechanical aspects of teaching (clear writing on the board, collecting homework efficiently, etc.). While these are certainly important aspects of learning to teach, there is a lot for graduate students to learn about how students think about mathematical ideas and most of that kind of learning happens once they set foot in the classroom and begin interacting with students and with the mathematics that their students produce. The suggested activities described below are modeled after elements of the professional development programs connected with the research on teachers' knowledge that was summarized earlier in this chapter. Using activities designed for K–12 teachers, however, is not always appropriate or feasible in the undergraduate teaching context. The suggestions below come from modifying such activities in ways to make them well-suited for graduate students and to take advantage of the rich learning environment that graduate students are in as they teach.

Focusing on students' thinking improves teachers' teaching practices

The first theme in the research discussed in this chapter is that knowing mathematics is a necessary but not sufficient condition for teachers to create good learning opportunities for their students. Findings indicate that students learn more when teachers have extensive knowledge of student strategies, student difficulties, and the mathematics tied to the teaching of particular ideas. Teachers develop this knowledge in various ways, but doing so requires focusing on issues of student thinking (in addition to other aspects of teaching).

As noted above, most pre-semester programs for mathematics graduate students do not focus on issues of student learning. This is, in some ways, very natural—graduate students are there to learn to teach and so time is spent discussing issues of teaching. This approach, however, may not help graduate students recognize the role that knowledge of student thinking plays in teaching and may not give them opportunities to begin to develop that knowledge.

Reasons why this is well-suited to the undergraduate mathematics context. There are several reasons why focusing on issues of student thinking might be a productive approach to professional development for graduate students. When designing professional development for K–12 teachers, one has to take into consideration the depth and breadth of the teachers' mathematical knowledge. Some K–12 teachers do not have as strong and deep knowledge of mathematical content as we might wish. As a result, preparation and professional development programs for K–12 teachers are often designed to address issues of mathematical content in addition to other issues (knowledge of student thinking, etc.). Professional development for graduate students, on the other hand, can take advantage of their extensive background and interests in mathematics. For example, some knowledge of student thinking can come from "unpacking" the mathematics contained in problems or topics. This unpacking involves thinking about all of the mathematical ideas that are connected to the particular problem. Then, one can envision a variety of ways someone might approach the problem if for some reason they chose to pursue a solution path that was not identical to the one first thought of by the graduate student.

Ways this can be accomplished. There are several ways that issues of student thinking could be added to or emphasized in professional development for graduate students.

- Talk about examples of student work. As part of graduate student instructor meetings, the professor who supervises those graduate students could orchestrate a discussion of examples of student work. These examples could relate to topics that everyone is going to be teaching soon. The professor (or experienced graduate students) could bring samples of work students produced when answering test or quiz problems on the topic. The discussion could be focused on trying to figure out why what the student did made sense to him or her and what the student might have been thinking as they worked on the problem. Faculty and experienced graduate students are apt to be able to provide quite a bit of insight into students' thinking. In addition to sharing their knowledge with less experienced graduate students, such discussions also may help new graduate students appreciate the role that knowledge of student thinking plays in teaching well. Professors and graduate students could build up libraries of these examples of student work for each course that could then be used in the future for discussions with new graduate students. Some examples can be found in the Boston College Case Studies materials (S. Friedberg et al., 2001a; S. Friedberg et al., 2001b).

- Use resources such as this volume. Graduate students and professors could read the chapter relevant to topics that are coming up in a course. Above, the suggestion was made that graduate students could examine examples of student work before teaching the related topics—in addition, graduate students can collect and examine student work from their students, read a summary of related research, and use the ideas from the research to analyze what their students did. As is done in the Cognitively Guided Instruction model, graduate students could also be given the task of interviewing a couple of students about how they solved a particular problem from class or a test. If there are several graduate students teaching the same course, each could select a different kind of problem and then report back to the group about what they learned about the students' difficulties and strategies. While this activity might be most meaningful if it used problems from the courses the graduate students are teaching, research articles on student thinking could also be used as a source for problems. In addition to this volume, the Research in Collegiate Mathematics Education (RCME) series provides collections of articles (Dubinsky, Schoenfeld, & Kaput, 1994; Dubinsky, Schoenfeld, & Kaput, 2000; Kaput, Schoenfeld, & Dubinsky, 1996; Schoenfeld, Kaput, & Dubinsky, 1998; Selden, Dubinsky, Harel, & Hitt, 2003).

Experienced Teachers Have Rich Knowledge of Student Thinking

Among the best resources for learning about student thinking are professors and advanced graduate students who have already acquired some of this knowledge. Providing TAs with access to the authentic practices of professors can give less experienced graduate student instructors ideas about what professors do when planning for teaching and examining student work.

Reasons why this is well-suited to the undergraduate mathematics context. For many graduate students, their teaching experiences include times when they are responsible for one section of a course or a recitation section associated with a lecture of a course. In many of these cases, there is a professor who supervises the graduate students who are all teaching some part of the same course. These professors have the responsibility for ensuring that various aspects of the course are coordinated among the graduate students (topics for the coming week, giving quizzes, grading exams, handling homework, etc.). To accomplish these things, often professors hold periodic meetings with the graduate students who are instructors for a course.

This kind of joint planning time is typically not found in K–12 teaching contexts, but having such time to discuss students and their thinking is often considered key to helping teachers improve their practice. In such meetings among graduate students and a professor, the focus is often on what is coming up next in the course—but such discussions could be focused on student thinking in addition to mechanics/administrative issues. Specific suggestions for how time during these meetings might be used are given below.

Ways this can be accomplished

- Have graduate students and professors discuss the examples that are used in particular sections of the textbook and what it is about those examples that makes them well-suited as an illustration of the particular ideas in the chapter.

- In staff meetings for large courses, have professors talk about planning for a particular lecture and how examples were selected. Discuss what specific ideas the examples are illustrating and/or how certain "classic" examples illustrate difficult ideas in particularly useful ways.

- Have graduate students write quizzes or problems and predict what students are likely to do with the problems and what difficulties they will have. Then talk with an experienced professor or more advanced graduate student to get feedback about other strategies or difficulties students might have to ensure that the problems are actually going to assess understanding of the ideas that the graduate student has in mind.

Some Teaching Practices Create Many Opportunities to Acquire Knowledge of Student Thinking

While graduate students and other teachers gain knowledge about student understanding in many ways, one of the most productive is from interactions with students when they are explaining their thinking. These interactions might take place in office hours, during a class discussion, or on paper when a question asks students to explain their reasoning. These practices are useful for student learning and they provide graduate students with authentic access to student thinking and learning.

Ways this can be accomplished. Particular approaches or models of instruction focus extensively on having students explain and justify their reasoning. At the undergraduate level, one of the most successful has been the Emerging Scholars Program (known by other names at some institutions). The Emerging Scholars Program (ESP) was designed based on research conducted on how successful students learn calculus (Fullilove & Treisman, 1990; Treisman, 1985). In ESP classes, students work in collaborative groups on challenging problems and the graduate student instructor assists students and moves from group to group asking them to describe how they arrived at their answers. This provides graduate students with especially extensive and rich access to student thinking. There is some evidence that such teaching experiences enable people to acquire considerable knowledge of student thinking in the course of their graduate school teaching careers (Kung, in press). For more information about ESP classes and their influence on students' learning of mathematics, see (Hsu, Murphy, & Treisman, this volume).

There are also small changes one can make to one's teaching to get more information about student thinking. The simplest is to ask questions that require students to provide an explanation for their answers. After a student responds to a question, if a teacher asks, "How did you get that answer?" not only will other students have access to the thinking behind an answer, but there is the possibility that the teacher will learn something new about how students think about the ideas or about the kinds of difficulties they have when solving such problems.

Concluding Thoughts

Given the relatively rapid increase in research on undergraduate mathematics education over the past few decades, it is likely that the future will generate a rich research base about how graduate students and professors learn to teach as well as insights into how best to support the development of their teaching practices. Currently most research activity is focused on students and how they think about and learn mathematics. The undergraduate mathematics education research community, however, is in the fortunate position of having the option to utilize research on student thinking in the professional development of teachers. Researchers of K–12 mathematics education have also amassed a base of research on student thinking and more recently have discovered ways to use that base of research in improving teaching practices. As described above, researchers have shown that knowing how students are apt to approach particular problems, what their difficulties are likely to be, and why specific ways of thinking make sense to students, are the types of knowledge that are likely to influence how students learn mathematics. Professional development programs that create opportunities for teachers' to enrich and expand their knowledge of student thinking have been successful in inducing changes in teaching practices that lead to increases in student learning.

As the other chapters in this volume demonstrate, much is known about how students think about and learn mathematics at the undergraduate level. Since this body of research exists (and is expanding), the undergraduate mathematics education community may very well already possess an important key to improving undergraduate education. The findings from K–12 research on teachers can play important roles in creating effective professional development for graduate students and professors. Questions of how best to adapt aspects of K–12 professional development for the undergraduate setting are among the issues to consider as more and more researchers take on the challenge of understanding how best to help people learn to be effective teachers of undergraduate mathematics.

References

Ball, D. L., & Bass, H. (2000). Interweaving content and pedagogy in teaching and learning to teach: Knowing and using mathematics. In J. Boaler (Ed.), *Multiple perspectives on the teaching and learning of mathematics.* Westport, CT: Ablex.

Ball, D. L., Lubienski, S., & Mewborn, D. S. (2001). Research on teaching mathematics: The unsolved problem of teachers' mathematical knowledge. In V. Richardson (Ed.), *Handbook of Research on Teaching* (pp. 433–456). Washington, DC: American Educational Research Association.

Begle, E. G. (1979). *Critical variables in mathematics education: Findings from a survey of the empirical literature.* Washington, DC: Mathematical Association of America and the National Council of Teachers of Mathematics.

Behr, M., Erlwanger, S., & Nichols, E. (1980). How children view the equals sign. *Mathematics Teaching, 92,* 13–15.

Calderhead, J. (1996). Teachers: Beliefs and knowledge. In D. C. Berliner & R. C. Calfee (Eds.), *Handbook of Educational Psychology* (pp. 709–725). New York: Macmillan Library Reference USA: Simon & Schuster Macmillan.

Carpenter, T., Fennema, E., & Franke, M. (1996). Cognitively guided instruction: A knowledge base for reform in primary mathematics instruction. *The Elementary School Journal, 97*(1), 3–20.

Carpenter, T., Fennema, E., Peterson, P., & Carey, D. (1988). Teachers' pedagogical content knowledge of students' problem solving in elementary arithmetic. *Journal for Research in Mathematics Education, 19,* 385–401.

Carpenter, T., Fennema, E., Peterson, P., Chiang, C., & Loef, M. (1989). Using knowledge of children's mathematics thinking in classroom teaching: An experimental study. *American Educational Research Journal, 26*(4), 499–531.

Darling-Hammond, L. (1999). *Teacher quality and student achievement: A review of state policy evidence.* Seattle: University of Washington, Center for Teaching and Policy.

Dubinsky, E., Schoenfeld, A. H., & Kaput, J. (Eds.). (1994). *Research in Collegiate Mathematics Education I* (Vol. 4). Providence, RI: American Mathematical Society.

Dubinsky, E., Schoenfeld, A. H., & Kaput, J. (Eds.). (2000). *Research in Collegiate Mathematics Education IV* (Vol. 8). Providence, RI: American Mathematical Society.

Fennema, E., Carpenter, T. P., Franke, M. L., Levi, L., Jacobs, V. R., & Empson, S. B. (1996). A longitudinal study of learning to use children's thinking in mathematics instruction. *Journal for Research in Mathematics Education, 27*(4), 403–434.

Fennema, E., & Scott Nelson, B. (Eds.). (1997). *Mathematics Teachers in Transition.* Mahwah, New Jersey: Lawrence Erlbaum Associates.

Ferrini-Mundy, J., Burrill, G., Floden, R., & Sandow, D. (2003). *Teacher knowledge for teaching school algebra: Challenges in developing an analytical framework.* Paper presented at the Annual meeting of the American Educational Research Association, Chicago, IL.

Friedberg, S., Ash, A., Brown, E., Hughes Hallett, D., Kasman, R., Kenney, M., et al. (2001a). *Teaching mathematics in colleges and universities: Case studies for today's classroom: Faculty edition.* Providence, RI: American Mathematical Society.

Friedberg, S., Ash, A., Brown, E., Hughes Hallett, D., Kasman, R., Kenney, M., et al. (2001b). *Teaching mathematics in colleges and universities: Case studies for today's classroom: Graduate student edition.* Providence, RI: American Mathematical Society.

Fullilove, R. E., & Treisman, P. U. (1990). Mathematics achievement among African American undergraduates at the University of California, Berkeley: An evaluation of the Mathematics Workshop Program. *Journal of Negro Education, 59*(3), 463–478.

Grossman, P. L., Wilson, S., & Shulman, L. S. (1989). Teachers of substance: Subject matter knowledge for teaching. In M. C. Reynolds (Ed.), *Knowledge base for the beginning teacher* (pp. 23–36). Oxford: Pergamon Press.

Hill, H., Rowan, B., & Ball, D. (2004). *Effects of teachers' mathematical knowledge for teaching on student achievement.* Paper presented at the Annual meeting of the American Educational Research Association, San Diego, CA.

Hill, H., Rowan, B., & Ball, D. (2005). Effects of teachers' mathematical knowledge for teaching on student achievement. *American Educational Research Journal, 42*(2), 371–406.

Hill, H., Schilling, S., & Ball, D. (2004). Developing measures of teachers' mathematics knowledge for teaching. *The Elementary School Journal, 105*(1), 11–0.

D. Holton (Ed.). (2001). *The teaching and learning of mathematics at university level: An ICMI study*. Dordrecht & Boston: Kluwer Academic Publishers.

Hsu, E., Murphy, T. J., & Treisman, P. U. (this volume). The Emerging Scholars Program turns 30: Supporting high minority achievement in introductory collegiate mathematics courses. In M. Carlson & C. Rasmussen (Eds.), *Making the Connection: Research and Practice in Undergraduate Mathematics Education*: Mathematical Association of America.

Kaput, J., Schoenfeld, A. H., & Dubinsky, E. (Eds.). (1996). *Research in collegiate mathematics education II* (Vol. 6). Providence, RI: American Mathematical Society.

Kieran, C. (1981). Concepts associated with the equality symbol. *Educational Studies in Mathematics, 12*(3), 317–326.

Kung, D. (in press). Teaching assistants learning how students think. *Research in Collegiate Mathematics Education*.

Ma, L. (1999). *Knowing and teaching elementary mathematics: Teachers' understanding of fundamental mathematics in China and the United States*. Mahwah, NJ: Lawrence Erlbaum Assoc.

Mason, J. (2001). Mathematical teaching practices at the tertiary level: Working group report. In D. Holton (Ed.), *The teaching and learning of mathematics at university level: An ICMI study* (Vol. 7, pp. 71–86). Dordrecht & Boston: Kluwer Academic Publishers.

Mewborn, D. S. (2003). Teaching, teachers' knowledge, and their professional development. In J. Kilpartick, W. G. Martin & D. Schifter (Eds.), *A research companion to Principles and Standards for School Mathematics* (pp. 45–52). Reston, VA: National Council of Teachers of Mathematics.

Monk, D. (1994). Subject area preparation of secondary mathematics and science teachers and student achievement. *Economics of Education Review, 13*(2), 125–145.

NSF (1992). *America's academic future: A report of the Presidential Young Investigator Colloquium on U.S. engineering, mathematics, and science education for the year 2010 and beyond*. Washington, DC: Directorate for Education and Human Resources, National Science Foundation.

Saenz-Ludlow, A., & Walgamuth, C. (1998). Third graders' interpretations of equality and the equal symbol. *Educational Studies in Mathematics, 35*(2), 153–187.

Schoenfeld, A. H., Kaput, J., & Dubinsky, E. (Eds.). (1998). *Research in Collegiate Mathematics Education III* (Vol. 7). Providence, RI: American Mathematical Society.

Selden, A., Dubinsky, E., Harel, G., & Hitt, F. (Eds.). (2003). *Research in Collegiate Mathematics Education V* (Vol. 12). Providence, RI: American Mathematical Society.

Shulman, L. (1986). Those who understand: Knowledge growth in teaching. *Educational Researcher, 15*(2), 4–14.

Speer, N., Gutmann, T., & Murphy, T. J. (2005). Mathematics teaching assistant preparation and development. *College Teaching, 53*(2), 75–80.

Speer, N., Strickland, S., & Johnson, N. (2005). Teaching assistants' knowledge and beliefs related to student learning of calculus. In G. Lloyd, M. Wilson, J. Wilkins & S. Behm (Eds.), *Proceedings of the twenty-seventh annual meeting of the North American Chapter of the International Group for the Psychology of Mathematics Education*. Blacksburg, VA: Virginia Tech University.

Treisman, P. U. (1985). A study of the mathematics performance of black students at the University of California, Berkeley: University of California, Berkeley.

About the Editors

Marilyn Carlson was awarded her PhD (in mathematics education) from The University of Kansas. Dr. Carlson joined the ASU faculty in 1995. She has served first as interim director and then director of CRESMET since 2003. Her teaching and research career began as a lecturer of mathematics on the Haskell Indian Nations in 1978, and she has also taught high school mathematics and served on the mathematics and computer science faculty at the University of Kansas. At both Kansas and ASU she has been Director of First-Year Mathematics, and at ASU she led the development of a Ph.D. concentration in mathematics education. Dr. Carlson is a frequent invited speaker and the author of more than 60 published and presented research papers. She received a National Science Foundation CAREER award, was a member of the Eisenhower Advisory Board for the State of Arizona, served as coordinator of the Special Interest Group for Research in Undergraduate Mathematics Education, served on a National Research Council panel investigating advanced mathematics and science programs in U.S. high schools, and has participated in policy deliberations at state and national levels. Dr. Carlson has been an investigator on a dozen projects funded by NSF. Currently, she is the principal investigator and co-principal investigator on two NSF grants of nearly $20 million that are funding research-based professional development to support teachers of math and science in Arizona secondary schools (Math and Science Partnership and Teacher Professional Continuum). The projects are conducting intensive research into coursework and professional learning communities for secondary mathematics and science teachers, seeking the structures and practices that help teachers to deepen their knowledge and improve their classroom practice. Dr. Carlson is currently leading an effort to reform the ASU program of college algebra. Dr. Carlson has been an invited speaker at numerous international conferences, is a consultant to peer universities seeking to improve undergraduate programs in mathematics, has traveled to study school mathematics as practiced in China, Japan, and Singapore, and is engaged in national efforts to improve U.S. programs that educate and support teachers, including the Teachers for a New Era project led by the Carnegie Corporation of New York. Under her leadership CRESMET has increased its supported research projects from less than $5 million to more than $30 million.

Chris Rasmussen earned a BS (in mechanical engineering) from the University of Maryland in 1985. He was awarded an MA in mathematics (1993) and then a PhD in mathematics education (1997) from the same institution. In 2006 Dr. Rasmussen was honored by the University of Maryland with their Distinguished Alumni Award: Outstanding New Scholar and also by San Diego State University with their Most Influential Teacher Award. He received the 2006 (inaugural) Annie and John Selden Award for Research in Undergraduate Mathematics Education. In 2002 Dr. Rasmussen won the Outstanding Faculty Scholar Award from Purdue University Calumet. He is active in the mathematics education community and is a member of the American Educational Research Association (AERA) and the AERA Special Interest Group for Research in Mathematics Education, the Association of Mathematics Teacher Educators, the Mathematical Association of America (MAA) and the Special Interest Group of the MAA on Research in Undergraduate Mathematics Education, the National Council of Teachers of Mathematics (NCTM), the International Group for the Psychology of Mathematics Education (PME), and the North American Chapter of the International group for the PME.